D1071457

HAZARDOUS WASTE CONTROL
in
RESEARCH and EDUCATION

Edited by
TAKASHI KORENAGA
HIROSHI TSUKUBE
SUMIO SHINODA
ISEI NAKAMURA

LEWIS PUBLISHERS
Boca Raton   Ann Arbor   London   Tokyo

Library of Congress Cataloging-in-Publication Data

Hazardous waste control in research and education / edited by Takashi
  Korenaga . . . [et al.].
     p.   cm.
  Includes bibliographical references and index.
  ISBN 0-87371-682-5
  1. Hazardous wastes. 2. Laboratories—Waste disposal.
  3. Universities and colleges—Waste disposal. I. Korenaga,
  Takashi.
  TD1030.H384   1994
  628.4'2--dc20

93-2503
CIP

© 1994 by CRC Press, Inc.
Lewis Publishers is an imprint of CRC Press

No claim to original U.S. Government works
International Standard Book Number 0-87371-682-5
Library of Congress Card Number 93-2503
Printed in the United States of America   1 2 3 4 5 6 7 8 9 0
Printed on acid-free paper

PREFACE

Research and educational activities often generate various wastes that may have severe environmental consequences. These wastes differ greatly from industrial wastes in the following ways: multiplicities in material, small quantities, and no periodical regularity. It is important that these wastes be properly controlled by various methods, such as recycling, conversion, and landfill.

Within the last decade much information has been available on the techniques of hazardous waste control in universities, research laboratories, and medical facilities, but a suitable book for the management of the wastes does not exist at the comprehensive level. This is what motivated us to prepare such a book at this time.

The main purpose of this book is to provide a balanced coverage concerning hazardous waste control in universities and research laboratories. After academic activity for global environment and environmental education are discussed, an overview on hazardous wastes, including chemical hazards and biohazards, is introduced in Section 1. The next section focuses on systems for minimizing waste by means of inventory control. Sections 3 and 4 describe off-site handling and on-site treatment systems for laboratory wastes. The monitoring and analysis of effluents and wastes are outlined in Section 5. Sections 6 and 7 are devoted to practical technology and new trends in hazardous waste treatment. Much of the subject matter in the book is about the activities of members in the Japanese Association for Laboratory Waste Management Facilities of Universities.

This book is appropriate not only for practicing engineers and managers who are responsible for management of hazardous wastes, but also for students involved in environmental science and engineering courses.

We express our thanks to the authors for their contributions to this book. The contribution made by Lewis Publishers, Inc. in preparing the book for press is also gratefully acknowledged.

Isei Nakamura, Ph.D.

EDITORS

Associate Professor Takashi Korenaga, Ph.D.

Takashi Korenaga was born in Okayama in 1949. He attended Okayama University, graduating with a B.S. and M.S. degree in analytical chemistry. He obtained a Ph.D. degree from Kyoto University in 1985 for work on automation of chemical oxygen demand measurement for flow injection analysis. He worked as a process analytical chemistry researcher at Japan Exlan Co., Ltd. from 1974 to 1977. In 1978, he joined the Center for Environmental Science and Technology at Okayama University and has served as Chief of the R&D Laboratory from 1983 to 1993. In 1987 he spent a few months at Virginia Polytechnic Institute and State University as a visiting professor. He planned the First International Symposium on Academic Activity for Waste Treatment in 1992.

His research interests include the design of new environmental analytical instruments and related chemical engineering problems, and the optimization of a treatment process or system for hazardous waste. He was the recipient of The Japan Society for Analytical Chemistry Award for Younger Researchers in 1984 and The Academic Award of The Sanyo Association for Advancement of Science and Technology in 1986.

Associate Professor Hiroshi Tsukube, Ph.D.

Hiroshi Tsukube was born in Osaka in 1953. He attended Osaka University, graduating with a B.S. degree in polymer chemistry in 1975. He obtained a Ph.D. degree from Kyoto University for work in organic chemistry including liquid membrane transport. In 1981, he was appointed a lecturer of chemistry at the College of Liberal Arts & Science at Okayama University and was promoted to associate professor in 1984. In 1990, he spent a sabbatical leave at the Department of Chemistry, University of California, Berkeley.

His research interests include the design of synthetic host molecules, new sensory devices, and carrier-mediated liquid membranes. He is also concerned with the development of laboratory methods for disposal of hazardous organic and inorganic chemicals. He received The Chemical Society of Japan Award for Young Chemists in 1987.

Professor Sumio Shinoda, Ph.D.

Sumio Shinoda was born in Hyogo prefecture in 1939. He attended Osaka University, graduating with a B.S. and M.S. degree in pharmaceutical science in 1964. He obtained a Ph.D. degree in 1968 for his work on the effect of thiamine derivatives on microbial cells. He worked as a research associate in the Research Institute for Microbial Disease of Osaka University in 1967 where he handled biohazardous organisms. After he spent a year in the Department of Microbiology at Case Western Reserve University, he was appointed associate professor in the Faculty of Pharmaceutical Sciences, Okayama University in 1971, and was promoted to professor in 1978. He was a director of the Center for Environmental Science and Technology, Okayama University from 1990 to the present date.

His research interests include pathogenic mechanism of vibrios, ecology of pathogenic bacteria in a natural water environment, treatment of infectious waste, and biodegradation of environmental pollutants. He organized the "Symposium on Environmental Pollutants and Toxicology" in 1982, *Vibrio parahaemolyticus* Symposium in 1988, and Annual Meeting of the Society of Antibacterial and Antifungal Agents, Japan in 1991.

Professor Isei Nakamura, Ph.D.

Isei Nakamura graduated from Tokyo University of Education with a B.S. degree in agricultural chemistry in 1953 and obtained a Ph.D. degree from Osaka University in 1965. He was a professor of the Institute of Applied Biochemistry, University of Tsukuba from 1978 to the present date and also took over as a director of the wastewater treatment facility of that university. His specialty and research interests include application of biochemical engineering.

He is President of the Japanese Association for Laboratory Waste Treatment Facilities of Universities and successfully hosted the First Asian Symposium on Academic Activity for Waste Treatment in 1992.

CONTRIBUTORS

Toshikazu Akita
Technical Services Division
NEC Environmental Engineering,
  Ltd.
Tokyo, Japan

Margaret-Ann Armour, Ph.D.
Department of Chemistry
University of Alberta
Edmonton, Alberta, Canada

Peter C. Ashbrook
Division of Environmental Health
  and Safety
University of Illinois
Urbana, Illinois

Tetsuji Chohji, Dr. Eng.
Department of Metallurgical
  Engineering
Toyama National College of
  Technology
Toyama, Japan

Kazuaki Isomura, Ph.D.
Department of Chemical
  Engineering
Kitakyushu National College of
  Technology
Kitakyushu, Japan

Miyoko Izawa, Ph.D.
Department of Bioengineering
  Science
Okayama University
Okayama, Japan

Satoru Kaseno, Ph.D.
Center for Environmental Science
  and Technology
Okayama University
Okayama, Japan

Mitsunobu Kitamura, Ph.D.
Environment Preservation Center
Kyoto University
Kyoto, Japan

Takashi Korenaga, Ph.D.
Department of Applied Chemistry
Okayama University
Okayama, Japan

Kengo Kurahasi, M.D.
Department of Chemistry
Kagawa Medical School
Kagawa, Japan

Edison Munaf, Ph.D.
Department of Chemistry
Andalas University
Padang, West Sumatra, Indonesia

Isei Nakamura, Ph.D.
Institute of Applied Biochemistry
University of Tsukuba
Tsukuba, Japan

Junko Nakanishi, Ph.D.
Environmental Science Center
University of Tokyo
Tokyo, Japan

Hiroyuki Ojima
Environmental Safety Center
Waseda University
Tokyo, Japan

Taneaki Okuda
Research and Development Group
NEC Corporation
Kawasaki, Japan

Peter A. Reinhardt
Safety Department
University of Wisconsin
Madison, Wisconsin

Shin-ichi Sakai, Ph.D.
Environment Preservation Center
Kyoto University
Kyoto, Japan

Harvey A. Shapiro
Faculty of Environmental Planning
Osaka Geijutsu University
Osaka, Japan

Sumio Shinoda, Ph.D.
Department of Pharmaceutical
  Sciences
Okayama University
Okayama, Japan

Izuru Sugano
NEC Environment Engineering,
  Ltd.
Tokyo, Japan

Yoshimitsu Suzuki
Environmental Science Center
University of Tokyo
Tokyo, Japan

Hiroshi Takatsuki, Ph.D.
Environment Preservation Center
Kyoto University
Kyoto, Japan

Toyohide Takeuchi, Ph.D.
Faculty of Engineering
Gifu University
Gifu, Japan

Yutaka Tamaura, Ph.D.
Research Center for Carbon
  Recycling and Utilization
Department of Chemistry
Tokyo Institute of Technology
Tokyo, Japan

Hideki Tatsumoto, Ph.D.
Department of Applied Chemistry
University of Chiba
Chiba, Japan

Hiroshi Tsukube, Ph.D.
Department of Chemistry
College of Liberal Arts & Science
Okayama University
Okayama, Japan

Etsu Yamada, Ph.D.
Center for Environmental Science
Kyoto Institute of Technology
Kyoto, Japan

Xiaojing Zhou, Ph.D.
Department of Applied Chemistry
Okayama University
Okayama, Japan

CONTENTS

Preface
I. Nakamura

**Section
1**

Introduction: Chemical Hazards, Biohazards, and Hazardous Wastes

CHAPTER 1.1

An Overview on Hazardous Waste Control

Yutaka Tamaura

TABLE OF CONTENTS

0-8493-682-5/94/$0.00 + $.50
© 1994 by CRC Press, Inc.

1.1.1 INTRODUCTION

The increase in the world population and the expansion of human activity has brought about global environmental problems. The laboratory wastes generated in experiments in the laboratories in universities and research institutes should be considered in relation to the global environmental problems. Laboratory wastes contain various kinds of hazardous materials; therefore, we must treat the laboratory wastes carefully when such wastes are generated. The laboratory researchers should have not only knowledge of the toxicity of the chemicals used in the experiments but also knowledge of the means for stabilizing the toxic substances to prevent environmental pollution.

Environmental education for handling laboratory wastes is one of the promising ways to insure that students have comprehensive knowledge of global environmental trends. A great consensus on global environmental issues and their seriousness is urgently required for human beings all over the world. There are many ways to create this consensus, and we have to pay attention wherever we can find the possibility of providing such environmental education. No major problems will be resolved in global environmental issues without a long-term educational effort. Today's university students are tomorrow's voters, taxpayers, and leaders.

1.1.2 ENVIRONMENTAL EDUCATION FOR HANDLING LABORATORY WASTES

After the United Nations Conference on the Human Environment held in Stockholm in June 1972, the recognition of global environmental problems has been promoted, and the United Nations Environment Program (UNEP) was established in Nairobi in 1972 by a resolution of the United Nations General Assembly. UNEP activities consist of earthwatch (observation and evaluation of the present conditions and future trends in the environment), environmental management, and supporting operations. As for the supporting operations, UNEP supports environmental education and environmental training programs. UNEP has played a central role in dealing with various global environmental problems. Environmental education ranks high as an indispensable action for human beings to solve global environmental problems.

In general, an effective environmental education requires training for raising awareness of environmental issues. In the chemical laboratories in universities, students use various toxic substances. This situation provides an ideal setting for initiating such an environmental education. We can train the students through the treatment of the hazardous materials used in the chemical experiments.

In Japan, the effluent discharged from the universities is regulated by law (Basic Law for Environmental Pollution Control) when the effluent is discharged into public water areas. The concentrations of the hazardous materials in the effluent discharged from the universities should be lower than those regulated by the law. The amounts of mercury and polychlorinated biphenyls (PCBs) are set, taking into consideration the possibility of health hazards due to their accumulation and concentration in fish or shellfish. In the law, a set of standards including biochemical oxygen demand (BOD), chemical oxygen demand (COD), phosphate (total phosphorus), and nitrate (total nitrogen) provides for preservation of the living environment and prevention of eutrophication (effluent standard). Therefore, in Japanese universities, the studenst are trained to carefully treat the laboratory wastes containing toxic substances. Also, during this decade, treatment facilities for laboratory wastes have been established in most every national university in Japan. The treatment facility in the universities provides an important role in educating the students to be aware of the environmental issues through practical treatment of the laboratory wastes in the universities.

1.1.3 WHAT HAPPENS TO DISPOSED LABORATORY WASTES

Disposing of chemically hazardous waste causes a threat to human health or to the environmental eventually. In the 1950s, some 700 people were killed and 9000 others were crippled in the city of Minamata, Japan. The disaster came from only a single cause, the dumping of a mercury-based compound into the bay by a chemical company. Often such health problems are caused by the hazardous chemicals released into the air, ground, or water and inhaled, eaten, or drunk by people. When chemically hazardous substances are once released into the environment, they contaminate the air, soil, ocean, bay, or groundwater. This may harm people and ecosystems. Hazardous substances cause a wide range of harmful effects on human health. Also they cause long-term or permanent damage to the ecosystem, such as Minamata Bay and PCB contamination. The number of chemicals in existence is increasing, with 70,000 chemicals now used regularly and 500 to 1000 new ones added each year. No data are available on the toxic effects of 79% of all chemicals, and complete data exist for only 2%. Under these circumstances, we have to render at least the toxic substances less hazardous before discharging them. This is the key principle for handling chemically hazardous materials. Even in university laboratories, the laboratory wastes containing toxic substances should be treated. The best strategy for dealing with toxic substances is to reduce the quantity of chemicals used in the experiments. However, in a chemistry laboratory, there is a limit to the reduction of the quantity of the chemicals used in the chemical experiments. In the universities, we have to educate the students to learn the necessity of treating chemically hazardous materials used in the experiments and practice how to render the chemicals less toxic or stabilize them.

The environmental education for handling laboratory wastes should involve two main points: (1) the meaning of the treatment of the laboratory waste and (2) practical training using the treatment. Starting from these two points, the environmental education will definitely attain a final goal, a great consensus on the global environmental issues and their seriousness.

1.1.4 RAISING AWARENESS THROUGH ENVIRONMENTAL EDUCATION FOR HANDLING LABORATORY WASTES

Acquiring knowledge concerning the global environmental issues should be a requirement in the educational institutes of the world. A more important factor in such environmental education is the practical training in solving these issues concerning global environmental problems. Especially in the universities, training the students is very effective for such environmental education. In Japan, the treatment facility in the universities assumes an important role in the environmental education through the practical treatment of laboratory wastes (laboratory wastewaters).

The laboratory wastewaters generated in the experiments are stored in bottles (20 liters) in each laboratory in most Japanese universities. The laboratory wastewaters collected in the bottles are carried to the treatment facility on the campus and treated to stabilize the chemically hazardous materials such as heavy metal ions (Cd, Mn, Cr, et al.) and cyanide. Organic solvent waste is generally treated by incineration. In most Japanese universities, the sink in the laboratory is directly connected to the river, ocean, or municipal sewage. Therefore, collecting the laboratory wastewaters in bottles is the key step in preventing the release of pollutants from this source. A very important and effective way to make this step adequate is to increase the students' awareness and to encourage the means of collection of laboratory wastewaters to prevent environmental pollution.

In the Tokyo Institute of Technology, a lecture concerning the treatment of laboratory wastes is especially given to the freshmen to increase the students' awareness for prevention of release of pollutants from the university. The lecture is given by a professor associated with the campus treatment facility. The student's awareness for protection of the environment increases after hearing the lecture. The number of students who want to advise a person who is discharging laboratory wastewater in the sink increases after hearing the lecture. This increase in the student's awareness resulting from the lecture proves that environmental education for handling laboratory wastes is very effective. In this lecture, global environmental issues are addressed, such as global warming, depletion of the stratospheric ozone layer, problems for marine ecosystems, tropical deforestation, desertification, extinction of wildlife species, acid rain, and transboundary movement of hazardous wastes. The background of these global environmental problems lies in the continued growth of the population and economic activities neglecting the necessary consideration of the environment and lacking sufficient environmental management. We have to do something to solve the global environmental problem. Environmental education provides an important role in increasing human awareness. By giving students a lecture concerning the laboratory wastes in relation to the background of the global environmental problem, students can readily understand what they have to do after the lecture. The practice of using the treatment of the laboratory wastes is thus closely linked with the environmental education for the global environmental problem. Thus, the treatment facility in the universities in Japan assumes a role in environmental education regarding global environmental issues, as well as handling laboratory wastes.

In the Kyto University in Japan, the laboratory researchers operate the treatment system in the campus facility by themselves. This is an effective way to make the researcher understand the importance of the treatment of laboratory wastes.

1.1.5 INFORMATION CENTER FOR LABORATORY WASTES

Most laboratory wastes are chemically hazardous, and they are in the form of gases, solids, and liquids. This means that laboratory wastes have various faces which can cause a threat to the global environment. Thus, through the treatment of laboratory wastes, we can effectively provide environmental education concerning general wastes in relation to the treatment of laboratory wastes.

Poor disposal methods for wastes will cause a serious environmental problem. From this point of view, we have to consider a suitable method for disposing of laboratory wastes. Chemically hazardous waste disposal in landfills will cause pollution of groundwater and surface waters when rainfall leaches the hazardous substance. When laboratory wastes are incinerated, effluent gases may contain dioxins and other hazardous air pollutants. Concerning these points, the students should be well informed of a suitable way to treat the laboratory wastes, as well as their toxicity. The information about toxic chemicals used in the experiment should be readily available in the universities. Some of the campus treatment facilities in Japanese universities provide such information. This kind of system will assume an important role in environmental education for handling laboratory waste. The Organization for Economic Cooperation and Development (OECD) activities relating to chemical substances include investigating the safety of existing chemical substances and handling environmental problems. Since information concerning the safety of chemical substances in OECD activities, is useful for consideration of the treatment of laboratory wastes, we need to establish an information center where students and researchers can obtain information concerning the safety of chemical substances found in OECD activities. Moreover, this kind of information center can also collect information concerning suitable treatment methods of chemically hazardous materials. Providing information to university students is a practical way to provide environmental education for handling laboratory wastes.

1.1.6 CHEMICAL ANALYSIS OF LABORATORY WASTES

Usually the chemical contents of laboratory wastes are not specified in detail. Very often, they are treated by a conventional means after classification into two categories of organic and inorganic wastes. In the chemistry laboratory, various toxic chemicals are used; therefore, their respective composition and contents in the laboratory wastes should be carefully monitored until they are treated. Thus, the chemical analysis before and after the treatment at the facility in Japanese universities is considered important among the jobs in the treatment facility. The analytical results are related to the researchers who request treatment of the laboratory waste. The laboratory researchers should know the chemical content and the amount of the laboratory waste. When they request treatment of the laboratory wastes, the chemical composition and amount are relayed to the facility to make treatment easy. If the actual chemical composition and content are largely different from those which are relayed to the treatment facility, they are warned to provide an accurate report on the chemical content. This feedback system is very useful for environmental education to facilitate raising the awareness of the students for environmental protection, since knowing the character of the waste is the primary obligation of the laboratory researcher. A final purpose of the research in the laboratory is to make human life better. Research that causes pollution through laboratory wastes cannot be said to be true research. Environmental ethics must remain alive in the minds of human beings to preserve human society in the future.

Treatment facilities in Japanese universities are equipped with several analytical instruments such as atomic absorption spectrometers, gas chromatographs (GC), and GC-MS (mass spectrometry); therefore, students visiting the facility can study practical ways to chemically analyze hazardous materials using those instruments. Moreover, we can train the students to operate the instruments and to practically analyze the chemically hazardous materials. In this respect, the treatment facility has the function of providing a practical environmental education through chemical analysis. This kind of technical training is required for environmental professionals, planners, and managers in industry and government.

1.1.7 SOLID WASTE MANAGEMENT

Solid waste management is also becoming a big consideration in global environmental problems. The solid waste volume is exceeding the disposal capacity in many cities. This increase is being accelerated by both growing population size and waste volumes. Wastestream reduction is the answer to this problem. Reduction of wastes can be achieved by minimizing the quantity of the materials used in daily life. Most of the waste can be recovered and reused as secondary resources; therefore, establishing a recycling system will be another answer. The term recycling is commonly applied to the processing of materials into new products that may or may not resemble the original material. Recycling reduces waste volume and lowers the energy requirements. To reduce the wastes, we must change lifestyles which cause the production of too much waste material. Training the students to reduce the waste will become a key educational feature in the treatment facilities of the universities.

The growing shortage of landfill sites has focused attention on incineration. Since most of the hazardous components of incinerator ash are metallic elements, removal of metals from the processes can largely eliminate their presence in ash. A better understanding is needed regarding the presence of metallic elements in the wastes which will be treated using an incinerator. Environmental education at the treatment facility in the university should include attention to these points. In the chemical laboratory, filter papers used for the separation of the metal ions are usually discarded together with other general waste papers. The need for clear separation of the filter papers used for chemical experiments should be well understood by the researcher in the chemical laboratory in the university.

1.1.8 TECHNICAL COOPERATION WITH DEVELOPING COUNTRIES

The problem of hazardous waste is not confined to a single country. The students should learn about the situation of toxic waste in foreign countries, especially in developing countries. A mutual understanding among countries is required for cooperation with recent international efforts to develop uniform regulations to control the export of hazardous substances. The developed countries must help developing countries establish their own technological capabilities and regulations for the use and disposal of hazardous substances. The nongovernmental organizations (NGOs) and their networks are conducting various activities. The treatment facility in the universities in the Asian area can form a unique network to communicate information concerning the treatment methods and the know-how for handling chemically hazardous substances.

CHAPTER 1.2

Chemical Hazards

Sumio Shinoda

TABLE OF CONTENTS

1.2.1 INTRODUCTION

Various chemicals are used in research laboratories, some of which are hazardous to humans. The chemicals are extremely diverse and indefinite, although the quantity used is much less than that in industries. The risks, therefore, are not clearly understood by the users in most cases. Hazardous chemicals used without recognition of the risks have the potential of impairing the health of the user and of resulting in environmental pollution if they are

0-8493-682-5/94/$0.00 + $.50

discharged without proper prior treatment. Handling of toxic chemicals is controlled by the "Poisonous and Deleterious Substances Control Law" in Japan, and the health precautions for handlers are prescribed by the "Industrial Safety and Health Law" and the "Ordinance of the National Personnel Authority". Wastewater is controlled by the "Water Pollution Control Law". The chemicals listed in these laws are controlled because they are frequently used for industrial purposes and have high incidence of causing health damage to laborers. Therefore, it is a matter of course that great care should be taken in handling such chemicals in research laboratories. There are, however, other uncontrolled or unregulated chemicals than those listed which are also used in laobratory research. These uncontrolled chemicals may be even more hazardous because of the lack of understanding of their risks by users.

1.2.2 LEGALLY REGULATED CHEMICALS

The Poisonous and Deleterious Substances Control Law sets the rules for health control in the manufacture and sales of poisonous and deleterious chemicals. Examples of the substances listed in the law are as follows: (1) the poisonous substances: 27 chemicals and their derivatives, including hydrogen cyanide, sodium cyanide, mercury, selenium, arsenic, diethyl-*p*-nitrophenyl phosphorothioate (parathion), and octamethyldiphosphoroamide (schradan); (2) deleterious substances: 93 chemicals and their derivatives, including acrylonitrile, aniline, hydrogen chloride, hydrogen peroxide, chloroform, sodium hydroxide, nitrobenzene, and pentachlorophenol (CP); (3) specified poisonous substances: 9 chemicals and their derivatives, including tetraalkyl lead and fluoroacetic acid. Only registered manufacturers are allowed to produce, import, and sell these listed chemicals. However, use of such chemicals for research purposes is not restricted, although the specified poisonous substances are allowed to be used only by "specified poisonous substance researchers and handlers" who have permission from the Japanese Ministry of Welfare. Laboratory researchers are thus able to purchase deleterious chemicals with a special slip, and there is no legal restriction in their use, although their storage and disposal are restricted by law. Excitants, hallucinogens, and anesthetics should be used with great care in research since the law prohibits the ingestion, aspiration, and possession of these chemicals without proper reason. Handling of explosive chemicals is also restricted. However, it is believed that in some cases researchers or students handle such chemicals without understanding how hazardous they are. A rule must be made to check the risk of a chemical before using it, especially for a new chemical being used for the first time. The Poisonous and Deleterious Substances Control Law prescribes that the bottles of poisonous and deleterious chemicals are to be marked by red labels with white letters and white labels

Table 1 Legal Regulation of Chemicals in Japan

(a) Hazardous chemicals prohibited for production
     1. Yellow phosphor match
     2. Benzidine and its salts
     3. 4-Aminodiphenyl and its salts
     4. 4-Nitrodiphenyl and its salts
     5. Bis(chloromethyl) ether
     6. β-Naphthylamine and its salts
     7. Gum containing benzene (5% or more)
     8. Products containing compounds 2 to 6 (1% or more)
(b) Hazardous chemicals requiring permission from authorities for production
     1. Dichlorobenzidine and its salts
     2. α-Naphthylamine and its salts
     3. Polychlorinated biphenyl
     4. *o*-Tolidine and its salts
     5. Dianisidine and its salts
     6. Beryllium and its compounds
     7. Benzotrichloride
     8. Products containing compounds 1 to 6 (1% or more) or
        compound 7 (0.5% or more)

From "Industrial Safety and Health Law" and "Ordinance for National Personnel Authority."

with red letters, respectively. Indications of small bottles, however, are often difficult to recognize, and the marking of a reagent taken out in a small quantity is often forgotten.

The Industrial Safety and Health Law prohibits the production, import, or usage of the chemicals shown in Table 1(a), and the production of chemicals in Table 1(b) is restricted because of the risks of health disorders to laborers. One who uses the chemicals in Table 1(a) for research must obtain permission from the Labor Standard Bureau and must undergo a periodical physical checkup. Similar regulations in the Ordinance of the National Personnel Authority apply to government employees. The regulation of these chemicals is mainly due to their carcinogenicity rather than their acute or subacute toxicity.

Furthermore, the Industrial Safety and Health Law regulates the methods of handling the specified chemical substances shown in Table 2. Handling of the chemicals in Class 1 is the most strictly regulated. This is based on their carcinogenicity and the risk of skin, nervous, and other disorders. Employers are obliged to take appropriate measures, such as preventing vapor generation, to protect the workers from health disorders. Although this regulation does not apply to users of these chemicals in a research laboratory, it is only common sense that researchers should themselves take proper precautions in using such hazardous chemicals.

In addition to the above laws, several ordinances such as the "Ordinance on the Prevention of Organic Solvent Poisoning (OPOSP)", the "Ordinance

Table 2 Legal Regulation of Chemicals in Japan — Specified Chemical Substances

Class 1
  Chemicals listed in Table 1(b)

Class 2
  1. Acrylamide
  2. Acrylonitrile
  3. Alkyl mercury
  4. Asbestos
  5. Ethyleneimine
  6. Vinyl chloride
  7. Chlorine
  8. Auramine
  9. o-Phthalodinitryl
  10. Cadmium and its compounds
  11. Chromates
  12. Monochloromethyl ether
  13. Vanadium pentaoxide
  14. Coal tar
  15. Arsenic trioxide
  16. Potassium cyanide
  17. Hydrogen cyanide
  18. Sodium cyanide
  19. 3,3'-Dichloro-4,4'-diaminodi-
      phenylmethane

  20. Methylbromide
  21. Dichromates
  22. Inorganic mercuries
  23. Trilene diisocyanate
  24. Nickel carbonyl
  25. Nitroglycol
  26. p-Dimethylaminobenzoate
  27. p-Nitrochlorobenzoate
  28. Hydrogen fluoride
  29. Propionolactone
  30. Benzene
  31. Pentachlorophenol (PCP) and its
      sodium salt
  32. Magenta
  33. Manganese and its compounds
  34. Methyl iodide
  35. Hydrogen sulfide
  36. Dimethyl sulfate
  37. Products containing compounds
      1 to 36

Class 3
  1. Ammonia
  2. Carbon monoxide
  3. Hydrogen chloride
  4. Nitric acid
  5. Sulfur dioxide

  6. Phenol
  7. Phosgene
  8. Formaldehyde
  9. Sulfuric acid
  10. Products containing compounds 1 to 9

From "Industrial Safety and Health Law" in an ordinance entitled "Prevention of the Hazards due to Specified Chemical Substances" established in accordance with the Industrial Safety and Health Law.

on the Prevention of Lead Poisoning", or the "Ordinance on the Prevention of Tetraalkyl Lead Poisoning" have been established to protect the workers. The solvents regulated by OPOSP are shown in Table 3. Most of those in Class 1 are chlorinated hydrocarbons, which have anesthetic toxicity and carcinogenicity. Trihalomethanes, such as chloroform, have attracted attention as a carcinogen in tap water which is produced by the reaction of organic pollutants with chlorine used for disinfection. Hepatic toxicity of halogenated hydrocarbons is also well known. For example, carbon tetrachloride is used to induce hepatic dysfunction in experimental animals. Benzene is a highly toxic solvent causing disorder of hematopoietic function. It is not listed in

Table 3 Legal Regulation of Chemicals in Japan — Hazardous Organic Solvents

Class 1
1. Chloroform
2. Carbon tetrachloride
3. 1,2-Dichloroethane
4. 1,2-Dichloroethylene
5. 1,1,2,2-Tetrachloroethane
6. Trichloroethylene
7. Carbon disulfide
8. Mixture of compounds 1 to 7 or products containing these compounds (5% or more)

Class 2
Acetone, Isobutyl alcohol, Isopropyl alcohol, Isoamyl alcohol, Ethyl ether, Ethylene glycol monoethyl ether (cellosolve), Ethylene glycol monoethyl ether acetate (cellosolve acetate), Ethylene glycol monobutyl ether (butyl cellosolve), Ethylene glycol monomethyl ether (methyl cellosolve), o-Dichlorobenzene, Xylene, Cresol, Chlorobenzene, Isobutyl acetate, Isopropyl acetate, Isoamyl acetate, Ethyl acetate, Butyl acetate, Propyl acetate, Amyl acetate, Methyl Acetate, Cyclohexanol, Cyclohexanone, 1,4-Dioxane, Dichloromethane, N,N-Dimethyl-formamide, Styrene, Tetrachloroethylene, Tetrahydrofuran, 1,1,1-Trichloroethane, Toluene, n-Hexane, 1-Butyl alcohol, 2-Butyl alcohol, Methyl alcohol, Methyl isobutyl ketone, Methyl ethyl ketone, Methyl cyclohexanol, Methyl cyclohexanone, Methyl butyl ketone, Mixture of the above compounds or products containing these compounds (5% or more)

Class 3
Organic solvents other than Classes 1 and 2

From "Ordinance on the Prevention of Organic Solvent Poisoning."

the ordinance, but is in the Industrial Safety and Health Law as shown in Table 1.

1.2.3 UNREGULATED CHEMICALS

As stated, in Japan, various laws and ordinances have been established to regulate the handling of generally distributed chemicals and to protect users from risk. Although regulation in the laboratory is not as strict, researchers should be careful in using recognized, legally regulated chemicals. In research laboratories of academic facilities, however, extremely diversified chemicals are used to carry out basic research, the main role of academic activities. Many such chemicals are legally unregulated, but may include those of unrevealed toxicity which could potentially be more hazardous than the legally regulated ones. Handling of such chemicals is therefore more troublesome because of the lack of information about them. For instance, for various reagents prepared by a manufacturer and used in diagnostic laboratories, there is little or no information on chemicals because of the industrial secrecy in which they were developed. Laboratory technicians tend to give no attention to the contents of these reagents. Some reagents may contain legally regulated chemicals or unregulated toxic chemicals. The manufacture must call this to the user's attention, and the user also must make it a rule to take note of the toxicity of the reagent and its potential as an environmental pollutant. It is very important to establish awareness among students to preinvestigate the toxic or hazardous properties of a chemical which they are about to use.

Of the unregulated chemicals, those which are carcinogenic or mutagenic require the greatest attention. An enormous number of chemicals exist on the earth, and their number is growing daily. The number of which the carcinogenicity has already been examined is extremely low in contrast with their total number. Tests for carcinogenicity in laboratory animals is very expensive and require much time, as animals administered various amounts of a test chemical through various routes are observed for a long period. Cancer is a condition in which normal somatic cells are transformed into abnormal (cancer) cells by mutation of the genes, so that the normal control of cell division no longer takes place. That is to say, the carcinogen is a mutagen. On the basis of this idea, tests for mutagenicity are carried out for the primary screening of a carcinogen. Various methods to test for mutagenicity have been developed. The Ames test[1] is used extensively because results can be obtained within a few days by this microbiological method. As microbial cells are different from animal cells in many points such as metabolism, membrane permeability, etc., various improvements have been made to adapt the results of the microbiological methods so that the effect on mammalian cells can be evaluated. Chemicals suspected in preliminary screening tests using microbiological methods are then subjected to carcinogen tests on laboratory animals. This saves times and money and allows information on a carcinogenic substance to be obtained sooner.

The Industrial Safety and Health Law obliges a company owner to carry out the mutagenicity tests of a chemical which is newly produced or imported. Many substances still remain to be examined for their possible mutagenicity; however, we have to keep in mind that we may be using a chemical of which the mutagenicity is still unknown. In studies involving the synthesis of new chemical substances, the handling of the newly synthesized substance should be done very carefully, especially if the mutagenicity or other toxicity of the analogues of the substance has been suggested.

Extensive investigation is ongoing on the correlation between the structure of the chemicals and the mutagenicity or carcinogenicity. For example, benzidine and some of its analogues (Structure 1) are listed in Table 1 because

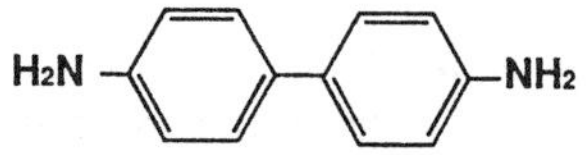

Benzidine

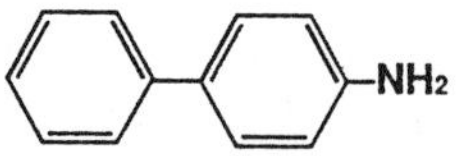

4-Aminodiphenyl

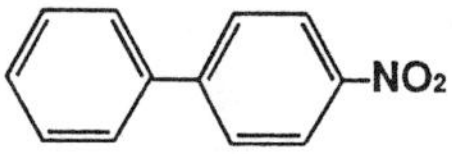

4-Nitrodiphenyl

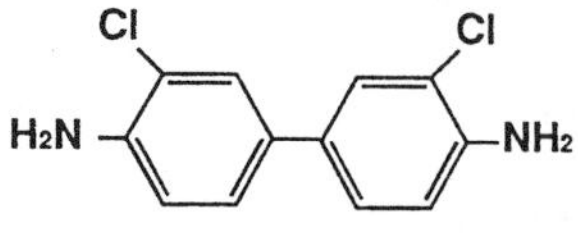

3,3'-Dichlorobenzidine   **1**   *o* - Tolidine

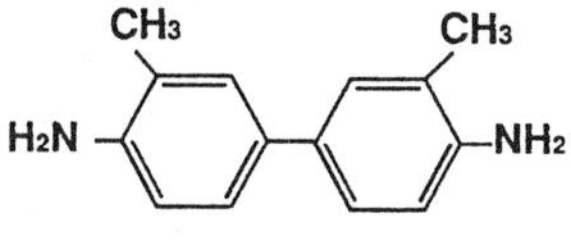

of their carcinogenicity. Actually, many analogous aromatic amines are known to show carcinogenicity. However, β-naphthylamine (Structure 2) is known as a potent carcinogen whereas the carcinogenicity of α-naphthylamine is not confirmed. It has been suggested that the carcinogenicity of contaminated β-naphthylamine, which is a by-product of α-naphthylamine production, contributes more effectively than that of α-naphthylamine itself. This bring us to two points regarding chemical hazards: the possibility of a significant change in toxicity by even a slight modification of the structure and the possibility of danger caused by impurities.

Carcinogenicity of the pyrolysates of amino acids (Structure 3) which are found on charred meat has been suggested. Polyaromatic hydrocarbons, such as benzo[*a*]pyrene which are produced by the combustion of various fuels, are also known as carcinogenic air pollutants. So-called "dioxin" (especially 2,3,7,8-tetrachlorinated di-benzo-*p*-dioxin), a teratogen, is known to be found in incinerator ashes of wastes. Thus, we have to recognize the possibility of the production of unexpected carcinogenic or other toxic substances through the thermal reaction of organic compounds.

Mutagens are frequently used in laboratories, since the induction of mutants of microorganisms is an important technique in biotechnology. N-methyl-N'-nitro-N-nitrosoguanidine[2,3] is the most frequently used mutagen in laboratory research and a well-known carcinogen. The remarkable progress being made

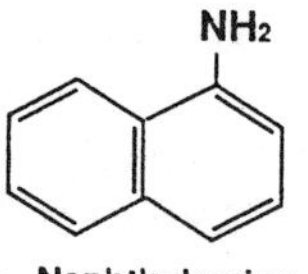

α - Naphthylamine   **2**   β - Naphthylamine

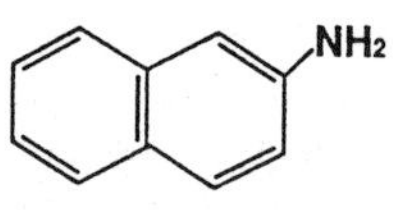

in biotechnology will further stimulate the use of such mutagens. It is also noteworthy that certain antitumor drugs are also carcinogenic. This is a problem not only for the patients and medical employees but also for the oncologist.

Trp - P - 1   3   Glu - P - 1

Benzo [a] pyrene

However, it is still unclear whether all medical employees or researchers understand the hazardous properties of the chemicals they use.

In addition to mutagens, there are many potentially hazardous chemicals which are toxic but unregulated because of their narrow distribution and the fact that the toxicity of the chemical remains to be determined. In principle, therefore, direct contact with any laboratory chemicals should be avoided as much as possible. Oral ingestion is out of the question. Handling of explosives and flammables requires the greatest possible care. Users of laboratory chemicals must develop the habit of learning and remembering the property of a chemical before using it.

For further information on the hazard of each chemical including its method of disposal, the reader should refer to technical books on these subjects.[4-6]

REFERENCES

1. Ames, B.N., Gurney, E.G., Miller, J.A., and Bartsch, H., Carcinogens as frameshift mutagens: metabolites and derivatives of 2-acetylaminofluorene and other aromatic amine carcinogens, *Proc. Natl. Acad. Sci. U.S.A.,* 69, 3129, 1972.
2. Holle, S., 5-Azacytidine as a mutagen for arboviruses, *J. Virol.,* 2, 1228, 1968.
3. Szybalskyi, W., Observations on chemical mutagenesis in microorganisms, *Ann. N.Y. Acad. Sci.,* 76, 475, 1958.
4. Committee on Biologic Effects of Atmospheric Pollutants, National Research Council, Medical and Biologic Effects of Environmental Pollutants, National Academy of Sciences, Washington, D.C., 1978.
5. Armour, M.A., Bacovsky, R.A., Browne, L.M., McKenzie, P.A., and Renecker, D.M., Potentially Carcinogenic Chemicals, Information and Disposal Guide, University of Alberta, Edmonton, Canada, 1986.
6. Armour, M.A., Browne, L.M., and Weir, G.L., Hazardous Chemicals, Information and Disposal Guide, University of Alberta, Edmonton, Canada, 1987.

CHAPTER 1.3

Biohazards

Sumio Shinoda

TABLE OF CONTENTS

1.3.1 INTRODUCTION

The term "biohazard" has primarily been used to refer to something which carries the risk of laboratory-associated infection, although in a broad sense it includes hazards or risks due to all biological species. Laboratory infections caused by potent pathogens occurred frequently in the late 19th to early 20th century when pathogens of various contagious diseases were just being

0-8493-682-5/94/$0.00 + $.50

© 1994 by CRC Press, Inc.

recognized, and some famous researchers were reported as victims.[1] In these cases, however, the researchers had well understood the risks involved because they were in a war against dangerous diseases. Since the introduction of recombinant DNA techniques in the 1970s, genetically engineered microorganisms have come to be regarded as new biohazards because of the possibility of creating unexpected and dangerous recombinant organisms. Genetic engineering is one of the most widely used biological techniques at present, and progress in modern molecular biology is opening the door on the mystery of life itself which cannot be done without recombinant DNA techniques. Furthermore, studies on pathogenic factors of infectious microorganisms are still in progress; in addition to known pathogens, new ones, such as HIV (human immunodeficiency virus, the pathogen of acquired immunodeficiency syndrome, AIDS), have been recognized. Laboratory animal-borne infections are also an issue requiring care. In view of the importance of biohazards in academic activities, this section deals with laboratory biohazards and ways of controlling them.

1.3.2 HANDLING OF PATHOGENIC MICROORGANISMS IN LABORATORIES

The handling of pathogenic organisms should always be done with utmost care. Pathogens are usually believed to be treated with adequate care when used in studies on infectious diseases because specialists in the field conduct their experiments with full understanding of the risks of what they are using. If pathogens are treated without adequate understanding of their pathogenic potential, however, problems may result. The use of pathogenic microorganisms is not confined to studies on infectious diseases, for genetic engineering is being further developed and various microorganisms including pathogenic ones are being studied and occasionally even cultured in large quantity. Some bacterial toxins such as cholera toxin, pertussis toxin, diphtheria toxin, or *Pseudomonas aeruginosa* exotoxin A are already commercially available as biochemical reagents because they are useful tools in the study of metabolic regulations. Those toxins are factors concerned with infections because of their physiological and pharmacological activity. Namely, toxins are potent, physiologically active agents which are usable reagents in basic research and are possible candidates for medical drugs. Streptokinase, streptodornase, and some proteases are, in fact, medical drugs which have been produced from pathogenic bacteria. The use of such physiologically active agents produced by pathogenic organisms is expected to grow more, and biohazards due to the use of such organisms without knowledge of infectious microbiology is feared. Although, generally, the same techniques are used in the handling of both pathogenic and nonpathogenic organisms, additional caution is called for the former. In microbiological study, the following precautions are essential: (1) avoid contamination of a culture by other organisms and (2) block escape of the subjected organisms into the environment. The later precaution is of the greatest importance to prevent infection when pathogenic organisms are being handled. Knowledge of the pathogen and adequate equipment is needed. The Japanese Society for Bacteriology enacted the "Guidelines for Biosafety on Distribution of Pathogens" in 1990.[2] They stipulate that culture collection facilities should examine the recipient of the culture to make sure equipment is appropriate and to learn the individual's microbiological education background.

To prevent biohazards, equipment for physical containment and sterilization or disinfection are necessary. The World Health Organization started a "Special Programme on Safety Measures in Microbiology" in 1976 and published a "Laboratory Biosafety Manual" in 1983,[3] in which pathogens are classified into the four risk groups shown in Table 1. The Centers for Disease Control (CDC) and the National Institutes of Health (NIH) in the U.S. published "Biosafety in Microbiological and Biomedical Laboratories" in 1984 and 1988[4] and recommended practices, techniques, safety equipment, and facilities for 4 levels of biosafety when using infectious agents (Table 2). These laboratory biosafety criteria are as follows:

Biosafety Level 1 is suitable for work involving agents of no known or of minimal potential hazard to laboratory personnel and the environment. **Biosafety Level 2** is similar to level 1 and is suitable for work involving agents of moderate potential hazard to personnel and the environment. **Biosafety Level 3** is applicable to clinical, diagnostic, teaching, research, or production facilities in which work is done with indigenous or exotic agents which may cause serious or potentially lethal disease as a result of exposure by the inhalation route. **Biosafety Level 4** is required for work with dangerous and exotic agents which pose a high individual risk of life-threatening disease.

The NIH of Japan also established "Rules for Biosafety Control of Pathogenic Organisms" with similar biosafety level classifications.[5] Classifications of the pathogens are shown in Table 3 and resemble those of the CDC/NIH.[4] Only bacterial names are listed in the table because of limited space; readers should refer to the original papers[4,5] for names of other pathogens such as virus or fungi. Examples of viruses listed are as follows:

Level 1. Live vaccine viruses (except vaccinia)
Level 2. Adeno-associated, corona, Coxackie (A, B), dengue, ECHO, hepatitis (A, B, C, D, E), herpes simplex (1, 2), HTLV (1, 2), influenza (A, B, C), Japanese encephalitis, mumps, papova, polio, rhino, rubella, vaccinia, etc.
Level 3. Colorado tick fever, Hantaan, human immunodeficiency (HIV 1, 2), rabies, etc.
Level 4. Congo-Crimean hemorrhagic fever, Ebola, Lassa fever, Marburg disease, variola minor, yellow fever, etc.

Table 1 WHO Risk Groups in Relation to Category of Laboratory[3]

Risk Group	Laboratory Classification	Examples of Laboratories	Examples of Organisms
1. (Low individual, low community risk)	Basic	Basic teaching	*Bacillus subtilis*, *Escherichia coli*, K12
2. (Moderate individual, limited community risk)	Basic (with biosafety cabinets or other appropriate personal protective or physical containment devices when required)	Primary health services: primary level hospital, doctors' offices, diagnostic laboratories, university teaching, public health laboratories	*Salmonella typhi*, Hepatitis virus B, *Mycobacterium tuberculosis*,[a] LCM virus[b]
3. (High individual, low community risk)	Containment	Special diagnostic laboratories	*Brucella* spp., Lassa fever virus, *Histoplasma capslatum*
4. (High individual, high community risk)	Maximum containment	Dangerous pathogen units	Marburg virus, Foot-and-mouth disease virus

[a] When larger volumes or high concentrations are used or when techniques may involve aerosol production, these and other agents should be promoted to Risk Group III.

[b] Includes research laboratories at appropriate risk group level.

From WHO, Laboratory Biosafety Manual, World Health Organization, Geneva, 1983.

Table 2 Summary of Recommended Biosafety Levels for Infectious Agents

Biosafety Levels	Practice and Techniques	Safety Equipment	Facilities
1	Standard microbiological practices	None: primary containment provided by adherence to standard laboratory practices during open bench operations	Basic
2	Level 1 practices plus laboratory coats; decontamination of all infectious waste; limited access; protective gloves and biohazard warning signs as indicated	Partial containment equipment (i.e., Class I or II biological safety cabinets) used to conduct mechanical manipulative procedures that have high aerosol potential which may increase the risk of exposure to personnel	Basic
3	Level 2 practices plus special laboratory clothing and controlled access	Partial containment equipment used for all manipulations of infectious material	Containment
4	Level 3 practices plus entrance through change room where street clothing is removed and laboratory clothing is put on; shower on exit; all wastes to be decontaminated before removal from the facilities	Maximum containment equipment (i.e., Class III biological safety cabinet or partial containment equipment in combination with full-body, air-supplied, positive-pressure personnel suit) used for all procedures and activities	Maximum containment

From CDC/NIH: Biosafety in Microbiological and Medical Laboratories, The Centers for Disease Control, Nat. Inst. of Health, U.S. Government Printing Office, Washington, D.C., 1988.

Table 3 Classification of Pathogens in Each Biosafety Level[5]

1. Virus, *Chlamydia* and *Rickettsia*[a]

2. Mycoplasma and bacteria

Level 1
  Bacteria other than Levels 2 and 3

Level 2
  Actinobacillus actinomycetemcomitans
  Aeromonas hydrophila, A. sobria
  Actinomadura madurae, A. pelletieri
  Actinomyces bovis, A. israeli, A. pyogenes, A. viscous
  Bacillus cereus
  Bordetella bronchiseptica, B. parapertussis, B. pertussis
  Borellia (all species)
  Branhamella catarrhalis
  Calymmatobacterium granulomatis
  Campylobacter coli, C. jejuni
  Clostridium botulinum, C. difficile, C. haemolyticum, C. histolyticum, C.
    perfringens, C. novyi, C. septicum, C. sordelli, C. sporogenes, C. tetani
  Corynebacterium diphtheriae, C. jeikeium, C. pseudodiphtheriticum
  Erysipelothrix rhusiopathiae
  Escherichia coli (except K12, B and their variants)
  Francisella novicida
  Fusobacterium necrophorum
  Haemophilus iducreyi, H. influenzae
  Helicobacter pylori
  Klebsiella pneumoniae, K. oxytoca
  Legionella (all species)
  Leptospira interrogans
  Listeria monocytogenes
  Mycobacterium avium, M. chelonei, M. fortuitum, M. haemophilum, M.
    intracellulare, M. kansaii, M. leprae, M. lepraemurium, M. malmoense,
    M. marinum, M. scrofulaceum, M. paratuberculosis, M. simiae, M. szulgai,
    M. ulceras, M. xenopi
  Mycoplasma fermentans, M. pneumoniae, M. hominis
  Neisseria gonorhoeae, N. meningitidis
  Nocardia asteroides, N. brasiliensis, N. caviae, N. farcinica
  Pasteurella multocida, P. neumotropica, P. urea
  Pleisomonas shigelloides
  Pseudomonas aeruginosa, P. cepacia
  Salomonella (all serovars except Level 3)
  Serratia marcescens
  Shigella (all species)
  Staphylococcus aureus
  Streptococcus pneumoniae, S. pyogenes
  Treponema carateum, T. pallidum, T. pertenue

Table 3 (continued)

Vibrio cholerae, V. fluvialis, V. mimicus, V. parahaemolyticus, V. vulnificus
Yersinia enterocolitica, Y. pseudotuberculosis

Level 3

Bacillus anthracis
Brucella (all species)
Francisella tularensis
Mycobacterium africanum, M. bovis (except BCG), *M. tuberculosis*
Pseudomonas mallei, P. pseudomallei
Salmonella typhi, S. paratyphi
Yersinia pestis

3. Fungi[a]

4. Parasites[a]

---

[a] Names of viruses, *Chlamidia, Rickettsia,* fungi, and parasites are omitted.

In using Level 4 organisms, very strict physical containment, sterilization, and control of individuals working with them are necessary, whereas the handling of Level 1 organisms can be carried out in an ordinary microbiology laboratory. Biological safety cabinets recommended by CDC/NIH are as follows:

The Class I biological safety cabinet is an open-fronted, negative-pressure, ventilated cabinet with a minimum inward face velocity at the work opening of at least 75 feet per minute. The exhaust air from the cabinet is filtered by a high efficiency particulate air (HEPA) filter. This cabinet may be used in three operational modes: with a full-width open front, with an installed front closure panel not equipped with gloves, and with an installed front closure panel equipped with arm-length rubber gloves.

The Class II vertical laminar-low biological cabinet is an open-fronted, ventilated cabinet also with an average inward face velocity at the work opening of at least 75 ft/min. This cabinet provides a HEPA-filtered, recirculated mass airflow within the work space. The exhaust air from the cabinet is also filtered by HEPA filters.

The Class III cabinet is a totally enclosed ventilated cabinet of gas-tight construction. Operations within the Class III cabinet are conducted through attached rubber gloves. When in use, the Class III cabinet is maintained under negative air pressure of at least 0.5 inches water gauge. Supply air is drawn into the cabinet through HEPA filters. The cabinet exhaust air is filtered by two HEPA filters, installed in series, before discharged outside of the facility. The exhaust fan for the Class III cabinet is generally separate from the exhaust fans of the facility's ventilation system.

Sterilization or disinfection is necessary for the discharge or reuse of biohazardous materials. As a rule, any material used in connection with a pathogenic organism, such as culture media, apparatus, or infected animals, must be sterilized or disinfected at the site of the experiment. Sterilization means the elimination of all viable microbes including bacterial spores, while disinfection generally refers to the use of a somewhat milder method to destroy the potential infectivity of a material and does not necessarily imply the elimination of all viable microbes. Sterilization and disinfection can be done by either a physical or a chemical method. Physical methods include heat treatment, radiation (γ-ray, X-ray, or ultraviolet rays), and filtration, while the chemical methods are treatment with gaseous, liquid, or solid reagent. Although treatment with ethylene oxide gas is effective in sterilization, most chemical methods are effective for disinfection but not for sterilization. Immoderate use of chemical disinfectants may cause disorders in the biological wastewater treatment system or the environmental ecosystem. Ethylene oxide gas, which is an effective sterilizing agent against dry materials, has no effect on waste in which there is moisture. Ultraviolet irradiation is effective only to disinfect the surface of a material, and the use of ionizing radiation is restricted. Therefore, heat treatment, such as incineration or autoclaving, is recommended to treat biohazardous laboratory wastes. Incineration and autoclaving are perfect methods of sterilization. However, overconfidence of those methods may themselves cause a biohazard. The temperature distribution in an incinerator is not uniform, and overload may leave unburned residue.[6,7] In autoclaving, the flowing steam is allowed to displace the inside air before building up pressure. In steam mixed with air, the temperature is determined by the partial pressure of water vapor. Thus, saturated steam is necessary for perfect sterilization. In this case also, overload may cause the effect to be less than complete because of the insufficient displacement of air. Although biohazardous laboratory wastes should, as a rule, be sterilized, the proper method of disinfection for each individual organism is available if the sensitivity of the organism to a disinfectant is well known. The reader should refer to Chapter 7.5 of this book on the sterilization and disinfection of biohazardous wastes.

In developed countries, classical contagious diseases decreased after World War II, but opportunistic infections to compromised hosts have become a serious problem. A compromised host refers to a person who is susceptible to various pathogens because of some underlying disease, such as a malignant tumor, immunodeficiency, liver or kidney dysfunction, superinfection due to overuse of antibiotics, immaturity, or aging. Even a weak pathogen, or one which would be nonpathogenic to a healthy individual, can cause severe symptoms in a compromised host. A patient suffering from AIDS is a compromised host due to the infection of HIV; this infection creates an acquired immunodeficiency in which pneumonia due to *Pneumocystis carinii,* mycosis, or another opportunistic infection is easily established. Many individuals work and study in academic laboratories; these include young and aged persons, and some of them may have underlying diseases. Various types of microorganisms are cultured on a large scale in laboratories. It should be recognized that even the most common of these can cause severe infection in a susceptible person.

1.3.3 BIOHAZARDS IN RECOMBINANT DNA TECHNIQUE

At the early stage of recombinant DNA research, the probability of a hazardous effect of unexpected recombinants to humans and the ecosystem of the natural environment was not well elucidated. Because of anxiety about creating an unexpected hazardous transformant, guidelines or rules regulating recombinant DNA experiments were legislated by the ministries or departments concerned. With growing knowledge about recombinant DNA technology, it has been demonstrated that the probability to create such transformants is extremely low. In 1987,[8] the National Research Council of the U.S. published "Introduction of Recombinant DNA-Engineered Organisms into the Environment: Key Issue". The basic point in assessing the hazards in genetic engineering is that the risk of recombinant DNA technology should be evaluated by the degree of risk of genes being studied and not by the technology itself. Thus, with time, genetic engineering has tended to become more liberated, although regulations remain necessary. Revision of the rules to ease the restrictions is proceeding, although the basic policy prescribing containment of the recombinant DNA organisms remains unchanged. In academic institutions of Japan, research with recombinant DNA technology must be performed according to the "Guidelines for Recombinant DNA Experiments in Monbusho" (Ministry of Education, Science and Culture of Japan), which was issued by the Ministry. The guide was first established in 1979 and has been revised several times, most recently in 1991.[9] The purpose of the publication is to prevent the escape of transformants bearing artificial recombinant DNA into the environment.

There are two methods of containment: physical and biological. Biological containment refers to the use of specified host-vector systems (B1 and B2) constructed with combinations of host-dependent vectors with ecologically infirm hosts which are able to grow in culture medium when sufficient nutrients are present, but not in a natural environment. Hosts in the B2 system are more infirm in the environment. Physical containment by which engineered or hazardous organisms are confined in a closed system is the required method to safeguard against biohazard. Types of physical containment (P1 to P4) are the same as those used in preventing infectious disease. Containment at the P1 level can be carried out in laboratories which have no special equipment but are approved to conduct recombinant DNA experiments. P2 level laboratories must have a safety cabinet and an autoclave. A P3 area is an isolated room equipped with a ventilation-filter system, a safety cabinet, an autoclave, and other pieces. P4 is the most strictly restricted system, and many pieces of specialized equipment are needed.

At the beginning of a recombinant DNA experiment, an outline of the experimental plan including the biological and physical containments must be presented to the responsible university committee. In the 1991 revised Monbusho guidelines, the recombinant DNA experiments are divided into three categories: (1) those requiring the permission of the Ministry of Education, Science and Culture, (2) those requiring permission of the head of the institution (the president of the university), and (3) those requiring only notification to the head of the institution. Experiments (Category 2) are examined by a committee of the institution. Depending on the subject matter, the Ministry of Education, Science and Culture may be notified to examine the outline (Category 1). Category 1 includes experiments involving the use of an uncertified host-vector system, DNA from newly recognized procaryotic and lower eucaryotic pathogens or viral pathogens, or genes of a protein toxin for vertebrates. The stipulated procedure is necessary for an experiment on recombination between biospecies in which gene transformation cannot take place in the natural environment, but not for one on naturally transformable genes. Although these restrictions are critical for experiments on recombinants having unidentified heterogeneous DNA fragments, an experiment on the cultivation of transformant with identified DNA can be carried out with a reduced level of physical containment. Thus, the rules for recombinant DNA experiments are established, although there are still some problems. The methods for containment and sterilization or disinfection to prevent the escape of recombinants into the environment are essentially the same as those for preventing biohazards due to pathogenic organisms.

Restrictions on recombinant DNA experiments are being somewhat relaxed. It was initially impossible to introduce recombinant DNA into the environment. As shown above, however, the U.S. National Research Council has established a basic policy allowing the application of genetically engineered organisms to agricultural fields. The policy is entitled ''Introduction of Recombinant DNA Engineered Organisms into the Environment: Key Issue, 1987[8]'' and was formulated by the Committee on the Introduction of Genetically Engineered Organisms into the Environment. The committee concluded that the recombinant DNA technique is a useful and safe method for the improvement of biospecies and approved the introduction of genetically engineered organisms into the environment subject to various restrictions. Such applications have already been started in various places. This is the tendency at this time, but strict surveillance should be made.

1.3.4 BIOHAZARDS DUE TO EXPERIMENTAL ANIMALS

Biohazards caused by experimental animals include: (1) infection from an animal artificially infected to study an infectious disease and (2) infection from an animal used in a general experiment and which may potentially carry zoonotic pathogens. It is very important to keep infected animals caged because they are a moving reservoir. More strict control is necessary when animals have been artificially infected, because they harbor a large number of the injected pathogens. Equipment commensurable with the degree of pathogenic hazard (Tables 1 and 2) must be used. A safety cabinet or a specified area with an HEPA filter from the exhaust is necessary for the hazardous experiments.

Generally, infectious experiments are carried out on the particular disease by specialists on the particular disease who have enough awareness to take the precautions needed to prevent an accident or injury. On the other hand, we are defenseless against infections from laboratory animals which naturally harbor zoonotic pathogens. In 1967, an outbreak of a viral disease carried by African green monkeys imported from Uganda to West Germany occurred. There were 20, 5, and 1 patients in Marburg, Frankfurt, and Belgrad (Yugoslavia), respectively, and 7 of them died. The disease was called "Marburg disease".[10] The tracing survey showed that the monkeys were not a natural reservoir for the virus, and it was suspected that they were infected from an unknown wild animal along their transport route. Marburg disease was designated as an international contagious disease, but we have had no case since that first outbreak in 1967. The possibilities of similar unexpected zoonosis from laboratory animals is still not unlikely, however, so caution is necessary in using imported animals.

Hemorrhagic fever with renal syndrome (HFRS, formerly called epidemic hemorrhagic fever or Korean hemorrhagic fever) is one example of a biohazard due to laboratory rats in Japan.[11,12] HFRS was originally known as an endemic disease carried by a field rat (*Apodemus*) on the Korean Peninsula and the northeastern part of China. The pathogen was called Hantaan virus after the place it was isolated, the Hantaan River. It is well known that several thousand soldiers of the Allied forces suffered from the disease during the Korean War.[11] In Japan, an endemic carried by water rats was reported in Osaka in the 1960s.[13] An infection from laboratory rats was revealed in the 1970s, and from 1970 to 1984, 124 HFRS cases of laboratory infection, including one fatal case, were reported.[12] In 1978 and 1981, 33 and 31 cases were reported, respectively. Ministry of Education, Science and Culture of Japan organized a project team to watch HFRS infection in academic laboratories in 1978, and we have had no laboratory cases in recent years. But continuous caution is still needed because the infected rat has no obvious symptoms; the only means of identifying the infection in rats is an immunological diagnosis of the animal's serum.[14] The symptoms in humans are fever with chill and shivering, cenesthopathia, hyposthenia, anorexia, nausea, diarrhea, and other discomfort.

In addition to the above examples, lymphocytic choriomeningitis in the mouse and hamster, brucellosis in dog, and infection with tubercle or dysentery bacilli in monkey are well-known zoonosis from laboratory animals; there are also others too numerous to mention, but they include psittacosis, leptospirosis, and anthrax. Thus, laboratory animals occasionally harbor zoonotic pathogens, but they are often healthy carriers without visible symptoms. There is no way to recognize the danger from the appearance of the animal in such cases. Infections can be avoided only if laboratory animals are properly handled at all times. First of all, we should remember that direct contact with the blood of an animal, especially through a wound in the skin, is the most hazardous, because blood is the reservoir for pathogens, not only in morbid animals but also in healthy ones. Animal experiments often involve the taking of blood or survey on the animal, and the blood and organs are sometimes treated without precaution against potential biohazards. A puncture wound made by a sharp used on the animal or handling of the blood or organs with bare (especially injured) hands should be avoided. The HFRS cases referred to above are suspected to have been caused by the improper handling of rat blood. Precautions should also be taken in the following: suction of aerosol, invasion of insects into laboratory, treatment of carcasses and excretions, against bites and scratches, as well as use of a mask, gloves, and proper laboratory clothing.

REFERENCES

1. Pike, R.M., Laboratory-associated infections: incidence, fatalities, causes, and prevention, *Annu. Rev. Microbiol.*, 33, 41, 1979.
2. The Japanese Society for Bacteriology, Guidelines for biosafety on distribution of pathogens, *Jpn. J. Bacteriol.*, 45, 865, 1990.
3. WHO, Laboratory Biosafety Manual, World Health Organization, Geneva, 1983.
4. CDC/NIH: Biosafety in Microbiological and Biomedical Laboratories, The Centers for Disease Control/National Institute of Health, U.S. Government Printing Office, Washington, D.C., 1988.
5. National Institute of Health of Japan, Rules for Biosafety Control of Pathogenic Organisms, 1992.
6. Peterson, M.L. and Stuzenberger, F.J., Microbiological evaluation of incinerator operation, *Appl. Env. Microbiol.*, 18, 8, 1969.

7. Shinoda, S., Miyoshi, S., Yamanaka, H., Takatsuki, H., and Sakai, S., Dynamics of microorganisms in heat-treatment of infectious waste, *Med. Wastes Res.*, 2, 3, 1989.

8. Committee on the Introduction of Genetically Engineered Organisms into the Environment, Introduction of Recombinant DNA-Engineered Organisms into the Environment: Key Issues, National Research Council, Washington, D.C., 1987.

9. Guidelines for Recombinant DNA Experiments in Monbusho, Ministry of Education, Science and Culture, Japan, 1991.

10. Smith, C.E.G., Simpson, D.I.H., Bowen, E.T.W., and Zoltnik, I., Fetal human disease from uervet monkeys, *Lancet,* 2, 1119, 1967.

11. Lee, H.W., Korean hemorrhagic fever, *Prog. Med. Virol.*, 28, 96, 1982.

12. Kawamata, J., Recent outbreaks of hemorrhagic fever with renal syndrome (Korean hemorrhagic fever) in Japan and its countermeasure, *J. Med. Technol.,* 26, 279, 1982 (in Japanese).

13. Tamura, M., Studies on epidemic hemorrhagic fever, *Jpn. J. Infect. Dis.,* 40, 286, 1966 (in Japanese).

14. Morita, C., Sugiyama, K., Matsuura, Y., Kitamura, T., Komatsu, T., Akao, Y., Jitsukawa, W., and Sakakibara, H., Detection of antibody against hemorrhagic fever with renal syndrome (HFRS) virus in sera of house rats captured in port areas of Japan, *Jpn. J. Med. Sci. Biol.,* 36, 55, 1983.

CHAPTER **1.4**

Hazardous Wastes in Research and Education

Junko Nakanishi and Takashi Korenaga

TABLE OF CONTENTS

1.4.1 INTRODUCTION

Wastes from university laboratories are of great concern as they include toxic chemicals and conventional pollutants. Also of great concern are new

0-8493-682-5/94/$0.00 + $.50

33

chemicals which are not yet regulated and have characteristics in wastes which are not known to even the researchers because the universities often play roles of technology pioneers. The wastes on campus are distinguished by a variety of constituents, some of which are hazardous or potentially hazardous.

Wastes from universities would degrade water quality and endanger aquatic life and human beings if they were directly discharged to the environment carelessly. Even if the universities were indirect dischargers, the careless handling of wastes would corrode sewer equipments, generate hazardous gases, cause treatment plant malfunction, and make sludge disposal harder.

Therefore, the wastes on campus must be handled very carefully with or without the service of a sewerage system. The overall effluent treatment system, such as biological treatment or chemical sedimentation treatment of all the effluents, is effective to remove conventional pollutants, but ineffective to eliminate hazardous chemicals. Therefore, we must provide a system to control campus wastes in which all the users of the laboratory are involved in handling wastes at source. The complete treatment at source, however, is not economical and imposes heavy burdens on the users. The system adopted at the University of Tokyo and Okayama University is a meeting point of treatment at source and central treatment. Here, handling of hazardous wastes in the system is introduced, taking mainly the University of Tokyo as an example.

1.4.2 DEFINITION OF HAZARDOUS WASTE

The term "hazardous waste" refers to solid and liquid wastes which contain hazardous chemicals. Hazardous chemicals include not only those regulated by state laws or prefectural ordinances, but also those which are not yet confirmed to be safe.

The term hazardous waste, however, does not include radioactive waste, which is managed by an isotope center on most campuses and is shipped to and treated at special facilities in the private sector outside urban areas. Pathogenic and infectious medical waste is also shipped and treated off-campus. Biohazardous waste relating to recombinant DNA experiments is pasteurized by researchers themselves and then segregated into hazardous or non-hazardous waste. Test animal carcasses are shipped to private handlers.

At the University of Tokyo, the chemicals are divided into eight groups from the perspective of handling, as follows:

1. Mercury, six valent chromium, trichloroethylene, and tetrachloroethylene
2. Chemicals that are identified as toxic to human beings and the ecology or are flammable
3. Chemicals that are regulated by the Sewerage Law except those in 1 and 2 (Table 1)
4. Chemicals with unavailable toxicity information
5. Chemicals that are suspected to be toxic
6. Refractory chemicals to biodegradation
7. Volatile chemicals
8. Biodegradable chemicals

Table 1 Maximum Concentrations of Chemicals Discharged to the Sewer

Chemical	Allowable conc. (mg/l)
Biochemical oxygen demand (BOD)	600
Suspended solids (SS)	600
Iodine consumption value	220
Oils	30
Phenols	5
Fluoride (F^-)	15
Cyanide (CN^-)	1
Arsenic (As)	0.5
Total mercury (total-Hg)	0.005
Alkyl mercury (alkyl-Hg)	Not detected
Polychlorinated biphenyl (PCB)	0.003
Organophosphorus compounds (organic-P)	1
Copper (Cu)	3
Cadmium (Cd)	0.1
Soluble iron (Fe)	10
Soluble manganese	10
Lead (Pb)	1
Zinc (Zn)	5
Chromium (total-Cr)	3
Hexavalent chromium (Cr^{6+})	0.5
Trichloroethylene	0.3
Tetrachloroethylene	0.1

The higher the group of the above eight in which the chemicals are classified, the more carefully they must be handled. The chemicals in the eighth group are allowed to be dumped into the sewer by diluting their concentrations to less than 1%. The other chemicals are hazardous in nature.

When toxic chemicals are used, the beaker solution is considered hazardous waste and must be disposed of in a special plastic container. Naturally, the solution residue in the beaker is hazardous waste. Hence, the first and second water rinses are also hazardous and are disposed of in the same container. Water in subsequent rinses is not considered hazardous and can be disposed of in the sink (sewer). The maximum allowable concentrations (ppm) of chemicals for sewer discharge are shown in Table 1.

The treatment flow of hazardous wastes has been discussed in detail.[1] At the University of Tokyo, we have two small incinerators for liquid and solid wastes and one distinctive treatment system for inorganic waste called the ferrite process. Ferrite is magnetite (Fe_3O_4) and contains ferric oxide and ferrous oxide incorporated into one crystalline structure. It is referred to as the spinel structure. Ferrite is dumped at a landfill site after having been subjected to a leachate test. Table 2 shows the maximum legal concentrations in parts per million (ppm) of chemicals from sludge leachate. As indicated,

Table 2 Maximum Concentrations of Chemicals from Ferrite Sludge Leachate

Chemical	Allowable conc. (mg/l)
Total mercury (total-Hg)	0.005
Alkyl mercury (alkyl-Hg)	Not detected
Cadmium (Cd)	0.1
Lead (Pb)	1
Organophosphorus compounds (organic-P)	1
Hexavalent chromium (Cr^{6+})	0.5
Arsenic (As)	0.5
Cyanide (CN^-)	1
Polychlorinated biphenyl (PCB)	0.003
Trichloroethylene	0.3
Tetrachloroethylene	0.1

Table 3 Hazardous Waste Data at the ESC, University of Tokyo

• Number of university students	~21,000
• Number of university workers	~8,000 (incl. 2,000 professors)
• Number of ESC workers	16 (incl. 3 professors)
• Total ESC budget (1991)	~$1,500,000 (U.S.)
• Chemical waste handled at the ESC	
Inorganic	10 m³
Organic	110 m³
Washwater of the exhaust gas	80 m³
Solid	1.2 tons
Unused chemicals	~6000 bottles

only 11 items listed in Table 2 are designated "hazardous" according to Japanese law. This is in contrast to the U.S. where as many as 700 chemicals are so designated.

The Environmental Science Center (ESC) at the University of Tokyo is in charge of hazardous waste management and recycling of waste (see Table 3). The ESC is a complex consisting of an administrative office, a research laboratory, and a small mill (plant). The center has four major objectives:

1. Management and treatment of waste, especially hazardous waste
2. Study of risk management of chemicals and treatment technology of toxic chemicals
3. Education of students and workers in the management of chemicals in compliance with rules of environmental protection and prevention of accidents, such as explosions, in the laboratory

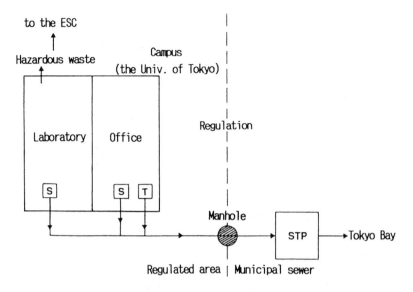

Figure 1. Sewer System at the University of Tokyo. STP: municipal sewage treatment plant, S: sink, T; toilet.

4. Proposal of policies or measures for implementation by the committee or the president of the university aimed at protecting the environment and human health; recent work in this category includes a proposal concerning the recycling of paper and cans

The first objective can be examined in greater detail with respect to the following two procedures: pick-up and treatment of hazardous waste from all laboratories and medical centers on campus, and monitoring of the sewage and air quality in laboratories.

1.4.3 SEWER SYSTEM ON CAMPUS

Figure 1 is a sketch of the sewer system on the campus of the University of Tokyo.[1] The wastewater from laboratories and sanitary sewage feeds into a common sewer, where they are mixed. The water quality at sewage manholes within campus limits is regulated by the Sewerage Law. It is very difficult to check and control the laboratory wastewater in this system; however, the system is beneficial for us because laboratory waste is diluted by sanitary sewage.

As shown in Table 1, the maximum concentrations (ppm) of chemicals discharged into the sewer are not critical. Because our waste is discharged

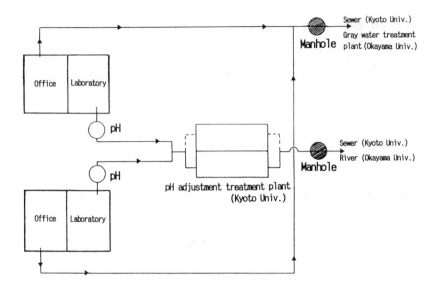

Figure 2. Sewer system at Kyoto University and Okayama University.

into the sewer, the standard biochemical oxygen demand (BOD) is 600 mg/l. If we were to discharge sewage directly into the river, the standard BOD would be only 20 or 10 mg/l. With respect to BOD and suspended solids (SS), we are thus subject to less stringent standards than using direct discharge. With regard to hazardous chemicals, such as mercury, cadmium, chromium, etc., however, those discharging indirectly and directly are subject to the same standards.

In the past, less stringent standards were imposed on indirect dischargers. However, in the 1970s, the great amounts of toxic compounds discharged by industry into the sewer system made the sewage treatment plant a serious source of pollution. Nakanishi's paper[2] which revealed this fact initiated a nationwide citizens' movement against the construction of sewage treatment plants as described. Therefore, in 1976, the government completely rewrote the Sewerage Law and set the same standards for indirect dischargers as for direct dischargers.

While the sewer system at the University of Tokyo requires substantial improvement, that at Kyoto University is excellent (Figure 2). An excellent system was also constructed at Okayama University over a 3-year period which was completed in 1984 at a cost of 8 million U.S. dollars.[3-5] Kyoto University and Okayama University previously had old systems identical to that of the University of Tokyo; however, these often did not comply with the standards, and as a result, students and teachers alike expressed their outrage to the university administration. Thus, the sewer system on campus was reconstructed so that laboratory waste and sanitary waste were separated, and pH autosensors were installed in all buildings with laboratories.

If the waste flowing through violates pH regulations, an alarm sounds inside the building and a speaker announces the violation, requesting that no further water be disposed. At the terminus of the laboratory sewer system, there are two tanks; laboratory wastewater usually flows in and out of the treatment tank and the other is empty. When a warning is given, the passage to the tank currently in use is closed and wastewater is guided to the empty one. Discharge from the campus is then stopped. The wastewater is checked, and if no problem exists, it is discharged. If the water quality is substandard, chemical coagulants are added to the tank for treatment.

1.4.4 WASTE SORTING AND RECYCLING ON CAMPUS

With the system described above, everyone engaged in laboratory work or medical activity at the University of Tokyo is required to sort their waste into 13 categories by chemical according to the guidelines of the ESC.[1,6] In addition, they must place the waste in specific containers which are distinguished by color and size. These plastic containers are picked up at specified collection sites once a week by workers at the ESC. Every department with laboratories or medical facilities has one or two collection sites.

In addition to experimental hazardous wastes, old and unused chemicals and medicine are collected upon request. Inorganic chemicals containing metals are sent to mining industries or treated with the ferrite process on campus; organic chemicals are incinerated at the facilities of the ESC or off-campus.

The first author (J.N.) will review recent operations from the perspective of recycling at the University of Tokyo. The ferrite process for hazardous inorganic waste treatment was adopted about 15 years ago because we expected that the final product, ferrite, would be in great demand. Magnetite, in its raw form, is now in common use in Japan. Many studies have been conducted on the usability of ferrite produced in the ESC, but unfortunately none has been successful; the final quality is greatly inferior to that of virgin magnetite. Thus, we were forced to dispense with plans to reuse the waste ferrite, which is now landfilled.

Currently, the focus is on recovery of selenide and fluorine from waste. Many kinds of organic fluorine compounds are used on campus. One of our colleagues is studying the technology required to treat and recover fluorine using titanium metal. Although he has not yet completed the study, one company recently announced that it had succeeded in recovering fluorine from the waste of semiconductor factories.

Twice a year, used dry batteries and fluorescent light tubes are also collected. They are sent, together with mercury-contaminated sludge and the resin used for mercury adsorption, to a former mercury mining company in Hokkaido, where mercury is recovered from the waste. Its purity is 99.9%, and it can be reused as pure mercury. Zinc is also recovered, but at a much lower purity level. We are currently grappling with the problem of collection of dry batteries. The current system is to collect and ship them to Hokkaido, because mercury is considered to be environmentally unsafe. In reality, mercury gas is reported to be discharged into the air from garbage incinerators. In response to this problem, and after concerted efforts, Japanese companies have begun to sell mercury-free, dry batteries. We are now discussing whether or not we should collect and recover metals from the mercury-free, dry batteries.

The University of Tokyo has recently begun collecting used paper and cans on the basis of a proposal by the ESC. Only part of the campus participates in this program at present, but within a couple of months, the entire campus will be involved. There is very advanced technology in Japan used in the production of recycled pulp from used paper. Twenty million tons of paper is produced nationwide every year, half of which is made from virgin pulp; the remaining half comes from recycled pulp. Recently, the government set the goal of a paper recycling ratio of 60% by the year 2000.

On the basis of recycling practices on part of the campus, we estimate that the amount of garbage can be reduced by 60% by recycling. The prospects are good for paper recycling; unfortunately, however, the outlook for recycling of steel cans is less optimistic.

As in the case of dry batteries, we have highly advanced technologies for recovering raw metals, though support of the social system in instigating a recycling program is lacking. Methods such as taxation, legislating deposits, and acknowledgment of responsibility on the producer's part are all taboo in Japan. Industries are dead set against these measures.

1.4.5 PROBLEM OF WASTE IN THE COMING BIOTECHNOLOGY AGE

Recognizing the breadth of past environmental damage and with concern for the far-reaching impacts on human health of chemicals introduced into commerce, the U.S. Congress passed the Toxic Substances Control Act which mandates the regulation of chemical substances, new or existing, that present an undesirable risk of injury to health or the environment.

Biotechnology is a new, rapidly growing industry with tremendous potential. The near future, from a decade to half a century, has been predicted to be the age of biotechnology. The future growth of biotechnology is almost a certainty, but the impacts of biotechnology on the environment and the means to mitigate them are still largely unknown. The deliberate or accidental release of biologically engineered organisms into the environment, and the subsequent exposure of humans and the environment to such organisms, could prevent the healthy growth of biotechnology and its possible positive contribution to the human good. Predictive capabilities are needed to foresee the release of and exposure to these organisms so that the risks can be estimated and the benefits can be assessed. The efficacy of cost-effective technology to contain or to destroy biotechnology wastes also needs to be evaluated to provide alternative approaches for risk management to the appropriate agency of the government.

In a series on global environmental education, we often tell students that ''Japan is a poor country. The only raw materials we have, the only resources we can utilize, are waste and our brain. You must develop technology to recover the elements from waste using your brain.'' However, we cannot control the structure of technology for recycling solely by creating the technology; the support of the social system is necessary for recycling to be successful and advantageous.

At the environment centers of universities, however, we are devoting concerted efforts to developing technologies to treat and recover chemicals. At the same time, we wish to propose a good social system on campus to support recycling, which can act as a model for the entire country and greater society.

ACKNOWLEDGMENT

This chapter has been rewritten in parts with permission from Water Report, Vol. 2, No. 2, P.1-5 (1992): Copyright 1992 by MYU, Tokyo.

REFERENCES

1. Nakanishi, J., The management and recycling of hazardous waste in Japan and at the University of Tokyo, *Water Rep.*, 2(2), 1, 1992.
2. Nakanishi, J., Some problems on the joint treatment of wastes and sewage in the UKIMA treatment plant, *Water Res.*, 7, 375, 1973.
3. *Instruction Manual for Laboratory and Ordinary Wastewater, in Center for Environmental Science and Technology,* Okayama University, 1985 (in Japanese).
4. *A Guidebook for Administration Center for Environmental Science and Technology,* Okayama University, 1985.
5. *Guidelines for Foreign Students: Campus Standards for Wastewater Disposal,* Okayama University, 1988.
6. Committee of Environment and Safety, *Guidelines for Environment and Safety,* Environmental Science Center, University of Tokyo, 1991 (in Japanese).

**Section
2**

Inventory Control and Waste Minimization

CHAPTER 2.1

Construction of an Inventory Control System

Satoru Kaseno

TABLE OF CONTENTS

0-8493-682-5/94/$0.00 + $.50

2.1.1 INTRODUCTION

Many types of chemical reagents are purchased and used for education, experiments, or research in academic research laboratories of an institution or university. These reagents are used in relatively small quantities, quite in contrast to amounts used in large-scale operations such as manufacturing.

In many instances, some reagents purchased for an experiment are left over and are stored in the laboratory as surplus or disposed of as chemical waste. Most of this surplus, except for general reagents such as acid or alkaline, is seldom used for other experiments and remains stored in the laboratory for a long time.

The fundamental principle of waste management should be waste abatement. An inventory control system is very important in assuring that leftover reagents are used up and should be set up in each academic research laboratory. This system would see that an adequate amount of a reagent was purchased for a planned experiment and would see to the use or the recycling of surplus reagents.

This chapter will begin by providing actual chemical waste management strategies and alternatives for academic research laboratories. The inventory system practiced by Okayama University will then be described.

2.1.2 CHEMICAL WASTE MANAGEMENT IN LABORATORIES

2.1.2.1 Chemical Waste Generation in Academic Research Laboratories

Shirasuka[1] randomly selected 1000 professors from among the 10,000 professors of Japanese national universities and institutions in fields of natural science (chemistry, biology, physics, geography, or medicine), and they were sent questionnaires concerning the handling and treatment of waste in their laboratories in 1983. From the 470 answers received, he totalled the amount of chemical reagents that had been treated in these laboratories during the year 1982 (Table 1).

"Reagents stored (I)" in Table 1 refers to the quantity in weight of reagents stored in these laboratories at the end of 1981. "Reagents purchased (II)" is the weight of reagents purchased by these laboratories during 1982. "Reagents held (III)" is the weight of those held at the end of 1982. Weight of "reagents used (IV)" is calculated by $I + II - III$. "Reagents disposed (V)" refers to those disposed of as waste by mixing with water or another solvent, changed to other matter, or reagents unwanted. The total weight of each category in Table 1 represents the sum of all questionnaire answers. As stated, the number of professors sent the questionnaire was about 10% of all those in natural science, and about 50% of those responded. The actual total weight in all academic research laboratories of Japanese national universities and institutions is therefore estimated to be about 20 times of the total weight shown in the table. Unit weight in each category is the average calculated by dividing the total weight by the number of answering professors.

Table 1 provides information on how reagents are diluted when they are disposed. Calculation of the dilution ratio, X_D, is made by dividing the "reagent disposed" by "reagents used": (V/IV). X_D is 13.8 for Category A, 14.9 for B, 6.7 for C, and 1.4 for D. Inorganic reagents purchased by the laboratories thus increase weight by 6 to 15 times when it is disposed of as chemical waste. For organic reagents, however, there is little weight change between purchased and disposed of states.

What portion of purchased reagents is actually used in the laboratories? Dividing the weight of "reagents used" by the weight of "reagents stored" and "reagents purchased" ($III/(I + II)$) would give the utilization ratio, X_U. X_U is 0.418 for Category A, 0.662 for B, 0.701 for C, and 0.817 for D. For inorganic reagents containing metals (A), the utilization ratio is low, and the weight of stored reagents is large. Because they do not easily change in quality, they tend to be stored for long periods; there is, therefore, a possibility that labels may deteriorate or become illegible on old containers. If this occurs, the nature of the reagent is unknown, it then becomes unwanted and is disposed of. The utilization ratio of inorganic reagents like alkaline or acid (C) and organic reagents (D), in contrast, is high, because these materials change in quality when stored for long periods, so there is a tendency to avoid purchasing too large an amount. Stored organic reagents are limited to those which are universally available solvents, so that little of these would fall in an unwanted category.

2.1.2.2 Chemical Waste Management in Academic Research Laboratories

All reagents purchased in academic research laboratories will finally be disposed of as waste. Experiments generate little in the way of a product,

Table 1 Total and Unit Weight of Reagents Treated in Natural Science Laboratories in Japanese National Universities or Institutions

Category	I. Reagents Stored	II. Reagents Purchased	III. Reagents Being Held	IV. Reagents Used	V. Reagents Disposed
(A) Inorganic reagents, heavy or other metals					
Total weight (kg/year)	1563	691	1200	1054	14550
Unit weight (kg/year/lab)	3.290	1.455	2.527	2.219	30.63
(B) Inorganic reagents, cyanogen, fluorine, or phosphorus anion					
Total weight (kg/year)	591	845	485	951	14200
Unit weight (kg/year/lab)	1.244	1.780	1.021	2.003	29.88
(C) Inorganic reagents, alkaline or acid					
Total weight (kg/year)	3713	5148	2653	6206	41660
Unit weight (kg/year/lab)	7.817	10.83	5.585	13.07	87.67
(D) Organic reagents					
Total weight (kg/year)	18600	49570	12490	55700	79170
Unit weight (kg/year/lab)	39.19	104.4	26.29	117.3	166.7

Waste Abatement	Do not make it
Waste Minimization	If it has to be made, one should minimize its volume and toxicity
Waste Reuse	See if someone else can use it
Waste Recycle	If it cannot be used as is, reclaim as much as possible that is useful
Waste Treatment	Treat what cannot be reclaimed to render it safe
Waste Disposal	Dispose of residue to air, water or land

Figure 1. Waste management hierarchy in the Ontario Waste Exchange.

and a reagent used would be changed in quality and quantity and become a waste material. Surplus reagents generated in academic research laboratories also become waste unless they are reused and recycled. Thus, the effective use of reagents is the key point in chemical waste management in laobratories. The establishment of an inventory control system in each laboratory is very important, and each university or institute should take the responsibility to see that this is implemented.

2.1.3 ALTERNATIVES FOR CHEMICAL WASTE MANAGEMENT IN LABORATORIES

2.1.3.1 Chemical Waste Management Hierarchy

L. Varangu and R. Laughlin[2] drew up the hierarchical principles for the Ontario Waste Exchange Program in Ontario, Canada (Figure 1). The management of chemical waste in academic research laboratories in Japan should follow these principles.

Although all reagents in laboratories eventually become waste, it is uneconomical if there are surplus reagents which languish, become unwanted, and are disposed of. The treatment and close control of reagents is an important aspect of the management of chemical waste.

Each laboratory should purchase only the amount of a reagent that it actually needs and will use. This is the first step in cutting down on waste. It is recognized, however, that reagents are sold in containers of set quantities, and this can make it difficult to purchase just the weight needed. A chemical reagent shop system supplying various laboratories in the university or institute would be one resolution to this problem.

Before beginning a project, experiments should be planned so that waste is kept to a minimum. Consideration should be given to using reagents safely, effectively, and economically in experiments; waste should be minimized. At the same time, reagents should be properly stored to assure there is no change in quality.

If a reagent is found to be surplus in a laboratory, a possible user in another laboratory should be sought before the reagent is put into the category of unwanted reagent. This type of exchange is a good procedure to avoid waste and to use surplus reagents. A good management policy is needed for this sort of exchange to implement a reagent control system.

If another user cannot be found for a surplus reagent so that the reagent becomes unwanted, another procedure must be sought, such as recycling.

When excess of a reagent cannot be avoided, it should be treated and disposed of properly, as done for products and by-products generated in an experiment.

As stated in Section 2.1.2.2, an inventory control system is an essential part of each laboratory's waste management system. Examples of inventory control systems will be introduced in the next section.

2.1.3.2 Inventory Control System

2.1.3.2.1 Effective Use System of Chemical Reagent

Though waste reduction is basic in inventory control, generation of a certain amount of reagents cannot be avoided. How should these surplus reagents be treated? Some examples of their effective use are as follows:

1. We can reuse surplus reagents in experiments or chemical analysis in other projects in our own laboratory. We can also use them for reagents of waste treatment.
2. Exchanges of reagents between other laboratories is another means. This will be explained later in detail.
3. Recycling is also a good way to handle excess supplies.

The distillation of an organic reagent that has undergone quality deterioration is one reclaiming procedure.[3] Some inorganic reagents which have changed have a quality that can be reclaimed by a combination of chemical reactions, such as neutralization, oxidation or reduction, precipitation, or complexation, and by physical operations, such as filtration, centrifugal separation, recrystallization, distillation, and evaporation.[4]

Inorganic reagents containing heavy metals can be scoured and recycled at a metal scouring plant, although a balance between the weight of reagents generated in academic research laboratories and the weight that can be accepted by a plant is a key factor in the feasibility of this process. Another very important feature in any such procedure is the necessity for a safe and dependable system of transport for chemical reagents from the research site to the recycling location.[5]

2.1.3.2.2 *Chemical Reagent Shop System*

Even though, ideally, only the minimum amount of reagent necessary for a given experiment should be purchased so that no surplus will remain, materials sold only in specified quantities sometimes make this impossible. This is where a chemical reagent shop can be particularly useful.[6] Such a system allows bulk purchasing, maintenance of an inventory, storage, and a record of use by various laboratories at the university or institution where the shop is located. Supplies of reagents can be ordered to the shop and received in the exact quantity needed by each laboratory. Details of this system are explained in Chapter 2.3 by H. Ojima.

2.1.3.2.3 *Chemical Reagent Exchange System*

Reagents considered surplus by a laboratory are often in usable form, many times in an unopened state, and certainly appropriate for other laboratories. Possible users should be sought and the reagents offered to them. The reagent exchange system is intended to further such exchanges between laboratories, thus assuring the reuse of assets and reducing the expense and disposal hazards, and making unnecessary a laboratory's retention of stock of a rare and possibly irreplaceable reagent.[2]

On the basis of safety and economy, the reagent exchange system should be constructed within each university or institute; however, a small institute could join with others nearby in setting up a common system.

The following section describes the construction of a reagent exchange system within Okayama University.

2.1.4 CONSTRUCTION OF A CHEMICAL REAGENT EXCHANGE SYSTEM

2.1.4.1 Purpose of Reagent Exchange System

The reagent exchange system in Okayama University (RESOU) was constructed in 1991 for the purpose of encouraging the internal exchange of surplus reagents between research laboratories. Prior to its establishment, the university had a water quality control rule which defined the options for the disposal of such material: the laboratory or individual who was responsible for the waste had to see to its proper treatment and disposal. This principle still prevails in RESOU, in both treatment and internal exchanges of reagents which are excess in a laboratory. At the time RESOU came into being, the

Administration Center for Environmental Science and Technology (ACE) was responsible for matters involving water on the Okayama University campus, and this center was active in the smooth introduction of the new system and its role.

RESOU functions in the following manner: a research laboratory presents information on surplus reagents generated to ACE. ACE then collects information from other laboratories on campus and creates a data base on surplus reagents. This information is then supplied to each laboratory. If a laboratory sees a reagent it needs on this supplied list, it advises ACE which arranges for exchange from the laboratory in possession of the reagent. Actual responsibility for the exchange falls to the donor and recipient laboratories involved, and the action is reported to ACE, which then updates its database.

2.1.4.2 Procedure in Reagent Exchange System

The numerical order in Figure 2 shows the reagent exchange procedure in RESOU, together with various information transfers.

1. When too much reagent has been ordered, there has been a change in a planned project or personnel, or a general cleaning out of inactive reagents has been done, surplus reagents may result. In this case, the laboratory must send prompt and accurate information on this to ACE. If the reagent is fresh, and detailed information on it is available, it is easy to find a new user in another laboratory. The person responsible for surplus reagents in the laboratory must be identified and must store these reagents carefully and safely. The advice to ACE is done by a specific document, on which information necessary for the data base can be indicated.
2. ACE collects information on surplus reagents from all university laboratories and inputs the information to a personal computer. The created database is then copied onto floppy disks which are made available to each laboratory.
3. From this disk, each laboratory can learn of surplus reagents available throughout the university. New floppy disks are periodically modified and distributed.
4. A laboratory user can research the reagents available on the surplus list through a personal computer and the floppy disk.
5. When a needed reagent is found on the list, the would-be user indicates his desire to ACE by submission of the stipulated document.
6. Upon receipt of the document from the requesting laboratory, ACE conveys it to the responsible person in the laboratory holding the reagent to the document. The document indicates the reagent desired, the name and telephone number of the requesting laboratory, and the date it is desired.
7. Upon receipt of the document in the donating laboratory, the responsible person contacts the requesting laboratory in writing or by telephone. The responsible persons in the two laboratories then arrange the detail of the transfer: confirming the packaging of the reagent and its weight, and deciding the place, date, time, and manner of the transfer of the reagent.

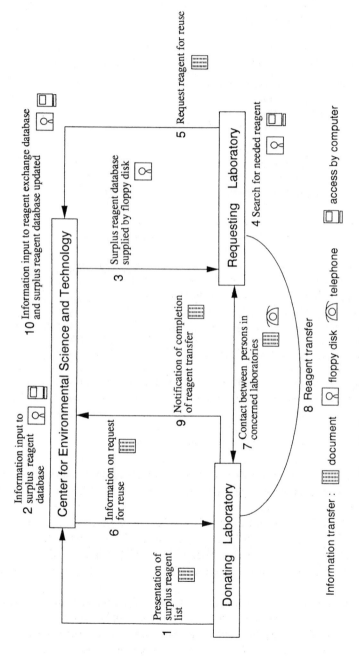

Figure 2. Reagent exchange system in Okayama University.

1. Name of Responsible Person in Donating Laboratory
2. Telephone Number of Responsible Person
3. Faculty Code Number (see Figure 5)
4. Department Name
5. Laboratory Name
6. Reagent Name
7. Reagent Category
   a) Inorganic Chemical   b) Organic Chemical   c) Other
8. Name of Reagent Manufacturer
9. Reagent Keywords
10. Reagent Grade
    a) GR(Guaranteed)   b) EP(Extra Pure)   c) TG(Technical Grade)
    d) Unknown
11. Reagent Volume or Weight
12. Reagent State
    a) Liquid   b) Solid
13. Opened or Unopened
14. Year Purchased

Figure 3. Information for surplus reagent database.

8. Proper execution of the transfer is the responsibility of the representatives of the two laboratories. The transfer should be conducted in a location which is well ventilated and designed to minimize fire and explosion hazards. All safety precautions should be taken in the transport procedure.
9. Following the transfer, the responsible individual in the donating laboratory immediately notifies ACE of its completion on the appropriate document.
10. Following receipt of this document, ACE inputs the information into the reagent exchange database by personal computer access and makes the necessary change to the surplus reagent database. The new database is copied onto a new floppy disk and this new information is sent to all laboratories.

2.1.4.3 Database for Reagent Exchange System

Thus there are two databases in RESOU: a surplus reagent database (Figure 3) and a reagent exchange database (Figure 4). Figure 5 shows the faculty code numbers used in these databases.

The databases are regarded as primary information, and the number of items used in them has been kept to a minimum. Actual reagent transfer is to be executed between the two laboratories, which share the ultimate responsibility. These databases are merely the procedural steps to be followed in the exchange.

The minimization of database subjects saves labor for the input and modification by ACE so that new information can be supplied quickly.

Keywords of the surplus reagent database show the element and radical of a chemical. The reagent keywords are only for inorganic chemicals.

1. Name of Responsible Person in Requesting Laboratory (RL)
2. Telephone Number of Responsible Person (RL)
3. Faculty Code Number (RL)   (see Figure 5)
4. Department Name (RL)
5. Laboratory Name (RL)
6. Name of Transfer   Reagent
7. Volume or Weight of Transfer   Reagent
8. Date of Reagent Transfer
9. Name of Responsible Person in Donating Laboratory (DL)
10. Telephone Number of Responsible Person (DL)
11. Faculty Code Number (DL)   (see Figure 5)
12. Department Name (DL)
13. Laboratory Name (DL)

Figure 4. Information for reagent exchange database.

1. Faculty of Teacher Education
2. Faculty of Science
3. Medical School
4. University Hospital Attached to Medical School
5. Dental School
6. University Hospital Attached to Dental School
7. Faculty of Pharmaceutical Sciences
8. Faculty of Engineering
9. Faculty of Agriculture
10. College of General Education
11. Graduate School of Natural Science and Technology
12. Institute for Agricultural and Biological Sciences
13. Junior College of Medical Science
14. Institute for Study of the Earth's Interior
15. Administration Center for Environmental Science and Technology
16. Others

Figure 5. Faculty code number.

These databases are accessible by personal computer. A popular access system and widely used software were selected, and the computer operation and research were quickly and easily done.

In accessing the surplus reagent database, a reagent can be sought by inputting the researched reagent name or its keyword (element or radical). A list of all surplus reagents can be seen on the CRT of the computer and can, of course, also be printed out by printer. Examples of CRT on access of the database are shown in Figure 6.

2.1.4.4 Future Action on Inventory Control System

As of 1992, 2000 reagents were listed in the surplus reagent database of RESOU. These, however, are only part of those on hand at Okayama

```
------------------------- SURPLUS REAGENT DATABASE -------------------------

RESPONSIBLE PERSON : Yoshinari Miura      TELEPHONE NUMBER : 529

FACULTY CODE : 8

DEPARTMENT : Applied Chemistry      LABORATORY : Inorganic Materials Science

CATEGORY : a    MANUFACTURER : Wako

REAGENT : Cadmium Oxide

KEYWORDS : 1  Cd        2  Oxide

GRADE : a    VOLUME or WEIGHT : 300    [ ml ]

STATE : b    OPENED ( Yes or No ) : Y    YEAR PURCHASED : 1990

                                             Record No. 714/2144
```

A

```
--------------------- REAGENT EXCHANGE DATABASE ---------------------

***** REQUESTING LABORATORY *****
RESPONSIBLE PERSON : Masami Inaba      TELEPHONE NUMBER : 8094

FACULTY CODE : 8

DEPARTMENT : Applied Chemistry      LABORATORY : Polymerization Chemistry

***** REAGENT *****
NAME : Mercury(II) Sulfate      VOLUME or WEIGHT : 20      [ g ]

DATE OF REAGENT TRANSFER : 09 / 28 / 92

***** DONATING LABORATORY *****
RESPONSIBLE PERSON : Yusuke Wataya      TELEPHONE NUMBER : 996

FACULTY CODE : 7

DEPARTMENT : Pharmaceutical Science    LABORATORY : Pharmaceutical Chemistry
                                             Record No. 138/435
```

B

Figure 6. Examples of CRT on access of database. (A) Information input to surplus reagent database, (B) information input to reagent exchange database, (C) research using reagent name, (D) research using keyword.

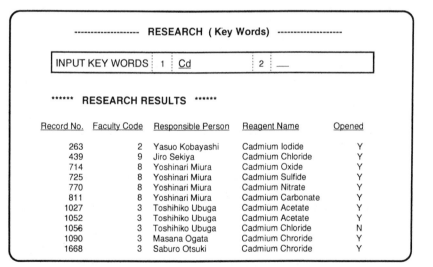

-------------------- RESEARCH (Reagent Name) --------------------

INPUT REAGENT NAME : Mercury Suifide

****** **RESEARCH RESULTS** ******

Record No.	Faculty Code	Responsible Person	Opened	Grade	Volume or Weight
290	7	Yusuke Wataya	Y	a	20 [g]
385	1	Yasuo Nakao	Y	b	10 [g]
432	9	Jiro Sekiya	Y	b	150 [g]
435	9	Jiro Sekiya	N	b	200 [g]
873	8	Ryuji Koga	Y	a	25 [g]
886	8	Sadao Tsuboi	Y	a	20 [g]
959	8	Seiki Saito	Y	b	25 [g]
1002	4	Yasuo Hori	Y	b	30 [g]
1003	4	Yasuo Hori	N	b	50 [g]
1083	3	Masana Ogata	Y	a	200 [g]
1194	10	Tadashi Iwachido	Y	b	40 [g]

C

-------------------- RESEARCH (Key Words) --------------------

INPUT KEY WORDS : 1 : Cd   : 2 : __

****** **RESEARCH RESULTS** ******

Record No.	Faculty Code	Responsible Person	Reagent Name	Opened
263	2	Yasuo Kobayashi	Cadmium Iodide	Y
439	9	Jiro Sekiya	Cadmium Chloride	Y
714	8	Yoshinari Miura	Cadmium Oxide	Y
725	8	Yoshinari Miura	Cadmium Sulfide	Y
770	8	Yoshinari Miura	Cadmium Nitrate	Y
811	8	Yoshinari Miura	Cadmium Carbonate	Y
1027	3	Toshihiko Ubuga	Cadmium Acetate	Y
1052	3	Toshihiko Ubuga	Cadmium Acetate	Y
1056	3	Toshihiko Ubuga	Cadmium Chloride	N
1090	3	Masana Ogata	Cadmium Chroride	Y
1668	3	Saburo Otsuki	Cadmium Chroride	Y

D

Figure 6 (continued).

University. We hope to enlarge the RESOU network to include all academic research laboratories of the university, and to do this, we must make students, teachers, and staff of the university aware of chemical waste management and promote the use of RESOU to learn of reagents which are surplus and readily available on the campus.

"RESOU" actually means ideal in Japanese, but there are still many problems requiring resolution in the system. First, RESOU must be improved so that it supplies new information more quickly and effectively. Also, linking of an internal surplus reagent exchange system like RESOU with a chemical reagent shop system would permit development of a new inventory control system.

2.1.5 CONCLUSION

Many types of chemical reagents are used by academic research laboratories engaged in teaching, research, or testing. Because these reagents, sooner or later, become chemical waste, their management is a very important part of chemical waste management. To abolish and minimize chemical waste, establishment of an inventory control system is required. We have introduced some examples of such a system here. Each university or institute should adopt the most practicable system among those available, considering safety and economics and based on the principles of waste management hierarchy.

An inventory control system cannot be effective without cooperation of all individuals handling reagents in laboratories. The importance of such a control system must, therefore, be well understood by all levels concerned.

REFERENCES

1. Shirasuka, K., Realities of disposal, in *Handbook for Laboratory Waste Treatment Facilities of Universities,* Japanese Association for Laboratory Waste Treatment Facilities of Universities, Ed., Nishinihon Hoki Syuppan, Okayama, 1988, chap. 2.
2. Varangu, L. and Laughlin, R., Chemical waste reduction and recycling in Canadian academic laboratories, in *Waste Disposal in Academic Institutions,* Kaufman, J. A., Ed., Lewis Publisher, Chelsea, MI, 1990, chap. 12.
3. Moriwake, T., Distillation for reuse of chlorocarbon solvents, *Annu. Rep. Jpn. Assoc. Lab. Waste Treatment Facilities of Universities,* 4, 76, 1987.
4. Ishii, D., Recycle of inorganic reagents waste, *Annu. Rep. Jpn. Assoc. Lab. Waste Treatment Facilities of Universities,* 4, 18, 1987.
5. Nakanishi, J., Recycle and reuse of unwanted chemicals, in *Report of Studies on Reuse of Waste in Academic Laboratories,* Higashimura, T., Ed., Kyoto University Press, Kyoto, Japan, 1992, chap. 6.
6. Koyama, T. and Katsura, T., Management policy of Waseda University Environmental Safety Center, *Annu. Rep. Jpn. Assoc. Lab. Waste Treatment Facilities of Universities,* 2, 8, 1985.

CHAPTER **2.2**

Disposal of Unused Chemicals

Yoshimitsu Suzuki

TABLE OF CONTENTS

0-8493-682-5/94/$0.00 + $.50
© 1994 by CRC Press, Inc.

2.2.1 INTRODUCTION

Disposal of unused chemicals is generally difficult. In many cases, unused chemicals to be disposed are not properly identified and are present in very small quantities. It is important to confirm the name of a chemical substance beforehand, as poisonous or explosive substances may exist in unused chemicals.

Presently, equipment for the disposal of undifferentiated refuse does not exist. This is similar to that of unused chemicals. Unused chemicals are disposed by various methods, and the selection of the treatment process is based on whatever equipment is available at each university.

The Environmental Science Center has been treating unused chemicals discharged from the University of Tokyo since 1978 and the treatment system for unused chemicals from the University of Tokyo is presented in this chapter.

2.2.2 THE COLLECTION SYSTEM

It is important to verify the identity of reagents in a collection of unused chemicals. The waste generator (a qualified trained chemist) is in charge of confirming the name of the reagent and the unused chemicals.

Figure 1 shows the outline of a collection system for unused chemicals.

1. First, the waste generator prepares a list of unused chemicals. At this stage, the weight and identity of each unused chemical are confirmed. A number for the reagent is assigned and listed.
2. The waste generator submits the list of unused chemicals to the Environmental Science Center (the Center).
3. The person in charge at the Center makes arrangements with the waste generator about the collection schedule of unused chemicals.
4. The person in charge at the Center comes to the laboratory of the waste generator and collects unused chemicals. In this collection work, the person in charge confirms the contents of unused chemicals.
5. The person in charge at the Center prepares a new list of the unused chemicals which have been collected.
6. This new list is sent from the Center to the waste generator and to the office of the person in charge of the department.
7. The department pays the disposal cost to the Center by way of the executive office.

We believe that this collection system has been successful to date. However, the occurrence of an unknown reagent is a problem which should not be ignored.

The analysis of unknown reagents is not an easy task, and this is the responsibility of the waste generator. The Center is teaching qualitative analytical methods or introducing private agencies to the waste generator.

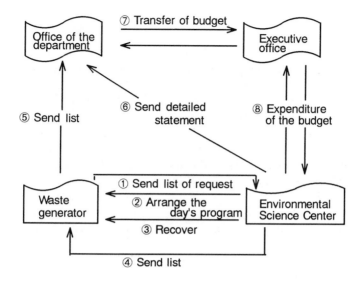

Figure 1. The collection system of unused chemicals for the University of Tokyo.

Unused chemicals are usually kept for a long period of time, and the labels eventually peel off. Thus, we must discharge unused chemicals as soon as possible. Most importantly, the occurrence of unused chemicals should be minimized or eliminated, if possible.

2.2.3 THE CLASSIFICATION AND STORAGE OF UNUSED CHEMICALS

Unused chemicals are classified under the dangerous substances by the Fire Services Act and others; Figure 2 shows the classification of unused chemicals at the Center. The dangerous substances of the Fire Services Act are classified into six classes:

1. Oxidative solids
2. Combustible solids
3. Self-igniting substances and substances which are not allowed to come into contact with water
4. Flammable liquids
5. Self-reactive substances
6. Oxidative liquids

Reagents, which are not classified under the Fire Services Act, are classified in accordance with a disposal method of the Center. These reagents include controlled substances under the Poisonous and Deleterious Substances Control Law and Law on Industrial Safety and Hygiene. In case of very

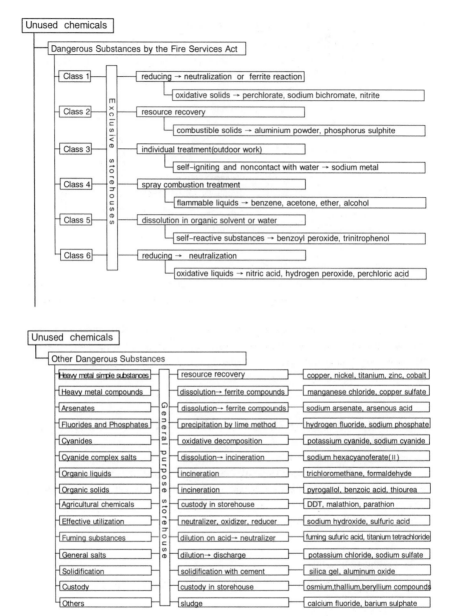

Figure 2. Classification of unused chemicals at the Environmental Science Center.

strongly poisonous and very dangerous substances, attention is noted directly on each reagent bottle. A new label is attached to the reagent when there is the possibility that a name disappears.

The Center has two storehouses built exclusively for unused chemicals. One of the storehouses is used for controlled substances classified under the

Fire Services Act, and the other is used for substances outside of classification. At any rate, custody of unused chemicals is temporary until their disposal. The arrangement of unused chemicals in the storehouse is profoundly connected with disposal plans. Management is difficult due to frequent coming and going of many reagents. Moreover, reagents which require custody of low temperatures are kept in an icebox.

2.2.4 TREATMENT PROCESS OF UNUSED CHEMICALS

The treatment of unused chemicals is basically done at a treatment plant of the university. A treatment process of unused chemicals is introduced below.

2.2.4.1 Organic Liquid Reagents

Treatment of combustible organic liquids is incineration disposal by means of the spray combustion plant. Inflammable liquids are treated by decomposition of inner heat in the incinerator. These two types of liquids are sprayed into the incinerator from different atomizers.

Maximum treatment capacity is 70 l/h for each liquid. Temperature within the incinerator is kept at about 900°C, where the exhaust gas that passes through a cooling tower, scrubbing tower, and mercuric adsorber is analyzed for SO_2, NO_x, HC, HCl, and CO concentrations by an exhaust gas monitor at the last stage and is finally discharged to the atmosphere from a stack.

Combustible and inflammable reagents are mixed exclusively in different pretreatment tanks. This mixture is transferred to the incinerator by means of a pipeline from the service tank. Precipitation occurs upon mixing of reagents; this is filtered off, and the filtrate is transported to each service tank. However, organic fluoride and organic phosphorus compounds should be subjected to a secondary treatment that treats wastewater from the wet gas scrubber as inorganic waste liquid. The compounds are rinsed by an alkali from a scrubbing tower of the gaseous chlorine, which is emitted by burning of chlorine compounds. If too much gaseous chlorine passes through a scrubbing tower at one time, rinsing will be inadequate and acid gas will be discharged to the atmosphere. It is necessary to control the amount of incineration of organic chlorine compounds to prevent the occurrence of too much gaseous chlorine in the incinerator. In actual practice, organic chlorine compounds are diluted by combustible organic solvents, such as benzene and toluene, and are disposed as combustible liquids.

The Center requires treatment conditions of the treatment plant to control chlorine concentration to less than 10%.

2.2.4.2 Organic Solid Reagents

Treatment of organic solid reagents is incineration disposal by an incinerator used exclusively for solid waste. Maximum treatment capacity is 30 kg/h for dead animals. Combustion temperature can be maintained at about 900°C. However, treatment capacity has dropped for reactive chemical substances.

Unused chemicals are divided according to inflammability or combustibility. The reagent taken from a reagent bottle is divided into small polyethylene pouches. These pouches are then individually tossed into the solid incinerator. However, this method is very dangerous for reactive substances. Reactive chemical substances (e.g., benzoyl peroxide and trinitro compounds) are initially diluted with either water or combustible organic solvent. As example of this is 0.5% solution of benzoyl peroxide in acetone, wherein the reactivity of benzoyl peroxide is lost, and it is disposed of as a general combustible waste liquid.

Many reactive chemical substances have been safely disposed of with this method.

2.2.4.3 Inorganic Liquid Reagents

Compared to organic substances, treatment of inorganic substances is more complicated. The treatment plant of inorganic waste liquid of the Center is classified into five kinds: mercury use, fluoride use, phosphate use, cyanide use, and heavy metal use. Pretreatment differs with treatment of general waste liquid.

Sulfuric acid and hydrochloric acid are discharged into public sewerage after neutralization by sodium hydroxide. Oxidizing and reducing agents that have no virulent properties are discharged into a sewer after oxidation-reduction treatment. Cyanide compounds are rarely discharged as a liquid, as they are, more often than not, discharged as a solid. The oxidative decomposition of cyanide is done by the addition of hypochlorite.

Hydrofluoric acid and phosphoric acid, respectively, are used as precipitating agents for obtaining calcium fluoride and calcium phosphate upon reaction with calcium salt. Potassium permanganate and sulfuric acid are added to mercurial compounds, and after oxidative decomposition, mercuric ion is adsorbed by means of chelate resin.

Reagents of heavy metals are treated by the ferrite method. Fuming sulfuric acid, fuming nitric acid, or sulfuric anhydride is treated by a similar method with sulfuric acid, after it is diluted by means of sulfuric acid, nitric acid, or others.

Fuming titanium tetrachloride is introduced into a fluorine treatment tank together with its reagent bottle which is wrapped in a basket of stainless steel after being cooled by dry ice. This reagent reacts strongly with water in the treatment tank, and as a result, the bottle within the basket is broken.

2.2.4.4 Inorganic Solid Reagents

We will look into a treatment method for cyanide. Cyanide is dissolved in water and is treated by an oxidative decomposition method. It is important to adjust the initial concentration of this solution to less than 1000 mg/l. At a concentration higher than this, a reactive intermediate, known as CNCl gas, is generated which is extremely dangerous. The cyanide ion concentration in the treated water should be less than 1 mg/l.

Solid cyanide can be disposed of by incineration. In cases where treatment is done by this method, measurement of cyanide content in the bottom ash is necessary. If decomposition of cyanide is insufficient, the bottom ash is reprocessed. However, the complete oxidative decomposition of a stable cyanide complex salt is impossible due to the presence of hypochlorite. The complex is directly charged into an incinerator and is subjected to pyrolysis by combustion. In this case, a second combustion of bottom ash is necessary. A stable cyanide complex salt can be subjected to incineration disposal as a liquid waste after being dissolved in water by means of atomized firing equipment.

Fluoride and phosphate, after being dissolved in water, react with calcium salt and are disposed of as precipitates of calcium fluoride and calcium phosphate, respectively. The liquid formed during precipitation is removed by means of filtration equipment. An elution test is performed on the precipitate. The precipitate which passes the elution test is disposed of by a land disposal. Fluoride and phosphate ion concentrations in the treated water should be less than 15 mg/l and less than 48 mg, respectively.

After both oxidizing agents and reducing agents are dissolved in water, an oxidation-reduction treatment is done. Treated water is discharged to a sewer after pH adjustment. Noxious substances are also subjected to incineration disposal by a solid incinerator.

Conditions of combustion for agricultural chemicals are confirmed for the purpose of treatment at a laboratory beforehand. When the conditions of combustion cannot be confirmed, treatment is suspended.

Carcinogenic and mutant substances are disposed by a solid incinerator. When conditions for proper incineration cannot be confirmed, such as those for agricultural chemicals, incineration disposal is cancelled. Confirmation of the reactivity of these substances is prior to incineration.

Unused chemicals decompose in spite of careful custody, and reactivity of matter changes. We should not rely on past experiences of incineration. The discharge amount of solid reagents to an incinerator differs depending on structure and size of the incinerator.

Heavy metal reagents are disposed by means of a ferrite method after dissolution in water or acid. At present, the reliability of the ferrite method is the highest as the treatment process of waste liquid which contains many kinds of heavy metal ions.

2.2.4.5 Mercury Reagents

In principle, disposal of mercury is not done in the university. Mercury reagents are turned over to a private agency, together with fluorescent lamps

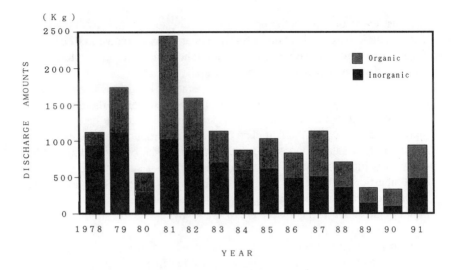

Figure 3. A change of discharge amounts of unused chemicals.

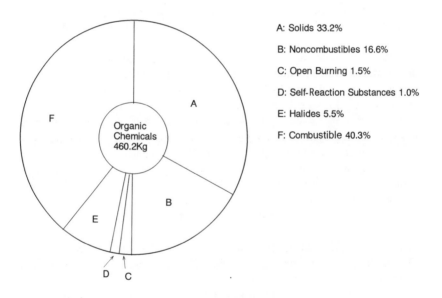

A: Solids 33.2%

B: Noncombustibles 16.6%

C: Open Burning 1.5%

D: Self-Reaction Substances 1.0%

E: Halides 5.5%

F: Combustible 40.3%

Figure 4. The rate of discharge of organic unused chemicals (1991).

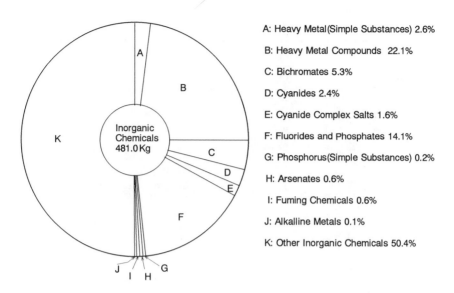

A: Heavy Metal(Simple Substances) 2.6%

B: Heavy Metal Compounds 22.1%

C: Bichromates 5.3%

D: Cyanides 2.4%

E: Cyanide Complex Salts 1.6%

F: Fluorides and Phosphates 14.1%

G: Phosphorus(Simple Substances) 0.2%

H: Arsenates 0.6%

I: Fuming Chemicals 0.6%

J: Alkalline Metals 0.1%

K: Other Inorganic Chemicals 50.4%

Figure 5. The rate of discharge of inorganic unused chemicals (1991).

and dry cells. The sole company in Japan dealing with the final disposal of mercury waste exists in Hokkaido.

2.2.5 DISCHARGE AMOUNTS FOR UNUSED CHEMICALS

Figure 3 shows a change of discharge amounts of unused chemicals. The total amount of discharge for 14 years was 14,821 kg. The ratio is organic reagents 55%, inorganic reagents 45%. Collections done in 1989 and 1990 were performed on a small scale due to shortage of storehouses.

Figures 4 and 5 show the rate of discharge of unused chemicals. As can be seen from these figures, unused chemicals of many kinds are discharged at various rates. Table 1 shows discharge amounts of unused chemicals of each department. It shows that unused chemicals are discharged from many departments.

Table 1 Discharge Amounts of Unused Chemicals of Each Department

Departments	1978	1979	1980	1981	1982	1983	1984	1985	1986	1987	1988	1989–90	1991	Total (kg)
Law	0	0	2.5	0	0	0	0	0	0	0	0	0	2.5	2.5
Medicine	2.5	167.1	0	161.8	157.5	109.3	20.3	403.0	0	0.1	0	0.6	231.7	1253.9
Hospital	263.0	469.7	46.8	144.7	247.3	274.9	258.3	167.4	13.4	571.6	278.2	142.3	271.5	3248.9
Engineering	589.4	293.5	46.5	443.2	190.5	328.2	11.1	201.9	162.8	166.7	153.0	232.8	139.3	2959.0
Letters	0	0.1	0	0	0	12.1	0	0	0	0	0	39.0	0	51.2
Science	6.2	180.4	55.8	25.0	54.2	93.8	54.6	14.7	73.5	157.0	50.6	105.1	77.2	948.1
Agriculture	0	469.6	52.8	258.7	91.4	71.5	7.0	0	37.2	0	0	0.1	11.9	990.2
College of Art and Science	6.4	0	3.0	0	24.7	0	23.3	26.3	0	9.0	0	0	77.2	169.9
Education	3.2	0	0	0	0	0	389.7	0	0	0	40.0	2.1	0	435.0
Pharmaceutical	1.0	0	3.3	28.3	0.6	11.1	0.6	7.5	33.9	84.9	1.6	71.2	0	244.0
Institute of Medical Science	198.3	0	126.7	384.0	13.9	0	13.0	170.4	169.2	12.7	2.2	0	37.0	1127.4
Earthquake Research	0	0	0	0	9.1	0	0	0	0	0	0.8	18.4	0	28.3
Institute of Applied Microbiology	0	0	0	17.0	5.3	8.5	34.8	0	16.2	2.9	5.4	20.5	0	110.6
Institute for Cosmic Ray Research	5.6	0	0	0	17.4	0	0	0	0	0	0	0	0	23.0

														Total
Institute for Nuclear Study	45.8	0	0	54.3	163.0	0	4.5	17.1	0	0	0	0	0	284.7
Institute for Solid-State Physics	0	0	0	43.6	0	0	0	0.2	202.9	0	59.1	8.9	29.6	344.3
Ocean Research Institute	0	3.5	0	0	0.6	0	0	0	0	0	9.5	0	0	37.1
Library	0	0	0	0	0	0	2.4	0	0	0	0	0	0	2.4
Health Service Center	0	9.7	68.8	0	17.9	0	0	6.4	0	0	4.8	0	8.7	116.3
Cryogenic Center	0	0	0	0	3.3	0	0	0	0	0	0	0	0	3.3
Museum	0	0	0	0	0	0	0	0	0	133.7	0	25.0	0	158.7
Research Center for Science and Technology	—	—	—	—	—	—	—	—	—	0	0	0	0.5	0.5
Bureau of Facilities	0	0	37.8	13.5	0	0	0	0	0	0	0	0	0	51.3
Bureau of Students	0	0	0	65.5	0	0	19.1	0	0	0	0	0	0	84.6
Institute of Cosmic Science Research	0.3	114.7	96.3	27.7	170.0	43.7	0.2	0	0	0	0	0	0	452.9
Total	1123.9	1744.4	557.8	2450.3	1593.5	1136.4	872.0	1027.0	836.2	1138.6	709.1	690.2	941.2	14820.6

CHAPTER 2.3

Chemical Reagent Shop System for Minimizing Waste

Hiroyuki Ojima

TABLE OF CONTENTS

0-8493-682-5/94/$0.00 + $.50

© 1994 by CRC Press, Inc.

2.3.1 INTRODUCTION

The social conditions centering on the waste issue are approaching a turning point characterized by such new trends as "streamlining and recycling", "preservation of resources", and "environmental protection".

The time has come to undertake full-scale implementation of the policy of replacing the idea of merely "rendering wastes harmless and stable" with that of "not producing wastes and not turning things into wastes".

Universities must deal in earnest with these issues, not only to meet their social responsibility as the highest educational institute, but also to fulfill their responsibility as business enterprises which emit large quantities of wastes. It is believed that since they use and discard great quantities of various types of chemicals, universities with experimental facilities, particularly those with departments of science and engineering, need to tackle the issues at hand immediately. However strong the social demand for reduction in the amount of wastes may be, under present circumstances, it is difficult to limit the purchase, use, and disposal of those chemical products that are indispensable to research and education.

The Waseda University (WU) is a case in point. Every year a great deal of laboratory waste is produced, and the amount of chemical waste in an unused state is increasing at a significant pace. Unused chemical waste is clearly unnecessary for research and education, and of all laboratory waste, the first to be targeted in any plan is to reduce the amount of wastes generated by the university.[1] The "Chemical Reagent Shop System" is a system that aims to reduce the amount of unused chemicals by improving the way chemicals are purchased and discarded.

Taking these circumstances into consideration, we will try to explain as concretely as possible the measures adopted — and their effectiveness — to reduce the amount of waste produced and the cost of disposing the waste under the system currently being implemented.

2.3.2 CHEMICAL REAGENT SHOP SYSTEM

2.3.2.1 Incentives for Waste Reduction

WU has a total enrollment of approximately 50,000 students and currently has 2700 faculty and administrative members. There are 6 campuses with a total area of 470,000 m², and among them, the campus of the School of Science and Engineering has been designated to improve the methods of purchase, use, and discard of chemicals.

The School of Science and Engineering in WU, as the oldest school of science and engineering in any of the private universities in Japan, was established in 1908. It consists of 14 departments — Mechanical Engineering, Electrical Engineering, Applied Chemistry, etc. — with an admission quota of about 10,000 students; over 900 members are employed on the teaching staff. Research and educational activities have been conducted in 200 laboratories, common laboratory sections, and academic research centers, within a campus area of 45,000 m².

A wide variety of chemicals had been ordered in each research laboratory and discarded as hazardous waste. For example, in 1984, the school generated about 420 kl of inorganic waste at the estimated concentration of 2000 ppm of metals and 150 kl of organic waste, some of which was over 1000 kg of unused chemical reagent waste. The quantity of the experimental waste generated in the School of Science and Engineering amounts to nearly 90% of waste produced at WU. Moreover, the total amount of chemical waste had been growing remarkably from year to year.

The Environmental Safety Center (ESC) was established in December 1979 to oversee the implementation of practical measures to eliminate (or reduce) environmentally harmful wastes produced in the course of university laboratory experiments.

2.3.2.2 The Environmental Safety Center (ESC)

The main purpose of the ESC is to plan and provide for the environmental safety of the faculty, students, and general public. The management systems in ESC are shown in Figure 1.

The ESC activities include the following: control and disposal of toxic substances, specifically, chemical reagents and organic solvents; environmental analysis for the purpose of ensuring environmental safety; and providing guidance and advice relating to environmental safety policies.[2] The services of the center are available not only to analysis consultations and requests from within the university, but also to the general public.

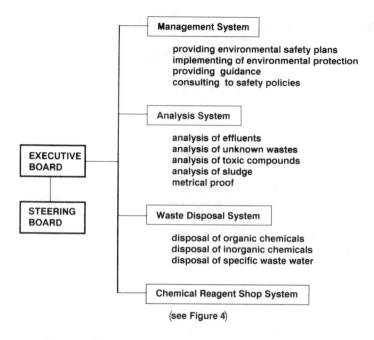

Figure 1. Systems of the Environmental Safety Center (ESC).

2.3.2.3 Waste Disposal System

The university usually disposes of the waste it generates in its experiments by separating it from ordinary waste such as paper trash. There are three ways to dispose of laboratory wastes. (1) Entrust it to a collecting firm that deals in waste disposal exclusively. (2) Undertake intermediate processing of waste by installing equipment for turning waste into harmless matter and entrust the final disposal to a collecting firm. (3) Adopt one of the previous ways, depending on the type of wastes being discarded. In recent years, because of lack of final dumping sites and surging labor costs, the cost in disposing of wastes, regardless of the method adopted, has tended to rise.

At our university, the third method is adopted now; that is, intermediate processing is undertaken for inorganic waste and a collecting firm is entrusted to undertake disposal of organic waste.[3] Figure 2 shows a disposal process for laboratory waste. The ESC distinguishes between inorganic and organic compounds in applying regulation requirements for hazardous waste generators. Table 1 lists disposal methods for laboratory waste, which are intermediately processed in the ESC (Figure 3).

We estimated the cost, under the existing system for disposal of 7 kg of copper sulfate (anhydrous) brought to the center as an unused chemical. Table 2 lists cost comparisons that would be incurred to carry out the disposal treatment. Here, a total cost includes the operation and labor costs; the cost

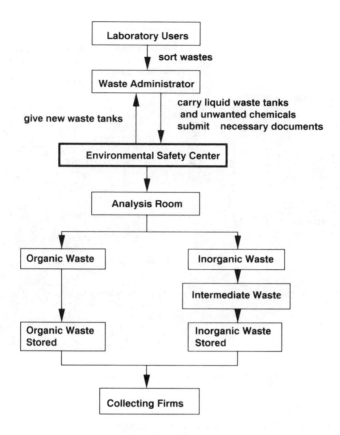

Figure 2. The disposal process for laboratory waste used at Waseda University.

Table 1 Disposal Methods for Chemical Wastes

	Inorganic Waste			Organic Waste
Disposal Method	**Ferrite Treatment**	**Iron Powder Treatment**	**Neutralization**	**Atomized Firing[a]**
Capacity (liters)	700/batch	50/batch	100/batch	20/batch
Disposal time (hours)	5/batch	3/batch	3/batch	Consecutive

[a] This method is not employed now because of surging disposal cost.

of ferrite disposal treatment is estimated at ¥3300 per batch to operate. Consequently, it costs about ¥22,000 to have a company dispose of the entire amount and about ¥31,000 if a ferrite treatment method is used. Thus, the cost of disposing wastes on campus is on the rise. The cost of disposing mercury, arsenic, and other chemical compounds is becoming even more

Figure 3. Installing equipment for waste disposal.

costly, with the cost rising to several tens of thousands yen in some cases. Also, there is a high probability that the range of chemicals designated as harmful will expand because of the Basel Treaty and revisions in the laws concerning the collection and treatment of wastes. Thus, the task of disposing laboratory waste is fraught with difficulties, such as its mounting cost and a host of technical problems that must be overcome.

2.3.2.4 Chemical Workshop Management

2.3.2.4.1 Purchasing Policy

Many of the discarded chemicals that were picked up from the campus were surplus and unneeded. To help researchers reduce the cost of purchasing new chemicals, there must be a minimizing of the purchase of unwanted ones. This will also cut down on disposal costs.

In April 1985, the Chemical Reagent Shop (CRS) was opened, and various chemicals have been handled collectively to improve safety and efficiency within the university. As shown in Figure 4, the major components of the CRS system include improvements in purchasing, taking inventory, storing, and using chemicals. This minimizes surplus agents and derived wastes gen-

Table 2 Cost Comparisons in 1987

Treatment cost conducted at WU

Ferrite disposal treatment	¥3300/batch × 2 batches	= ¥ 6,600
Sludge disposal treatment (off-site)	¥65/kg × 60 kg	= ¥ 3,900
Labor costs (2 persons)	¥10,000/batch × 2 batches	= ¥20,000
	WU subtotal	¥30,500
Commercial disposal cost	¥3200/kg × 7 kg	= ¥22,400
	Total difference in costs	¥ 8,100

Note: ¥120 = $1.

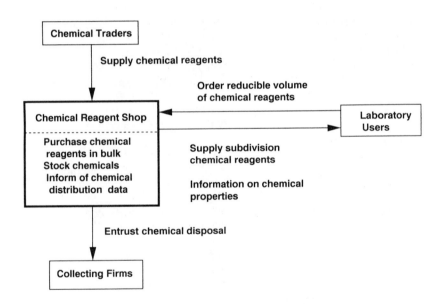

Figure 4. A chemical reagent shop system.

erated from chemical experiments. In other words, minimizing unused reagents in the laboratory helps to encourage reducing the likelihood of chemical wastes. Unfortunately, one barrier to the reduction of surplus chemicals is that it takes a few days before researchers can obtain new chemicals. In order for a university to implement chemical treatment methods without preventing the development of research, chemical reagents have to be purchased in bulk, stocked, and requirements met by all the users. The system is designed and written for users where small volumes of literally hundreds of different chem-

Table 3 Stocked Chemicals

Organic Reagent		Inorganic Reagent	
Code	Substance	Code	Substance
A00010	Acetic acid	B00010	Aluminium sulfate
A00020	Acetone	B00020	Ammonia water
A00030	Acetonitrile	B00030	Ammonium chloride
A00040	Benzene	B00040	Ammonium sulfate
A00050	Carbon tetrachloride	B00050	Calcium chloride (anhydrous)
A00060	Chloroform	B00060	Calcium nitrate
A00070	Cyclohexane	B00070	Copper sulfate
A00081	Dichloromethane	B00081	Distilled water
A00090	Diethyl ether	B00090	Hydrochloric acid
A00100	*N,N* Dimethylformamide	B00100	Hydrofluoric acid
A00110	1,4 Dioxane	B00110	Hydrogen peroxide
A00120	Ethyl alcohol	B00120	Iron (III) chloride
A00130	Ethyl Acetate	B00130	Magnesium carbonate basic
A00140	Ethylene glycol	B00140	Magnesium sulfate
A00150	Glycerin	B00150	Nitric acid
A00160	*n*-Hexane	B00160	Phosphoric acid
A00170	Isopropyl alcohol	B00171	pH standard solution (4.01, 6.86, 9.18)
A00181	Kerosene	B00180	Potassium chloride
A00190	Ligroin	B00190	Potassium hydroxide
A00200	Methyl alcohol	B00200	Potassium iodide
A00210	Methyl ethyl ketone	B00211	Silica gel blue
A00220	Paraffin liquid	B00220	Sodium carbonate (anhydrous)
A00230	Petroleum ether	B00230	Sodium chloride
A00240	Pyridine	B00240	Sodium hydrogen carbonate
A00250	Tetrahydrofuran	B00250	Sodium hydroxide
A00260	Toluene	B00260	Sodium nitrite
A00270	Trichloroethylene	B00270	Sodium sulfate
A00280	Triethanolamine	B00280	Sodium sulfite (anhydrous)
A00290	Xylene	B00290	Sodium tiosulfate
A00300	*n*-Pentane	B00300	Sulfuric acid
A00310	Formalin (formaldehyde)		

ical reagents are the norm, but it is more economical to purchase chemical reagents in bulk if they are interested in any of the stock chemicals.

2.3.2.4.2 *Implementing Chemical Treatment Methods*

Those employed by the university order chemicals necessary for research and education from the CRS. The CRS has about 60 kinds of chemicals in stock, and these are itemized and sold in any quantity desired. Table 3 lists the stocked chemical reagents in the CRS, which includes acids, ketones,

Figure 5. Delivery of chemicals.

alcohols, sulfates, chlorides, nitrites, etc. These chemicals can be subdivided into a relatively small volume of reagents.

The system treatment forms a simple routine that can be carried out by someone in a normal laboratory. The ESC checks the need for chemical reagents, and the shoproom coordinator orders ones of high frequency in bulk. The appropriate quantity of chemical reagents are provided for each laboratory at no charge. Data on the supply of chemicals on the campus are controlled by computer and used for information on the distribution of chemicals by the CRS and the orders of on-campus organizations (see Figures 5, 6, and 7). For a successful production, it is essential to have communication between academic researchers and the ESC.

2.3.2.4.3 Chemical Inventory Service

In addition to reducing the amount of wastes generated, the CRS provides effective guidance on ways to ensure safe handling and storage of chemical products. A wide variety of chemicals are stored, used, and disposed on the campus, some of which, being extremely toxic or highly inflammable or explosive, can cause serious damage to the human body or the natural environment. In other words, since they may trigger a disaster, an accident or an incident where they are stolen and abused is very serious; many of the chemicals available on the campus are potentially very dangerous. Thus, by introducing the CRS system, in addition to reducing unnecessary chemicals

Figure 6. Storage of chemicals.

Figure 7. The on-line system used in the Chemical Reagent Shop.

as much as possible, our university collects accurate data on the flow of chemicals into the campus and uses the data to ensure the safe handling of those chemicals. In other words, by providing information on various aspects of the chemicals involved, including their danger, the university provides guidance on how to handle the chemicals safely at the time of their delivery.

For instance, in case a researcher wishes to purchase a toxic substance for one's experiment, he or she must follow the instructions on the use, storage, and disposal of the substance at the time of its delivery by the campus organization in charge of controlling toxic substances. A sample form of an administration card for poisonous reagents is shown in Figure 8. The CRS supports the safe and sure control of chemicals by providing someone in charge of controlling toxic substances with data on the purchases of toxic substances.

2.3.3 EVALUATION OF THE SHOP SYSTEM

2.3.3.1 Minimizing Unused Chemical Reagent Wastes

Figure 9 shows the amounts of unused chemical reagent in the School of Science and Engineering between 1982 and 1988. We achieved a high level of reduction of unwanted chemicals in laboratories because of the introduction of the CRS system which allowed the users to obtain a better understanding of surplus laboratory chemicals. Options to reduce unused chemicals have been promoted since 1985 with the support of the ESC, and in 1988, an annual quantity rate of wastes had improved 59% for the 3 years.

$$\frac{1350 \text{ kg} - 550 \text{ kg}}{1350 \text{ kg}} \times 100 = 59\%$$

There are three main incentives for considering reduction: saving new chemicals, cleaning out of inactive chemicals, and good storage practices in each laboratory.

If we analyze that 70% of a loss weight of unused chemical wastes should be disposed by ferrite treatment, we can save a total weight of 12,000 kg of sludge (Table 4).

2.3.3.2 Minimizing Waste Formation

Figure 10 shows the amounts of laboratory waste. Unfortunately, there is little fluctuation in the amounts in spite of the achievement of unused agents reduction. There are a number of reasons why laboratory waste reduction has

Substance:	Total Quantity:
Purchase Date:	
Room:	Phone:

Head user's Name:_____

Head administor's Name:_____

User's Data

Date	Use's name	Quantity Used [g]	Quantity Remained [g]	Remarks

Waste Date:
Acknowledgement of Receipt of Materials
Signature:

Figure 8. An administration form sheet for control of poisonous reagents.

not occurred in the school: (1) both qualities and quantities of laboratory waste depend on variations of the content of research every year; (2) current users of chemical reagents are increasing in various departments, such as Mechanical Engineering, Electrical Engineering, and Physical Science and Engineering; (3) there is a need to consider how the ESC can provide technical assistance on waste reduction.

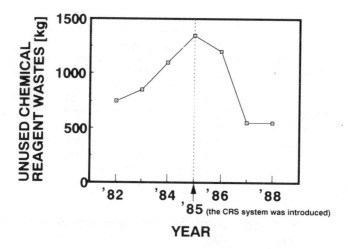

Figure 9. Variations in the amount of unused chemical reagent wastes in the years 1982 to 1988.

Table 4 Estimation of Sludge Saving

Disposal waste weight	800 kg $\times$ 0.7 = 560 kg
Rate of dilution	500
Disposal volumes	560 kg $\times$ 500 = 280,000 l
Ferrite disposal capacity	700 l/batch
Number of batch operation	280,000 l/700 l = 400
Sludge quantitation per batch	30 kg/batch
Total weight of sludge	30 kg/batch $\times$ 400 batches = 12,000 kg

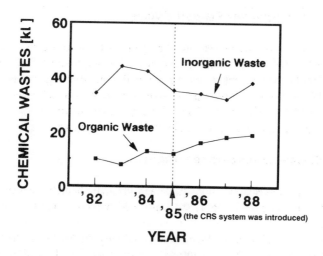

Figure 10. Variations in the amount of chemical wastes from 1982 to 1988.

Table 5 Comparison of Chemical Reagents Prices

Reagents (first class)	Unit Volume of Weight	Chemical Shop Price (¥)	Commercial Price (¥)	WU Net Saving (%)
Acetone	500 ml	135	500	73
Methyl alcohol	500 ml	70	400	82
Carbon tetrachloride	500 g	270	500	46
Hydrochloric acid	500 ml	163	400	59
Sodium chloride	500 g	252	520	51

Note: ¥ 120 = $1(1993).

Table 6 Cost Savings Due to an Implementation of the CRS System

Discount prices	¥ 8,330,000
Savings in purchases	800,000
Savings in disposals	¥ 4,000,000
Total	¥ 13,130,000

Note: ¥ 120 = $1 (1993).

2.3.3.3 Cost Effects

By setting up a centralized management of chemical reagents, two types of cost savings have been accomplished: discounts of reagent prices and savings in disposal costs.

2.3.3.3.1 Discounts of Reagent Prices

Table 5 lists the prices of chemical reagents. Chemical users can purchase reagents 40 to 80% off the commercial prices. If a reagent costs ¥ 12,500,000 ~$104,200 for the stocked chemicals annually under no regulations, and an average discount rate is approximately 40%, the total savings for the treatment program would amount to ¥ 8,330,000 per year.[4] Furthermore, the estimated value of saving new chemicals is about ¥ 800,000 per year.

2.3.3.3.2 Savings in Disposal Costs

Assuming an average cost for disposal of unused chemical reagents would be ¥ 5,000 per kilogram, a total cost saving of disposal is about ¥ 4,000,000 per year.

As mentioned above, the estimation of cost saving due to the introduction of the CRS system results in an annual reduction of ¥ 13.13 million in waste disposal, purchase, and other related costs (Table 6).

2.3.4 FUTURE ACTION FOR CHEMICAL MANAGEMENT SYSTEM

Seven years have elapsed since the CRS system was introduced, and its effectiveness has been fully demonstrated. However, there are still problems that have to be solved, such as reducing the operating costs of the system itself and making further contributions to the implementation of safety policies.

2.3.4.1 Efforts of the University to Maintain the System

As it stands now, while operating the CRS system results in an annual reduction of about ¥ 13 million in waste disposal, purchases, and other related costs, it also adds to operating costs by increasing labor and equipment costs. If these expenses lead to such added values as contributions to social welfare and increased safety by reducing the amount of surplus chemicals used and chemical wastes produced on the campus, operating the CRS system can be said to be fully profitable. However, a greater effort to further reduce operating costs in the future must be made before the large reduction in the cost of chemicals for the university can be regarded as a major motive for operating the CRS system or as a driving force behind the efforts of the university to maintain the system.

A possible solution to the problem of reducing labor costs, which accounts for most of the operating costs, is to have an employee on load to the CRS from the distributor of the chemical products to support part of the distributor's business. A number of legal problems must be cleared before such an arrangement can be introduced. Moreover, support provided by the distributor must be limited to such matters as itemizing and delivering the chemicals, while university personnel will have to endeavor to supply more information on chemical products and provide more useful safety tips.

2.3.4.2 Safety Improvement

With regard to safety policies, the CRS system, which at the present only serves the School of Science and Engineering, should be expanded into a system that serves the entire campus. To control the flow of chemical products on the campus as a whole, the computer software used in the CRS system should be linked to an on-line system. At the same time, a data base, separate from the currently available one on delivery and demand, should be established on the properties of chemicals sold and on disused chemicals. By making this data base available to the public at large, there should be a heightened awareness by those using the chemicals of the importance of safety, thereby securing the safety of chemical products when they are being handled as well as contributing to the reduction of the amount of chemicals used.

2.3.5 CONCLUSIONS

Although up to now we have encountered a number of difficulties in running the CRS system, we have solved them each times by repeatedly educating the user on the basic concepts of the system. We believe that as a result of such steady efforts by the operating staff and the enlightened understanding of the user, the amount of chemicals stored has been reduced substantially, thus greatly increasing the safety of our campus. Here it should not be forgotten that cutting costs is always a secondary effect, that is, a means for building consensus on the campus or an effect needed in adhering to the policy of the safe purchase, use, and disposal of chemicals. Since the CRS system is based on the concept of ensuring campus safety from various angles concerning the handling of chemicals, thereby protecting society and the environment, it should not be overly concerned with policies aimed at reducing operational costs. In the future, the CRS will have to double its efforts to enlighten and guide the user in reducing the amount of chemical wastes generated and in the proper handling of chemical products. At the same time, as an expression of our gratitude for the user's cooperation, we believe further efforts must be made to improve our service and avoid the pitfalls of adopting stereotypical policies.

REFERENCES

1. Ojima, H., Reuse of Unwanted Chemicals in an Academic Laboratory, reprints of the 4th Council of Academic Waste Disposal Facilities, Tokyo, Japan, 1986.
2. *User's Manual of the Environmental Safety Center,* Waseda University, Tokyo, Japan, 1990.
3. *Environment Safety Center News,* Waseda University, Tokyo, Japan, 1991.
4. Annual Reports of Waseda University, Waseda University, Tokyo, Japan, 1991.

**Section
3**

Off-Site Handling and Disposal Technology

CHAPTER 3.1

Laboratory Methods for Disposal of Toxic Inorganic and Organic Chemicals

Margaret-Ann Armour

TABLE OF CONTENTS

0-8493-682-5/94/$0.00 + $.50
© 1994 by CRC Press, Inc.

89

3.1.1 INTRODUCTION

Scientific and engineering research and teaching activities in academic institutions can result in the generation of relatively small quantities of a wide variety of waste and surplus chemicals. Many of these chemicals are hazardous to human health and to the environment and must be disposed of in an appropriate and responsible manner. However, the disposal of small quantities of chemicals can present a difficult and expensive problem.

Of course, it is highly desirable to minimize the waste at source, and several strategies for achieving this have been suggested.[1,2] For example, in many undergraduate chemistry laboratories the scale of the experiments has been drastically reduced;[3,4] in research laboratories, used solvents from operations such as chromatography are being distilled for reuse; where possible, chemicals no longer needed by their original purchaser are being made available to others who can use them, thus reducing the purchase of new supplies.

When the waste volume has been minimized as far as possible, a number of disposal methods are available to treat the remainder. These include incineration, landfilling, and physical treatment.[5,6] This requires that the chemicals be packaged and transported to a central location on campus or to a commercial disposal facility.

However, for some chemicals, it is possible to convert the waste to a nontoxic product, or at least to a product that is less hazardous than the original, at the laboratory bench.[7-14] This has several advantages. First, the researcher, teacher, or student who has been handling the material would be expected to be familiar with its properties and reactivity and, therefore, would know the appropriate precautions to take and personal protection to wear during its disposal. Second, the risks associated both with mixing the wastes for transportation and with the transportation itself to a central disposal facility are avoided. Third, and perhaps most importantly, this treatment method emphasizes the responsibility of the generator of the waste to ensure its safe disposal. Not only does this apply to researchers and laboratory supervisors, but students can also be made aware of the need to plan to be responsible for disposal of the hazardous waste material they generate by using the methods as a teaching tool.

For example, after completing a laboratory exercise synthesizing the insecticide methoxychlor, students can spend the next laboratory period following its conversion using bleach to the nonphysiologically active and environmentally acceptable product, 4,4'-dimethoxybenzophenone. Other examples of such applications of disposal procedures to university undergraduate laboratories have been described.[4]

In many institutions, it is common practice for researchers to expect their hazardous laboratory wastes to be collected and disposed of in a central facility at the institution or by an outside contractor. For this to change, a plan needs to be implemented to encourage treatment of the appropriate wastes by the researcher at the laboratory bench. The first requirement of such a plan is the availability of practical procedures which can be performed safely and which give reproducible and reliable results. In this way, the researcher does not have to spend an inordinate amount of time both finding information and then performing reactions to convert the hazardous wastes to nontoxic products.

Since 1980, it has been the objective of a research group at the University of Alberta to develop and examine in the laboratory such procedures. The disposal methods have been tested for the rate of disappearance of the toxic material, the degree of conversion to nontoxic products, the nature and identity of the products, the practicality of the method, and the ease of reproducibility. Methods were selected which were shown to be safe to the operator, reliable, and reproducible. For the method to be acceptable, greater than 99% of the starting material had to be destroyed in a reasonable length of time. As far as possible, the products of the reaction were identified; where this was not practical, for example, when a large number of fragments was produced, and especially when the chemical being treated was a suspect carcinogen, the reaction mixture was submitted for an Ames test which detected whether the material formed had excess mutagenicity over background. The disposal procedures that met the chosen criteria have been documented in detail[10-14] and, where possible, have been tested by workers in an independent laboratory.

Several types of reaction can be used in laboratory disposal procedures. Heavy metals are precipitated from aqueous solutions as insoluble salts, often as sulfides, and, if required, the solid immobilized.[4,5] In Japan, stabilization of the metal using the ferrite process is widely used.[16] Oxidation with aqueous sodium hypochlorite in the form of household bleach, with dilute hydrogen peroxide, or with potassium permanganate solutions will destroy a variety of hazardous organic chemicals. This method is often applicable to complex molecules such as chemotherapeutic drugs. In some cases, (e.g., picric acid) compounds can be converted to less hazardous and more easily transported materials by reducing agents such as granular tin in acidic aqueous solution. In still other instances, a reaction specific to a particular chemical is used to convert it to a nontoxic product.

In addition to disposal procedures, we have been developing methods for recycling materials from teaching laboratories. Many institutions collect used solvents from both undergraduate and graduate laboratories and distill the liquids for reuse. This is especially practical in large undergraduate classes where used solvent can be collected and distilled to be used in the same experiment the following year. Commercial semiautomatic and automatic stills are available for simple, fractional, and reduced pressure distillations. Silver, nickel, and cobalt ions can be recovered as silver nitrate and nickel and cobalt chlorides from aqueous solutions containing other impurities.

Examples of several of the specific procedures which we have developed are described in the following paragraphs.

While performing any of the techniques, appropriate personal protection must be worn. This includes adequate eye protection and gloves which are impermeable to the chemicals being used. A laboratory coat is highly recommended. In some instances, where additional protection is needed, this is noted in the text. Wherever possible, reactions should be run in the fume hood.

The disposal of hazardous chemicals, by whatever method, must be performed in accordance with local regulations.

3.1.2 RECOVERY OF METAL SALTS FOR REUSE

3.1.2.1 Solutions Containing Silver Salts

Solutions containing silver salts are often used in undergraduate introductory general or inorganic chemistry laboratories. The silver can be recovered from waste solutions as silver nitrate. All operations should be performed in the fume hood and the operator must wear goggles, gloves, and a laboratory coat. The direct procedure is applicable only to solutions which do not contain cyanide ion, since the first step is the addition of hydrochloric acid to precipitate the silver as silver chloride. If cyanide ions are present, highly toxic hydrogen cyanide gas could be liberated in this step. For solutions containing cyanide ion, an initial step in which the cyanide ion is oxidized to cyanate with sodium hypochlorite must be included.

3.1.2.1.1 Silver Solutions Not Containing Cyanide Ion

Sufficient 6 M hydrochloric acid is added to solutions containing silver ions to completely precipitate the silver as silver chloride. The solid is collected by filtration and boiled with fresh 6 M hydrochloric acid to dissolve impurities. The acid is decanted and the step repeated with fresh 6 M hydrochloric acid. The solid silver chloride is collected by filtration through a glass fiber filter and washed with water. The solid is combined with any solid silver wastes which have been collected (such as electrodes), and the mixture is boiled with about twice its volume of 40% sodium hydroxide solution for at least 30 min while stirring mechanically. This converts the silver salts to silver oxide. The precipitate is allowed to settle, the liquid decanted, and the solid reboiled with fresh 40% sodium hydroxide solution. Water is added to the mixture and the precipitate collected by filtration through a glass fiber or sintered glass filter. The solid is washed thoroughly with water, dried, and, in the fume hood, slowly added to stirred, concentrated nitric acid. Enough acid should be added to convert the dark oxide to white crystals of the nitrate. The solution is diluted with three times its volume of water, any precipitate is removed by filtration (the solid can be added to the next batch of silver waste to be recycled), and the filtrate is evaporated to dryness on a rotary evaporator (or allowed to evaporate to dryness in the fume hood in the dark) to yield crystals of silver nitrate.

$$Ag^+ \rightarrow AgCl \rightarrow Ag_2O \rightarrow AgNO_3$$

3.1.2.1.2 Silver Solutions Containing Cyanide Ion

To the solution containing both silver and cyanide ions is added sodium hypochlorite solution (5%, household bleach). The quantity of the latter required depends upon the concentration of cyanide ion in solution. Thus, 100 ml of a solution containing 10% silver cyanide would require 135 ml of 5% sodium hypochlorite solution. The Prussian blue test can be used to ensure that all of the cyanide has been oxidized to cyanate. To 1 ml of the test solution is added 2 drops of a freshly prepared 5% aqueous solution of ferrous sulfate. The mixture is boiled for 1 min, cooled to room temperature, and 2 drops of 1% aqueous ferric chloride solution added. The solution is made acid to litmus with 6 M hydrochloric acid. If the color of the solution is pale yellow, all of the cyanide has been oxidized to cyanate. On the other hand, a deep blue precipitate indicates continuing presence of cyanide ion, and more sodium hypochlorite must be added to the silver cyanide/bleach solution until the Prussian blue test is negative.

When it has been shown that no cyanide ions remain, the silver can be recovered from the solution by following the procedure described in Section 3.1.2.1.1.

3.1.2.2 Solutions Containing Nickel Salts

The nickel ions present in aqueous wastes are precipitated as nickel hydroxide and recovered as nickel chloride.

$$Ni^{2+} \rightarrow Ni(OH)_2 \rightarrow NiCl_2$$

The nickel-containing wastes are stirred mechanically and 20% sodium hydroxide solution added until there is no further precipitation of nickel hydroxide. The green precipitate is collected and washed with water until the washings are neutral to litmus. The precipitate is dissolved in 6 M hydrochloric acid, diluted to three times the original volume with water, filtered, and the liquid removed using a rotary evaporator or allowed to evaporate in the fume hood. The residue is recrystallized from ethanol to yield pure nickel chloride.

3.1.2.3 Solutions Containing Cobalt Salts

Cobalt carbonate is precipitated from solutions containing cobalt ion by the addition of sodium carbonate and converted to cobalt chloride for reuse.

$$Co^{2+} + Na_2CO_3 \rightarrow CoCO_3 \rightarrow CoCl_2$$

An estimate is made of the amount of cobalt present in the solution, and a two times molar excess of sodium carbonate is added. Sparingly soluble cobalt carbonate precipitates, is collected by filtration and washed with water. Cobalt carbonate easily dissolves in 6 M hydrochloric acid. When the liquid is evaporated, the product is cobalt chloride. Thus, if it is estimated that the waste solution contains 1 g of cobaltic ions, 2 g of anhydrous sodium carbonate are required to precipitate the cobalt carbonate, and the filtered precipitate is dissolved in 10 ml of 6 M hydrochloric acid (concentrated hydrochloric acid added to its own volume of water). The final solution can be allowed to evaporate in the fume hood, or the liquid can be removed by distillation under reduced pressure.

3.1.3 METHODS FOR THE DISPOSAL OF INDIVIDUAL CHEMICALS

3.1.3.1 Solutions Containing Heavy Metal Ions

Increasingly, the disposal into landfills of aqueous solutions containing heavy metal salts is banned. The metals can be precipitated as insoluble salts which are acceptable for disposal. The insoluble salt of choice has often been the sulfide. This requires the use of highly toxic reagents such as hydrogen sulfide, sodium sulfide, ammonium sulfide, or thioacetamide. To avoid the use of these reagents, a number of heavy metal ions can be precipitated as silicates. These salts show similar solubility properties to the sulfides in neutral, acidic, and basic aqueous solutions. Thus, the effect of acid rain on the leaching of silicates is comparable to that of the sulfides. Also, natural ores often contain metals in the form of silicates so that the metals are being returned to the ground in the same form as they were taken out. In some cases, for complete precipitation, the pH has to be controlled. As an example, the method is described in detail for solutions containing lead ions.

$$Pb^{2+} + Na_2SiO_3 \rightarrow PbSiO_3 + 2Na^+$$

To the solution containing lead ions is added, with stirring, an aqueous solution of sodium metasilicate ($Na_2SiO_3 \cdot 9H_2O$, 17 g in 100 ml of water) until there is no further precipitation. If the concentration of the lead salt is known, then 200 ml of sodium metasilicate solution is used for each 0.04 mol of lead ions. If the concentration is not known, it is helpful to allow the precipitate to settle, to withdraw a few milliliters of the supernatant liquid and to add to it several drops of sodium metasilicate solution to test whether or not precipitation is complete. The pH is adjusted to between 7 and 8 by the addition of 2 M aqueous sulfuric acid. For each 100 ml of solution, about 20 ml of 2 M sulfuric acid will be needed. The precipitate is collected by filtration, or the supernatant liquid can be allowed to evaporate in a large evaporating basin in the fume hood. The solid is allowed to dry, and then is packaged and labeled for disposal in a secure landfill, or in accordance with local regulations.

For dilute solutions of lead salts, the sodium metasilicate solution should be added until there is no further precipitation, the pH adjusted to between 7 and 8 by the addition of 2 M sulfuric acid, and the solution allowed to stand overnight before collecting the solid by filtration or allowing the liquid to evaporate.

Solutions of cadmium and antimony salts can be treated similarly to the lead salts, and several other heavy metal salts can be precipitated as silicates by this procedure. These include iron(II) at pH 12, iron(III) at pH 11, zinc(II) at pH 7 to 7.5, and aluminum(III) at pH 7.5 to 8. Copper(II), nickel(II), manganese(II), and cobalt(II) can be precipitated without adjustment of the pH from that after the addition of the solution of sodium metasilicate. This method for precipitation of heavy metals as insoluble silicates is particularly applicable to such wastes from high school laboratories.

3.1.3.2 Solutions Containing Mercury Ions

Mercury salts can be precipitated as the water-insoluble sulfide.

$$Hg^{2+} + Na_2S \rightarrow HgS + 2Na^+$$

Appropriate precautions must be taken to prevent inhalation of the highly toxic gas, hydrogen sulfide. The waste mercury salts are dissolved as far as possible in water (100 ml for each 10 g of waste). The pH of the solution is adjusted to 10 with 10% sodium hydroxide solution. In a fume hood, aqueous sodium sulfide solution (20%) is added with stirring until no further precipitation occurs. To check whether precipitation is complete, a small sample of supernatant liquid is withdrawn and a few drops of sodium sulfide solution are added. If a precipitate or cloudiness appears, more sodium sulfide solution is required. After the precipitate has settled, the liquid is removed by decantation or the solid collected by filtration. The liquid is washed into the drain, and the solid packaged and labeled as mercuric sulfide for appropriate disposal. Depending on regulations, this may be into a secure landfill or by encapsulation in a cement block.

3.1.3.3 Solutions Containing Chromium Ions

Acidic solutions of potassium dichromate are often used in reactions where a strong oxidizing agent is required. They also used to be widely employed for cleaning glassware. This practice has now been largely discontinued in favor of other types of cleaning solutions which do not contain chromium ions. Insoluble chromium hydroxide is formed by reduction of the dichromate with sodium thiosulfate solution. The efficiency of the reduction and the formation of the product as an easily handled flocculent precipitate rather than as a gel is dependent upon the pH of the solution. For this reason, the solution is first neutralized, then reacidified with a measured volume of acid.

$$Cr_2O_7^{2-} + 3S_2O_3^{2-} + 2H_2O \rightarrow 2Cr(OH)_3 + 3SO_4^{2-} + 3S$$

The method is illustrated for a specific quantity of acidic dichromate solution. Thus, to acidic dichromate solution (100 ml) is added solid soda ash

Table 1 Conditions for the Reduction of Oxidizing Agents with Sodium Metabisulfite

Oxidizing Agent in Waste Stream	Agent in Aqueous Solution (Quantity and % Conc.)	10% Aqueous Sodium Metabisulfite (Quantity)	Comments
Potassium permanganate	1 liter of 6%	1.3 liter	Solution becomes colorless
Sodium chlorate	1 liter of 10%	1.8 liter	50% Excess reducing agent added
Sodium periodate	1 liter of 9.5%	1.7 liter	Solution becomes pale yellow
Sodium persulfate	1 liter of 10%	0.5 liter	10% Excess reducing agent added

slowly and with stirring until the solution is neutral to litmus. About 108 g of soda ash will be required. The color of the solution changes from orange to green. It is reacidified to pH 1 by the careful addition of about 55 ml of 3 M sulfuric acid. The color of the solution returns to orange. While swirling, sodium thiosulfate (40 g of $Na_2S_2O_3 \cdot 5H_2O$) is added. The solution becomes blue-colored and cloudy and it is neutralized by the addition of soda ash (10 g). After a few minutes, a blue-gray flocculent precipitate forms. The mixture can be filtered immediately through Celite or allowed to stand for one week, when much of the supernatant liquid can be decanted. In the latter case, the remaining liquid is allowed to evaporate or filtered through Celite. Analysis by atomic absorption spectroscopy showed that the supernatant liquid contains less than 0.5 ppm of chromium. Most local regulations allow this solution to be washed into the drain. The solid residue should be packaged and labeled for appropriate disposal depending on local regulations. Aqueous solutions of other chromium salts, such as chromium trioxide, can be treated similarly.

3.1.3.4 Oxidizing Agents

Solutions of compounds such as potassium permanganate, sodium chlorate, sodium periodate, and sodium persulfate should be reduced before being discarded into the drain to avoid uncontrolled reactions in the sewer system. The reduction can be accomplished by treatment with a freshly prepared 10% aqueous solution of sodium bisulfite or metabisulfite. The use of sodium metabisulfite is preferred because of the greater stability of this salt. Precise quantities and conditions for these reactions are detailed in Table 1. If the concentration of the oxidizing agent to be destroyed is greater than that listed in Table 1, the solution is diluted with water until the stated concentration is reached.

3.1.3.5 Solutions of Picric Acid

When dry, solid picric acid is a powerful explosive. In dilute aqueous solution, the hazard arises from the possibility of evaporation of the solution on stoppers, caps, or lids of containers, since the explosive can be detonated by friction. Picric acid is smoothly reduced with tin and hydrochloric acid to triaminophenol, which is no longer explosive.

The reaction should be performed behind a shatter-proof screen.

The picric acid (1 g) is placed into a 3-neck round-bottomed flask fitted with a dropping funnel and condenser. Any traces of acid on glassware or equipment are rinsed into the flask using about 20 ml of water. To the solution is added 4 g of tin and the mixture stirred magnetically. Into the dropping funnel is placed 15 ml of concentrated hydrochloric acid and, while cooling the flask in an ice-water bath and with stirring, the hydrochloric acid is added dropwise. The first few ml of hydrochloric acid should be added slowly since the initial reaction is vigorous; the rate of addition may be increased as the reaction moderates. When all of the hydrochloric acid has been added, the mixture is allowed to warm to room temperature, then heated under reflux for 1 h to complete the reduction. The unreacted tin is filtered and the precipitate washed with 2 M hydrochloric acid (10 ml) (the residual tin can be reused). The filtrate is neutralized with 10% sodium hydroxide solution and refiltered. The tin chloride precipitate can be treated as normal garbage. The solution which contains 2,4,6-triaminophenol can be packaged, labeled, and sent for incineration or it can be treated with acidic potassium permanganate solution to decompose the triaminophenol. In the latter procedure, to the solution is added slowly and cautiously 50 ml of 3 M sulfuric acid containing 12 g of potassium permanganate. After standing at room temperature for 24 h, solid sodium bisulfite is added until a clear solution is obtained. The liquid is neutralized with 10% sodium hydroxide solution and poured into the drain. This method has been used to decompose batches of up to 8.5 g of picric acid at one time.

It has been found that very dilute solutions of picric acid (including large volumes, e.g., 45-gal drums) can be reduced by acidifying to pH 2 or less with concentrated hydrochloric acid and allowing the solution to stand in the presence of granulated tin for 14 d. The color gradually darkens from yellow to brown, and the complete disappearance of the picric acid can be determined by thin layer chromatography of the solution on silica gel plates using methanol:toluene:glacial acetic acid, 8:45:4, as eluant. Picric acid has an R_f

value of about 0.3 and the bright yellow spot is easily visible. The detection limit can be increased by developing the plate in iodine vapor. Thus, tin powder can be added to 45-gal drums of acidified dilute picric acid solutions and the acid is reduced to 2,4,6-triaminophenol. When it can be shown that no picric acid is present in the solution, contractors are willing to ship the material.

3.1.3.6 Cyanides

Highly toxic aqueous solutions of sodium or potassium cyanide are oxidized to nontoxic cyanates by reaction with household bleach (5% sodium hypochlorite solution).

$$CN^- + ClO^- \rightarrow CNO^- + Cl^-$$

Solutions of sodium or potassium cyanide should be diluted with water to a concentration not greater than 2%. (This dilution is advisable to avoid the possibility of a delayed rapid reaction of the cyanide with the bleach). To each 50 ml of solution is added 5 ml of 10% sodium hydroxide solution and household bleach (60 to 70 ml). The solution can be tested for the continued presence of cyanide as follows. About 1 ml of the solution is removed and placed in a test tube. Two drops of a freshly prepared 5% aqueous ferrous sulfate solution are added and the mixture is boiled for 30 s. After cooling to room temperature, 2 drops of 1% ferric chloride solution are added. The mixture is acidified to litmus with 6 M hydrochloric acid. If cyanide is still present, a deep blue precipitate forms. Concentrations of cyanide greater than 1 ppm can be detected. If the test is positive, more bleach is added to the cyanide solution, and the test repeated. When the blue precipitate no longer forms, the solution can be washed into the drain.

Note that although they also contain a cyano group, this method cannot be used for organic nitriles such as benzonitrile, which do not react with household bleach. Treatment methods for nitriles are described in the following paragraph.

3.1.3.7 Nitriles

Waste solutions containing organic nitriles are treated by hydrolyzing the nitrile to a nontoxic acid. For example, benzonitrile (1 g) is converted to benzoic acid by heating under reflux with 10% ethanolic potassium hydroxide solution (30 ml) for 3 h. The cold solution is neutralized with dilute hydrochloric acid and washed into the drain.

3.1.3.8 Inorganic and Organic Azides

Oxidation of inorganic azides by ceric ammonium nitrate is a useful method of disposal. However, for organic azides this is a slow and unsatisfactory process. Preferable is a reduction to the corresponding amine with tin and hydrochloric acid. The two procedures are detailed.

For inorganic azides, work in the fume hood and behind a safety shield. The concentration of the azide solution should be about 1 g/100 ml of water. Prepare a quantity of 5.5% ceric ammonium nitrate solution which is 4 times the volume of the azide solution. Slowly add the ceric ammonium nitrate solution to the azide solution and stir for 1 h. At the end of the reaction time, the solution should show the orange color of ceric ammonium nitrate. The solution can be washed into the drain.

For solutions of organic azides, work in the fume hood behind a safety shield.

$$C_6H_5N_3 + Sn + 2HCl \rightarrow C_6H_5NH_2 + SnCl_2 + N_2$$

Slowly add the azide (1g) to a stirred mixture of granular tin (6 g) in concentrated hydrochloric acid (100 ml). Continue stirring for 30 min. Cautiously add the solution to a pail of cold water. Wash the residual tin with water and reuse. Neutralize the aqueous solution in the pail with soda ash and wash it into the drain.

3.1.3.9 Metal Carbonyls

Metal carbonyls such as iron pentacarbonyl and nickel carbonyl are highly toxic and reactive materials; the latter is a suspect carcinogen. These compounds are destroyed by stirring solutions in the appropriate solvent with bleach. The choice of solvent is very important, and these are summarized in Table 2, together with quantities and reaction conditions.

3.1.4 DISPOSAL OF SOME POTENTIALLY CARCINOGENIC CHEMICALS

3.1.4.1 Aromatic Amines

Many aromatic amines have been shown to cause cancer in rodents and are confirmed or suspect human carcinogens. They can be rendered physiologically inactive by removal of the amino group. For example, 4-aminobiphenyl (Structure 1) is converted to biphenyl (Structure 2) by treatment with sodium nitrite and then hypophosphorous acid.

Table 2 Conditions of the Reaction of Metal Carbonyls with Bleach

Metal Carbonyls +	Bleach (NaOCl)	Reaction ($\rightarrow$)	Result
$Fe(CO)_5$ — iron pentacarbonyl solution in hexane (5 ml in 200 ml)	65 ml	Stirred 30 min under helium	$Fe(OH)_3$ — solid filtered, discard aqueous layer to drain, hexane layer recycled or incinerated
$Fe(CO)_9$ — diiron nonacarbonyl solution in toluene (1 g in 100 ml)	50 ml	Stirred 25 h	$Fe(OH)_3$ — solid filtered, discard aqueous layer to drain, toluene layer recycled or incinerated
$Cr(CO)_6$ — chromium hexacarbonyl solution in tetrahydrofuran (1 g in 200 ml)	30 ml	Stirred 15 min	$Cr(OH)_3$ — solid filtered, packaged for disposal, liquid incinerated
$Ni(CO)_4$ nickel carbonyl solution in tetrahydrofuran (5 g in 200 ml)	250 ml	Stirred 2 h under helium	$Ni(OH)_2$ — solid filtered, packaged for disposal, liquid incinerated

1   **2**

Thus, to 1.0 g of 4-aminobiphenyl in a 125-ml Erlenmeyer flask is added a mixture of 0.8 ml of water and 2.5 ml of concentrated hydrochloric acid. The mixture is stirred for 10 to 15 min until a homogeneous slurry is formed. The slurry is cooled to 0°C in an ice-salt bath, and a solution of 1.0 g of sodium nitrite in 2.5 ml of water is added dropwise at such a rate that the temperature of the mixture does not rise above 5°C. After stirring for 1 h, 13 ml of ice-cold 50% hypophosphorous acid is added slowly. Some foaming may occur. When addition of the acid is complete the mixture is stirred for 18 h at room temperature. The solid is collected by filtration, the filtrate washed into the drain with a large volume of water, and the solid (biphenyl) discarded with normal refuse or disposed of by burning. Other aromatic amines such as 2-aminofluorene (Structure 3) can be made harmless by a similar reaction. In this case, the product is fluorene (Structure 4).

3   **4**

3.1.4.2 Ethidium Bromide

Ethidium bromide (Structure 5) is a compound frequently used in genetic testing since it intercalates double-stranded DNA and RNA. It is known to be mutagenic and therefore must be handled with care. It is normally used in very dilute solution, and we have found that it can be destroyed in these solutions by treatment with household bleach. When a solution of 34 mg of ethidium bromide in 100 ml of water is stirred at room temperature with 300 ml of household bleach for 2 h, the ethidium bromide is converted to the physiologically inactive product 2-carboxybenzophenone (Structure 6). The product solution does not show excess mutagenicity over standards in the Ames test.

5   **6**

Table 3 Conditions for the Disposal of Drugs with Potassium Permanganate

Drug and Quantity (mg)	Volume of Water (ml)	Volume of H₂SO₄ (ml)	Wt. of KMnO₄ (g)	Time at Room Temp. (h)
Chlorambucil (50)	10	2	1.8	4
Daunorubicin (30)	10	2	1.0	2
Doxorubicin (30)	10	2	1.0	2
Dichloromethotrexate (10)	10	2	0.5	1
Methotrexate (50)	10	2	0.5	1
6-Mercaptopurine (18)	17	3	0.13	12
6-Thioguanine (18)	17	3	0.13	12
Streptozotocin (48)	10	2	2.0	12
Vincristine sulfate (10)	10	2	0.5	2
Vinblastine sulfate (10)	10	2	0.5	2

3.1.4.3 Chemotherapeutic Drugs

Frequently, the manufacturer's recommended method for the disposal of spills or waste quantities of hazardous pharmaceuticals is incineration at temperature above 1000°C. Research on the drugs in academic institutions results in small quantities of waste. Further, with chemotherapeutic drugs being dispensed in hospital pharmacies, clinics, and physician's offices, many users do not have access to such facilities, and an urgent need was recognized for the development of practical, safe, and environmentally acceptable methods for converting these drugs into nontoxic and nonmutagenic products. Disposal methods for a variety of antineoplastic drugs are described.

3.1.4.3.1 Chemotherapeutic Drugs That Can Be Treated with Acidic Potassium Permanganate

Acidic potassium permanganate is a powerful oxidizing medium and can often be used to break a molecule into nontoxic and nonmutagenic products which can be washed into the drain. In Table 3 are listed the drugs which are made harmless by treatment with this reagent, the quantities of potassium permanganate and concentrated sulfuric acid required to deactivate a given quantity of the drug, and the time for which the solutions should be allowed to stand. Conditions are given for both solid material and reconstituted aqueous solutions.

While performing these reactions, wear protective gloves, goggles, and a laboratory coat. As far as possible, work in a fume hood. Concentrated sulfuric acid is extremely corrosive and should be handled with care. Spills on the skin should be washed immediately with cold running water for 20 min.

Table 4 Conditions for Disposal of Drugs with Bleach or Calcium Hypochlorite

Drug and Quantity (mg)	Volume of Bleach or Calcium Hypochlorite Solution (ml)	Time at Room Temp. (h)
Amsacrine (10)	100 (bleach)	0.5
Dactinomycin (10)	20 (bleach)	1
Methotrexate (20)	25 (bleach)	0.5
Fluorouracil (500)	40 (calcium hypochlorite)	0.5
Procarbazine HCl (500)	60 (calcium hypochlorite)	12
Mercaptopurine (3.2)	25 (bleach)	1.5
Thioguanine (3.2)	25 (bleach)	1.5

Sulfuric acid must always be added to water, never the other way around. Heat is produced during the neutralization of the acidic solution with 10% sodium hydroxide solution. When sodium bisulfite is added, some sulfur dioxide may be produced; therefore, this addition should be made in a fume hood or in a well-ventilated area.

Solid material is dissolved in the water and the sulfuric acid added. Sulfuric acid is added to the reconstituted solutions. In both cases, solid potassium permanganate is slowly added to the stirred mixture. Should the purple color fade before reaction time is complete more potassium permanganate should be added. When the reaction is complete, the mixture is neutralized with 10% sodium hydroxide, then sodium bisulfite is added until a colorless solution is formed. This solution is washed into the drain with a large volume of water.

Glassware, including vials containing residues of the drugs, is immersed in a solution prepared by dissolving 4.7 g of potassium permanganate in 83 ml of water and 17 ml of concentrated sulfuric acid. After standing overnight, the glassware is rinsed thoroughly with water and discarded or reused. The permanganate solution is neutralized by the careful addition of 10% sodium hydroxide, then sodium bisulfite is stirred into the mixture until a colorless solution is formed. This solution is washed into the drain with ten times of its volume of water.

3.4.1.3.2 Chemotherapeutic Drugs That Can Be Treated with Sodium or Calcium Hypochlorite

Sodium hypochlorite solution (household bleach) and calcium hypochlorite solution, although not as powerful oxidizing agents as acidic potassium permanganate, can be used to decompose several hazardous pharmaceuticals. Conditions for the effective denaturation of these materials are listed in Table 4.

While performing these reactions, wear protective gloves, goggles, and a laboratory coat. As far as possible, work in the fume hood. Bleach solution is corrosive. Spills on the skin should be washed immediately with cold running water for 20 min. The vapors from bleach should not be inhaled.

Calcium hypochlorite solution is prepared by stirring 10 g of calcium hypochlorite with 100 ml of water for 2 h, then removing the undissolved material by decantation or filtration. The bleach or calcium hypochlorite solution is added to the solid or solution. When the reaction is complete, the solution is washed into the drain with ten times its volume of water. Vials or other glassware containing residues of the drugs are immersed in bleach and allowed to stand overnight. The glassware is rinsed thoroughly with water and reused or discarded. When the reaction is complete, the residual liquid is washed into the drain.

3.1.4.3.3 Lomustine and Carmustine

Lomustine (Structure 7) and carmustine (Structure 8) are decomposed in a synthetic mixture which is similar to a detergent solution. The reagent is prepared by dissolving 4.8 g of sodium pyrophosphate in 100 ml of water and adding 1 ml of Triton® X100, a surfactant. When 100 mg of lomustine or carmustine is added to 25 ml of the detergent solution and heated at 55°C for 90 min, nontoxic and noncarcinogenic products are formed. When the reaction is complete, the cooled solution is washed into the drain. The major product of the reaction with lomustine was shown to be dicyclohexylurea (Structure 9).

3.1.5 TREATMENT OF SPILLS

A practical and versatile spill mix has been developed consisting of a mixture of sodium carbonate to neutralize any acid present, clay cat litter (calcium bentonite) to absorb liquid rapidly, and sand to moderate any reaction.

To treat a spill of acid, acid derivative, reactive chemical, or solvent, wear nitrile rubber gloves, laboratory coat, and goggles. (Self-contained breathing apparatus may be necessary depending on the nature and size of the spill.) Isolate the area of the spill and cover it with a 1:1:1 mixture by weight of sodium carbonate, clay cat litter (calcium bentonite), and sand. When all of the liquid has been absorbed, scoop the mixture into a plastic pail, and in the fume hood, very slowly add the mixture to a pail of cold water. Allow to stand for 24 h. Test the pH of the solution and neutralize, if necessary, with sodium carbonate. Decant the solution to the drain. Treat the solid residue as normal garbage. Sodium carbonate can be replaced with calcium carbonate both in the spill mix and in the neutralization of the aqueous solution.

For spills of those chemotherapeutic drug solutions which are particularly hazardous to handle, it may be desirable to pour a solution of the appropriate deactivating agent, i.e., acidified potassium permanganate, bleach, or synthetic detergent, directly onto the spill. Solutions of potassium permanganate for treating spills are prepared by dissolving 4.7 g of potassium permanganate in 100 ml of 3 M sulfuric acid (17 ml of concentrated sulfuric acid added to 83 ml of water). This solution should be prepared fresh daily. If the purple color of the permanganate fades during treatment of the spill leaving a brown mixture, more acidic potassium permanganate solution should be added. The liquid resulting from treatment of the spill is adsorbed onto suitable material such as paper towels or absorbent cotton; the material is placed in a container, covered with more of the deactivating solution, and allowed to stand overnight. If the deactivating solution is acidic potassium permanganate, the solution is carefully neutralized with 10% aqueous sodium hydroxide, decolorized by the slow addition (with stirring) of solid sodium bisulfite, and decanted to the drain. Where the deactivating solution is bleach or synthetic detergent, the liquid is washed into the drain. In all cases, the residual absorbent material is discarded as normal refuse.

3.1.6 CONCLUSION

There are several general chemical techniques which can be used to reduce, reuse, recycle, or recover waste. Where the waste cannot be further minimized, the remaining material can often be safely and effectively converted to nontoxic and environmentally acceptable products. In combination with the strategies to minimize waste chemicals, the disposal procedures allow for the development of an integrated waste management program. We are continuing to develop, test, and document acceptable waste disposal procedures for individual chemicals, including azo dyes and pesticides, the latter with two goals in mind: firstly, to allow destruction of residual pesticides in containers and secondly, to find efficient ways of removing all pesticide residues from workers' clothing.

ACKNOWLEDGMENTS

The author gratefully acknowledges financial support from the Alberta Environmental Trust, Alberta Occupational Health and Safety, and the Alberta Heritage Foundation for Medical Research. Consultants in the project are Dr. Lois Browne, Dr. Gordon Weir, and Ms. Rosemary Bacovsky. The laboratory development and testing of the procedures has been done by Donna Renecker, Patricia McKenzie, Carmen Miller, Katherine Ayer, John Crerar, Paul Cumming, Richard Young, Girard Spytkowski, Mui Chang, and Dr. Roger Klemm, all of whom provided many excellent suggestions as the work progressed.

REFERENCES

1. Task Force on RCRA, *Less is Better,* American Chemical Society, Department of Government Relations and Science Policy, Washington, D.C., 1985.
2. Pine, S.H., Chemical management, a method for waste reduction, *J. Chem. Educ.,* 61, A95, 1984.
3. Pike, R., Szafran, Z., and Foster, J., *Microscale Laboratory Manual for General Chemistry,* John Wiley & Sons, New York, 1993.
4. Williamson, K., *Macroscale and Microscale Organic Experiments,* D.C. Health, Lexington, MA, 1989.
5. Phifer, R.W. and McTigue, W.R., *Handbook of Hazardous Waste Management for Small Quantity Generators,* Lewis Publishers, Chelsea, MI, 1988.
6. Connor, J.R., *Chemical Fixation and Solidification of Hazardous Wastes,* Van Nostrand Reinhold, New York, 1990.
7. Prudent Practices for the Disposal of Hazardous Chemicals in Laboratories, National Academy of Sciences, Washington, D.C., 1983.
8. Pitt, M.J. and Pitt, E., *Handbook of Laboratory Waste Disposal,* Wiley-Halstead, New York, 1985.
9. Lunn, G. and Sansone, E.B., *Destruction of Hazardous Chemicals in the Laboratory,* Wiley-Interscience, New York, 1990.
10. Armour, M.A., *Hazardous Chemicals Disposal Guide,* CRC Press, Boca Raton, FL, 1991.
11. Armour, M.A., Bacovsky, R.A., Browne, L.M., McKenzie, P.A., and Renecker, D.M., *Potentially Carcinogenic Chemicals, Information and Disposal Guide,* University of Alberta, Canada, 1986.
12. Armour, M.A., Browne, L.M., and Weir, G.L., Tested disposal methods for chemical wastes from academic laboratories, *J. Chem. Educ.,* 62, A93, 1985.
13. Armour, M.A., Chemical waste management and disposal, *J. Chem. Educ.,* 65, A64, 1988.
14. Castegnaro, M., Adams, J., Armour, M.A., Barek, J., Benvenuto, J., Confaloneri, C.C., Goff, C., Ludeman, S., Reed, D., Sansone, E.B., and Telling, G., *Laboratory Decontamination and Destruction of Carcinogens in Laboratory Wastes: Some Antineoplastic Agents,* International Agency for Research on Cancer, Lyons, 1985.
15. Tamaura, Y., M. Kitamura, and H. Takatsuki, *Ind. Water,* 367(4) 29, 1989.

CHAPTER 3.2

Technical Aspects and the Costs of Hazardous Waste Management

Peter C. Ashbrook and Peter A. Reinhardt

TABLE OF CONTENTS

0-8493-682-5/94/$0.00 + $.50
© 1994 by CRC Press, Inc.

3.2.1 OPTIONS FOR HANDLING AND DISPOSAL OF HAZARDOUS WASTES

When possible, treatment of wastes to render them nonhazardous in the laboratory where they are produced will be the safest and least costly disposal method. For various reasons, in-lab treatment may not be possible. At this point, the academic institution must look at other options.

One general management strategy an institution may use is to collect all its waste chemicals and contract for direct disposal. Most academic institutions use this strategy when they first develop a hazardous waste program. Now, however, economics and safety concerns have caused these institutions to assume a more active role in managing their wastes. A sample management strategy is presented in Figure 1.

3.2.1.1 Redistribution, Returns, Recycling, Reclamation

The first point to remember in managing laboratory wastes is that some of these materials are not really wastes. Before handling these materials as waste, one should examine the "R" options. These "R" options typically present little in the way of safety hazards, but can have large economic benefits.

These terms are sometimes used interchangeably. Although I will give brief definitions and examples of each, the important thing to remember is that one is trying to improve resource usage, minimize direct disposal costs, and avoid potential future liability. "Redistribution" applies to the use of those chemicals that have been disposed of solely because the owner accumulated more than was needed. A surprisingly large proportion of wastes from laboratories — as much as 10% — consist of excess chemicals. Of these excess chemicals, many are in the original unopened containers. Obviously, in many cases, these excess containers may have value to other individuals. In the U.S., "returns" have become a desirable option for certain kinds of wastes. Return of small compressed gas cylinders (lecture bottles) and excess pesticides to the manufacturers is often a viable option. Perhaps other types of returns will become possible as management of laboratory wastes develops. "Recycling" means that one takes a waste and finds another use for the waste without any reprocessing of it. For example, one might have waste xylene from a histology laboratory. This waste xylene would not be suitable for reuse in the histology laboratory, but might have a use for cleaning equipment in the paint shop. "Reclamation" means that the waste has some value, but only after the valuable parts are separated from the waste. Examples of reclamation would be: distillation of solvents, rerefining of used oil, recovery of mercury, and recovery of precious metals.

Several advantages to these "R" options are readily apparent. First of all, resource utilization becomes more efficient. Second, disposal costs are greatly

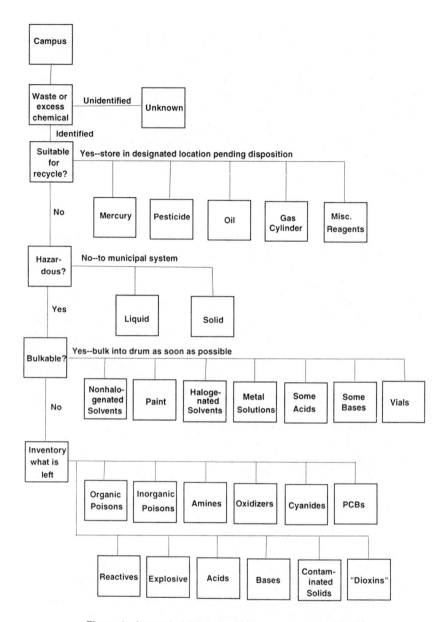

Figure 1. A sample laboratory waste management scheme.

reduced and often eliminated altogether. In the case of redistribution, savings arise from both avoidance of disposal costs and avoidance of purchase costs. Third, these options minimize or eliminate long-term liability concerns arising from disposal. Lastly, safety hazards to personnel are minimal because wastes are handled only in closed containers.

In pursuing these desirable management options, care must be taken not to create new hazards. For example, redistribution of suitable chemicals is clearly a desirable option. However, one must have suitable storage space which allows for segregation of chemicals by hazard class, adequate ventilation, and is closely supervised. Used chemical storerooms have a great potential to become dumping grounds. These areas must be inventoried on a regular basis to assure that chemicals and containers have not deteriorated and that all containers are properly labeled. Likewise, accumulation of wastes for recycling, reclamation, or returns can get out of hand if not closely monitored.

3.2.1.2 On-Site Treatment

An earlier section of this book addressed treatment of wastes in the labs where generated. Another management option is treatment of the wastes in a separate location on campus. The primary objective in on-site treatment is to create a nonhazardous waste that can be disposed of as such. In some cases, on-site treatment may be used to reduce the amount of hazardous waste requiring off-site disposal.

Relatively large quantities of wastes may be more appropriately treated in a central facility than in a lab. The basic concepts for treatment are the same as discussed in Chapter 3.1. One must be careful, however, in scaling up procedures because chemical reactions may occur differently when the reagents are present in larger quantities. In addition, the consequences of accidents or unexpected reactions are much more dangerous when larger quantities are involved. Examples of on-site waste treatment include neutralization of corrosive materials, oxidation of cyanides, deactivation of water reactives, and precipitation of heavy metals.

Economics are not the only reason for pursuing on-site treatment. Reduced liability and safety concerns are also the result of rendering wastes nonhazardous on-site. Treatment at a central location provides the institution greater control over such procedures when compared to allowing each laboratory researcher to treat wastes in the lab. There may be times when such control is desirable.

The obvious concern with on-site treatment is the potential for unexpected reactions. Treatment procedures must be conducted using the proper equipment, including adequate ventilation and suitable personal protective equipment. A significant barrier to larger institutions in the U.S. is that most treatment procedures require a permit from the state and/or federal government.

In determining the economic viability of pursuing in-house treatment, one must be careful to include staff time, potential for spills and accidents, and the cost of equipment. These costs of performing treatment can then be compared to the various alternatives to determine the cost/benefit ratio.

3.2.1.3 Bulk Wastes

The hazardous waste treatment and disposal industry has developed to treat wastes in large quantities. Thus, more options are available and prices are more competitive when one has wastes packaged in bulk form in 55-gal drums. Bulk wastes are formed when compatible wastes are commingled into larger containers. Fifty-five gallon drums are typically used for commingling, but smaller containers may be useful in some situations.

Cost is the primary advantage of commingling compatible wastes. In the U.S., the cost of disposal for waste solvents in bulk form is typically on the order of $1/kg — less than 10% of the cost when disposed of as labpacks (see Section 3.2.1.4). Many waste solvents can be used as supplemental fuels, cutting disposal costs by half again to about $0.50/kg. If one assumes a disposal cost of $20/kg for a labpack of solvents, the cost for disposal of 200 kg would be $4000. By contrast, disposal of these same solvents after bulking them into a 55-gal drum would cost only $200 for incineration — a savings of $3800. Because solvents typically account for over half of a university's waste, commingling of solvents is the most common bulking procedure. Other laboratory wastes that can be bulked under the proper conditions are: used oil, aqueous solutions, and some solids.

Safety is the primary disadvantage of commingling wastes. Whenever wastes are mixed together, there is the possibility of incompatible reactions. Great care must be exercised to ensure that such reactions do not occur. The actual bulking process must be carried out in suitable facilities and by persons wearing appropriate personal protective equipment. Good ventilation is essential. Most commingling of liquids should be conducted in a walk-in fume hood or equivalent device to provide additional control of vapors. When combining flammable solvents, the drums must be grounded. Respiratory protection may be required for workers, along with appropriate gloves and protective clothing in the event that chemicals splash. Handling of full 55-gal drums provides hazards not usually associated with labpacks just because of the much larger bulk involved. Most full 55-gal drums weigh at least 200 kg. Handling of these drums presents the potential hazards of back strains and crushed fingers or toes. If a full 55-gal drum leaks or catches on fire, the consequences will be much more severe than would be the case with a laboratory size container.

Commingling of wastes can be economically advantageous to the institution. The institution must be sure that it has developed adequate procedures to prevent accidents from happening. One additional caution before getting into commingling relates to analytical costs. Laboratory analysis of each container is rarely required for wastes disposed in labpacks. However, with bulk wastes, laboratory analysis is almost always required. With the wide variety of chemicals regulated as hazardous waste and the wide variety of chemicals used in research and teaching institutions, the analytical costs cannot

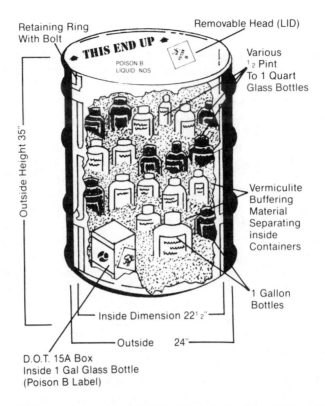

Retaining Ring With Bolt

THIS END UP

POISON B LIQUID NOS

Removable Head (LID)

Various ¹₂ Pint To 1 Quart Glass Bottles

Outside Height 35"

Vermiculite Buffering Material Separating inside Containers

1 Gallon Bottles

Inside Dimension 22¹ ₂"

Outside 24"

D.O.T. 15A Box
Inside 1 Gal Glass Bottle
(Poison B Label)

Figure 2. A labpack drum — typical D.O.T. 17H steel with removable head (LID). (Reprinted with permission from *Less Is Better: Laboratory Chemical Management,* copyright 1985, American Chemical Society.)

be considered trivial. Taking the hypothetical case of the waste solvents discussed above, analysis of the wastes could cost $1500 if a detailed solvent scan is required. These costs would decrease, but by no means eliminate, the economic benefit.

3.2.1.4 Labpacks

The simplest method for disposing of chemicals from laboratories is to collect them and ship them off for disposal in ''labpacks''. A labpack is prepared by using a suitable drum, placing compatible chemicals in the drum, and surrounding the containers with an absorbent material such as vermiculite (see Figure 2). The labpacks may be disposed directly into an incinerator or landfill (but see landfill discussion below), or may be unpacked as a special handling and treatment facility designed for laboratory wastes.

Advantages of labpacks are the simplicity of management for the institution generating the waste and minimal safety hazards. If labpacks are used exclusively, chemical waste management merely becomes an exercise in inventory control on the part of the academic institution. Except for container breakage, safety concerns are minimal because the containers are never opened, so that potentially incompatible chemicals are not mixed together. Worries about compatibilities can be transferred to the hazardous waste contractor, who usually provides the technical expertise to properly package the wastes.

Disadvantages of labpacks include the relatively high cost of disposal and the high use of resources. Incineration of labpacks in the U.S. has typically cost $15 to 30/kg net weight of chemicals disposed. Landfill of labpacks, when legal, has typically cost $10 to 20/kg net weight. These costs are often more than the purchase price of the chemicals. The high use of resources is not an issue that is discussed much. The simplest example would be the disposal of 200 l of nonhalogenated solvents that are in 4-l glass containers. Using labpacks for disposal involves 4 55-gal drums and about 6 20-kg bags of vermiculite. These four drums, containing the glass containers and vermiculite, would be incinerated, leaving the drums, glass, and vermiculite as ash requiring disposal. On the other hand, the 4-l containers could be emptied into a single 55-gal drum. This drum would be pumped out at the incinerator and available for reuse with cleaning, if necessary. Little ash would be formed from the incineration process. Even the glass containers could be washed and the glass recycled.

3.2.1.5 Miscellaneous Handling Techniques

As noted above, in the U.S., the hazardous waste disposal industry has developed to serve large generators of hazardous waste. Further, the types of wastes best handled are large quantities of a single waste type. At a large university, the total quantity of waste may be relatively large, but the waste is far from homogeneous. From a cost point of view, it is in the generating institution's interest to try to convert wastes either to a nonhazardous form or to a bulk form. A side benefit of many strategies for cost reduction is that safety to the institution's waste handlers and the public is enhanced. When cost reduction strategies present a significant potential for significant hazards, one must question whether the strategy should be pursued.

One technique not discussed elsewhere is segregation of chemical wastes as they arise at a storage facility. The first objective is to segregate wastes by hazard class. Prevention of reaction between incompatible chemicals in the event of an accident should receive high priority. Segregation by hazard class also reduces the labor required by a hazardous waste contractor when disposal services are procured. Once such segregation occurs, one can segregate the nonhazardous chemicals from the hazardous. In many places, nonhazardous chemicals may be disposed by the normal trash or via the sewer system. Disposal as normal trash generally involves little cost and minimal labor.

Another technique is dissolving organic solids in a suitable solvent and commingling the mixture into a bulk waste solvent drum. The basic concept is to increase the packing density of wastes being disposed. As with other commingling strategies, great care must be exercised at all times to avoid reactions between incompatible chemicals. One can extend this strategy of dissolution to inorganic chemicals and placing the resulting mixtures into appropriate drums of aqueous wastes.

3.2.1.6 Technical Aspects of Biohazardous Waste Management

Biohazardous waste disposal problems have grown with the awareness that human immunodeficiency virus (HIV) and other bloodborne pathogens (e.g., Hepatitis B Virus or HBV) post a risk to biological scientists and health care workers. In 1987, the U.S. Centers for Disease Control recommended that all human blood and body fluids be handled with "universal precautions". That is, because the pathogenic hazard of specimens are usually unknown, all specimens should be handled with the same extreme care. The physical hazards of biohazardous waste is an increasing concern. So is aesthetic degradation of the environment, such as the problem of medical waste washing up on the beaches of the U.S. east coast resorts in 1988. As a result, certain wastes in biological and medical facilities have been newly identified as biohazardous, and the volume of biohazardous waste has grown.

Waste from biological and medical facilities can include waste that is infectious, waste that is potentially infectious, waste that can cause physical harm (such as a puncture wound), and medical waste that is aesthetically offensive. The terms "biohazardous waste", "infectious waste", "medical waste", and "biomedical waste" are often used interchangeably. The choice of off-site disposal methods for biohazardous waste depends upon the nature of the waste. If infectiousness is a concern, safe handling and transport, containment prior to treatment or disposal, and destruction of the infectious agents are the criteria for the selection of the disposal route. Traditional methods for destroying infectious agents include chemical disinfection, heat and steam sterilization (e.g., autoclaving), and incineration. If physical hazards are present (e.g., needles and other sharps), containment and destruction are of primary importance. Destruction is also desirable when aesthetics are of concern. Shredding, encapsulation, and incineration have all been used to dispose of sharps and aesthetically offensive medial waste.

On-site treatment has been the preferred method for disposal of biohazardous waste because off-site transport and additional handling risks release of infectious agents. Prompt treatment prevents putrefaction if refrigerated storage is not provided. On-site autoclaving is the preferred disposal method for waste cultures and stocks and for contaminated media and labware from microbiology and bacteriology laboratories. Other biohazardous waste can be treated in small on-site pathological incinerators. In the U.S. today, however, air pollution control regulations make on-site incineration prohibitively expensive for all but the largest institutions. As a result, off-site disposal of some types of biohazardous waste is usually necessary.

3.2.1.7 Mixed Wastes

In the U.S., the term "mixed wastes" usually refers to wastes that are hazardous both because of radioactivity and because of chemical hazards. However, mixed wastes could also refer to wastes that exhibit a biohazardous characteristic as well as another hazardous characteristic.

Regardless of the hazard characteristics present, there are very few disposal sites in the U.S. willing to accept mixed wastes. Therefore, the best strategy is to prevent mixed wastes from occurring. In the case of radioactive materials, prevention of mixed wastes is relatively easy because most institutions carefully control the use of such items. Control of chemicals and biohazardous materials is far less common.

If mixed wastes are generated, the institution may often find it desirable to provide some management of the wastes on campus. For example, wastes that are both biohazardous and chemically hazardous could possibly be autoclaved to remove the biohazardous characteristic. Incineration of used liquid scintillation cocktail in an on-site power-plant boiler can destroy the hazardous chemical characteristic and release radioactivity in compliance with regulations. An alternative method for addressing liquid scintillation cocktail is to require researchers to use a biodegradable cocktail that may be disposed of down the drain, provided the radioactive content is within specified regulatory limits.

3.2.1.8 Nonlaboratory Wastes

Those responsible for hazardous waste management at academic institutions often neglect to give adequate attention to nonlaboratory sources of wastes. The Physical Plant can be a major source of chemical wastes, generating such items as paint thinner, used oil, degreasing solvents, excess or degraded pesticides, and miscellaneous other items. Print shops, photo labs, and other administrative units may also generate hazardous wastes.

3.2.2 ALTERNATIVE STAFFING SCHEMES

Academic institutions have two basic choices for staffing a hazardous waste management program. They can either hire personnel to operate a centralized program or they can allow laboratory workers to handle their own waste management needs. Some institutions will find that a combination of the two approaches is optimum. The pros and cons of these alternatives are summarized in Table 1.

Table 1 Comparison of Alternative Management Schemes for Hazardous Waste Management

Central Program	No Central Program
Possible advantages	
• Institution can exert greater control	• No cost for administrative staff or space
• Hazardous waste management staff will have better specialized training	• Responsibility for waste disposal rests with those who produce the waste
• Better opportunities for controlling disposal costs	• Wastes are accumulated in smaller quantities, which minimizes the consequences from accidents
• Relieves researchers of some responsibility for regulatory compliance	
• Central staff may better understand disposal options	
Possible disadvantages	
• Additional administrative costs for staff	• Difficulty to ensure total institutional compliance with regulations
• Space requirements for storage and management of wastes	• Researchers may have difficulty keeping current with regulations
• Large quantities of waste in storage increase the potential consequences from accidents	• Difficult to dispose wastes; may become almost impossible to dispose
Most appropriate for	
• Large schools	• Small schools

3.2.2.1 No Central Program

For many years, little concern was paid to disposal of waste chemicals from academic institutions. These wastes were typically disposed in the normal trash, down the sewer, or even directly into bodies of water. Over the past 20 years, most researchers have come to realize that the relatively small amounts of chemical wastes they produce must be disposed of responsibly.

In an academic environment, it would at first appear that each researcher or teacher should be capable of devising ways of treating or disposing of the chemical wastes generated. There is certainly merit to this approach. It is reasonable to expect that those producing hazardous wastes would be the ones most familiar with the potential hazards. By having those who generate the waste be responsible, one should also minimize potential hazards from transporting the wastes. Potential hazards are further minimized by keeping wastes in relatively small amounts and dispersed throughout the institution. Should an accident occur with hazardous wastes, the consequences to the institution are likely to be relatively small because only wastes from a single generating point should be involved.

There are, however, some potentially significant drawbacks to this decentralized approach. There are always some wastes that cannot be disposed of or rendered nonhazardous in the lab. These must receive special handling. Another drawback is that the institution has little control over hazardous waste management procedures. The potential for improper disposal of hazardous wastes through ignorance is greatly increased under this scenario. If an institution adopts this approach, it should examine very closely whether wastes are being properly handled or whether it is merely ignoring the problem.

3.2.2.2 Part-Time Oversight by Central Staff

Most academic institutions of any size eventually reach the conclusion that it is worthwhile assigning responsibility for hazardous waste management to some centrally housed person. This person is typically an employee of the Chemistry Department, the Safety Department, or the Physical Plant. This person has responsibility for developing policies for proper hazardous waste management, complying with government regulations, finding solutions to difficult disposal issues, and arranging for off-site disposal. Rather than directly providing services, this individual serves as a coordinator to see that needs are met. Thus, researchers are relied upon to take care of most of their hazardous waste disposal needs, but the institution has someone to provide some organization and oversight.

3.2.2.3 Active Central Program

Many of the larger academic institutions in the U.S. have decided that they wish to have hazardous waste management services provided primarily by an administrative support unit. This approach usually means that the administrative staff are responsible for collecting hazardous wastes, storing or handling it if necessary, and then arranging for proper disposal. The researcher still has a role in the system, because the researcher is the one actually generating the waste. In this scheme, however, researchers may only need to have responsibility for collecting the waste, placing it in suitable containers, and labeling it properly. The central management staff then has responsibility for collecting the waste, safely transporting the waste, developing management techniques, and arranging for proper disposal of the wastes.

This option greatly increases the institutional control over hazardous waste management. It also can relieve researchers of some of the responsibility for hazardous waste management, thereby freeing time for other pursuits. Central handling of an institution's hazardous wastes can potentially reduce unit disposal costs through some of the management techniques discussed in Section 3.2.1 above.

There are potential drawbacks to having an active central staff. First is the addition to the administrative staff. Ideally, these persons are hired to minimize disposal costs; however, as with any program, the institution must keep close watch over the program to make sure it is run as efficiently as possible. Sometimes, administrative personnel have difficulty working productively with academic personnel. It is essential that those hired to operate a central management program have good skills in interpersonal communications. Safety issues may also be of concern. When the relatively small quantities of wastes from each lab are aggregated together in one place, the consequences of an accident may be much more severe than if wastes are widely dispersed throughout the campus.

CHAPTER 3.3

Transportation of Hazardous Waste from the Laboratory

Peter C. Ashbrook

TABLE OF CONTENTS

0-8493-682-5/94/$0.00 + $.50

3.3.1 STORAGE SITUATION AT UNIVERSITY

The type and availability of storage areas within a university setting has obvious implications for developing a transportation strategy for moving hazardous wastes from the laboratory. Three general situations are discussed below; however, the reader may think of other possibilities. Table 1 summarizes the advantages and disadvantages of these options.

3.3.1.1 No Accumulation/Storage Area Available (Staging Area Only)

Hazardous waste management at many universities began by having periodic campus-wide cleanouts at which researchers or staff would bring hazardous wastes to a central location. The central location was a staging area where a hazardous waste contractor would separate wastes into the various hazard classes, package the wastes for transportation, and immediately remove these wastes for disposal.

This approach has several good points. Labor for the program is contracted out on a short-term basis to experts who handle the waste. The labor to move wastes to the staging area is split up among a large number of people from the various campus laboratories. There is no need to use resources to supply safe and adequate storage space for wastes.

Disadvantages of this approach are the possible inconvenience to researchers and relatively high unit disposal costs. If wastes can only be disposed on a periodic basis, each laboratory must make arrangements for storage of wastes between collections. In my experience, many labs do not have extra space for such storage. As a result, wastes are often stored unsafely. Disposal costs are relatively high because it is difficult to pursue cost-saving activities such as bulking of wastes at a temporary staging area. One last disadvantage of not having a storage area is unusual or emergency situations. In the case of a chemical spill, there is often a large quantity of spill cleanup materials that require storage pending disposal. When one does not have a storage area, these unusual items can be highly frustrating to deal with.

Having a storage area cannot always be justified. Periodic cleanouts are sometimes the most appropriate approach for a university. This approach may be best when a hazardous waste management program begins. It prevents unsafe conditions due to improper storage. It also enables the university to collect information for planning purposes so that good decisions can be made in planning for a more permanent solution. Small schools that do not generate much waste may also find it makes more sense to ship off wastes periodically in spite of the relatively high unit costs because the alternative of putting resources into operating a storage facility may be even higher. If this approach is pursued, the university should spend some time developing plans for dealing with chemical spills and other emergencies.

Table 1 Comparison of Alternatives for Hazardous Waste Storage

Type of Storage	Possible Advantages	Possible Disadvantages
None — staging area only	• No costs or maintenance required for storage space	• No space available when special situations arise
	• No dumping area for wastes	• Limited options for cost control; relatively high unit disposal costs
	• Less handling of waste minimizes possible accidents	• Labs may not have adequate storage space
Short-term accumulation	• Space available for emergencies	• Careful oversight required
	• Cost-control management options	• Additional regulatory concerns
		• Possible dumping ground
Long-term storage/handling/ treatment facility	• If properly designed, adequate storage space for all campus needs	• High cost of facility
		• Greater fixed costs
	• Allows full-range cost-control options	• Many regulatory requirements

3.3.1.2 Short-Term Accumulation Area

Short-term accumulation areas can mitigate many of the disadvantages that occur when one does not have a storage area. A facility of some kind exists where wastes can be stored for short periods of time.

The presence of an accumulation area allows hazardous wastes to be removed from the laboratory in a timely fashion. It also allows for possibilities of managing wastes through waste minimization techniques, such as redistribution, and through improved handling, such as bulking. In the U.S., storage areas where wastes are stored for less than 90 d (for large quantity generators) or less than 180 d (for small quantity generators) usually do not require permits from regulatory agencies.

Good management practices for short-term accumulation areas are:

1. Proper labeling of all containers
2. Segregating hazard classes
3. Maintaining an inventory of wastes in storage
4. Planning for emergencies

5. Restricting access to the area
6. Making sure one person has responsibility for the area

One potential drawback with any storage area is that it may turn into a dumping ground. Often, responsibility for oversight of short-term accumulation areas is shared among many people. This is a mistake. When everyone is in charge, no one is in charge. Universities with short-term storage areas must be careful that the area is used solely for short-term storage. With the very wide variety of wastes found at universities, problem wastes appear. It is often tempting to put problem wastes to the side to take care of them later. All too often "later" never comes.

3.3.1.3 Long-Term Storage/Handling/Treatment Facility

If a university generates a large quantity of waste or wants to conduct a wide range of management options, it will probably want to develop a more substantial facility than would be the case for short-term storage. In the U.S., such facilities are called treatment, storage, or disposal (TSD) facilities.

In the U.S., facilities that store wastes for more than 90 d (large quantity generators) or more than 180 d (small quantity generators) must have a permit to do so. Similarly, if a facility wishes to treat wastes, no matter how small the amount, at a central facility, a permit must be obtained. Major components of these permits are:

1. Waste analysis plan
2. Preparation and maintenance of an operating log
3. Secondary containment
4. Prevention of accidents
5. Emergency planning
6. Site cleanup at the end of its useful life
7. Training of site personnel

There is no doubt that life is less complicated for a university if it can operate its hazardous waste facility without a permit. Most large universities in the U.S. have or will end up obtaining a permit, however. Most have at least small quantities of hazardous wastes that are difficult or impossible to dispose of. Some find that they can treat wastes more economically on their campus than can off-site facilities. Without having the pressure to ship wastes frequently, a university may find more options available for managing wastes that are cheaper or otherwise more desirable.

As with the short-term accumulation area, there are several precautions. There must be clear and well-designated control over the facility. Personnel operating the facility must make sure that wastes which are hard to handle are not allowed to accumulate for excessive periods of time.

Although the funds required to construct a long-term storage/handling/treatment facility can be very large, the university will usually find that these

Table 2 Comparison of Options to Collect Wastes from the Laboratory

Method	Possible Advantages	Possible Disadvantages
Directly from laboratory by central staff	• Ensure proper labeling of wastes	• Cost of staff and equipment
Accumulation areas at each building	• Minimizes transport requirements for researchers • Increases efficiency of central management staff	• Labor intensive for central management staff • Areas require careful oversight; potential for dumping
	• May minimize amount of wastes stored in labs	• Responsibility for improper labeling and packaging may be difficult to trace
No collection — laboratory staff transports waste	• No cost for central staff or equipment • Researchers required to take greater responsibility for wastes generated	• Wastes may be allowed to accumulate in labs too long • Wastes more likely to be transported in unsafe manner

funds are well spent. If the facility is properly planned and funded, it can be designed to meet the unique needs of the university. Having a properly designed facility makes it possible for staff to work efficiently and pursue the most economical management options. A good facility enhances safety, enhances morale of those working in it, and puts forth a positive public image for the university.

3.3.2 OPTIONS FOR MOVEMENT OF WASTES FROM THE LABORATORY

Three general options for movement of wastes from the laboratory to a collection facility are presented below. Each of these has both positive and negative attributes (see Table 2). The system actually used may involve any one or a combination of methods depending on the unique situation of the institution.

3.3.2.1 Collection at Individual Laboratories by Central Staff

The most organized approach to the collection of wastes is to have the university's hazardous waste management staff go to each individual laboratory to collect the wastes. This approach gives the university a high degree of control. The central management staff can make sure that all wastes are properly identified and can easily request any additional information about the hazards of the wastes. By giving responsibility for on-campus transport of wastes to the central staff, the university can focus its training on a small number of people. Some training of laboratory staff would be required, but would be much less than if the researchers had more responsibility for transporting wastes themselves.

One implication of this approach is that the time required for laboratory staff to spend on hazardous waste management is minimized. Some would argue that this is a positive situation since laboratory staff are hired to conduct teaching and research. Others would argue the opposite — that laboratory staff should play a more active role in managing their wastes to encourage them to adopt waste minimization practices.

3.3.2.2 Collection at Accumulation Areas in Each Building

One way to increase the efficiency of central management staff is to have central accumulation areas in each building where hazardous wastes are generated. These areas could be laboratories, storage rooms, or, in some cases, locked cabinets. While making the central staff's time more efficient, the additional burden placed on laboratory workers would be minimal because wastes only have to be transported to another location within the building. In fact, laboratory workers might prefer this approach as a way of minimizing accumulation of wastes within their work areas.

This approach clearly has some advantages. However, there are potential pitfalls. Any place where wastes are allowed to accumulate must be closely monitored. Even if accumulation areas are carefully set up with good oversight, over a period of time these controls may become lax. Once the controls start to slip, wastes are found to be poorly labeled, and other safety hazards may arise. Once wastes are improperly labeled, responsibility for wastes can be difficult to determine.

3.3.2.3 No Collection, Laboratory Staff Transports Waste

As an alternative to having central administration staff collect waste chemicals from the laboratories, the institution could require laboratory workers to transport the wastes to a collection site. This approach would minimize administrative staff requirements.

There are several concerns that arise from this approach. First, because laboratory workers may see waste transportation as inconvenient, they may be tempted to allow wastes to accumulate too long in their laboratories. Worse, they may be tempted to dispose of the wastes improperly down the drain or in the trash. Some training of laboratory workers will be required to address issues such as proper labeling of containers, proper precautions for safe transport, and hours of operation for the collection site. When operating a collection site, one must take precautions to make sure wastes are not left during times when the site is closed.

In spite of the potential drawbacks mentioned above, the central collection site is a good option in many situations. Certainly for small schools where waste generation is minimal, a periodic campus-wide collection program may be the best alternative. Larger institutions may also find that allowing laboratory workers to transport their own wastes may be desirable in some situations.

3.3.3 SAFE PACKAGING AND TRANSPORT REQUIREMENTS

3.3.3.1 Safety Issues

The first safety rule in transporting wastes from the laboratory is to recognize that there are potential safety hazards. Those working with chemicals on a daily basis often have a tendency to become lax on safety issues because little ever goes wrong.

When evaluating the potential hazards from using chemicals in laboratories, one analyzes the hazards inherent in each chemical and tries to construct scenarios under which hazards might arise. Too frequently, this kind of analysis stops before hazardous wastes are considered.

Laboratory chemical wastes rarely present potential hazards to the general public. However, the potential hazards should not be underestimated by those handling the wastes. When it comes to the hazards in transportation, the potential hazard of most concern is container breakage. In the case of liquids, some are highly flammable. Others are corrosive to skin and materials. Many will give off toxic vapors. Some chemicals are air- or water-reactive, which present potential fire and explosion hazards. Still other chemicals are relatively safe, unless they come into contact with incompatible chemicals thereby creating a potential for fire or explosion.

Whenever hazardous wastes are handled, appropriate personal protective equipment should be worn. In most cases, gloves of some kind will be required. In some cases, eye protection and protective clothing may also be desirable.

3.3.3.2 Labeling of Wastes

All containers with chemicals must be labeled at all times. At the very least, the labels should include the idnetity of both the major and minor constituents of the material and the relative amounts of each constituent. In addition, it is very helpful to have information about what hazards (e.g., flammable, corrosive, oxidizer, reactive, and others) might be present with the material.

Good labels are also essential in determining optimum handling methods. Wastes that are properly identified are more likely to be recycled or disposed of by a cost effective method than wastes that are unlabeled. When unlabeled containers are encountered, someone must make a determination of what the material is before it can be disposed. It is almost always more desirable to have this determination made by the persons generating the material than by someone outside the lab.

Placing dates on containers when they are opened or first used is useful for two reasons. First, the date helps determine whether the material has become old and possibly degraded. From the safety point of view, some chemicals, such as peroxide formers, become more hazardous while in storage.

3.3.3.3 Packaging of Wastes for Short-Term Transit

Requirements for packaging of wastes for on-campus transport depend on how far the waste will be transported and how it will be transported. The first issue is the selection of suitable containers. To prevent spills, containers with screw cap tops are recommended. Some institutions may decide that glass containers present too great a hazard for breakage and that plastic containers should be used in some or all cases.

For short transport distances, such as from a laboratory down to the building loading area, one may be tempted to just carry small quantities of wastes. In general, this practice should be discouraged. A much better alternative is to put the waste into a poly tray or box and use a cart that can be pushed. Using a cart keeps one from attempting to carry too heavy a load and allows one to respond more quickly in the event of an accident.

Frequently, when one is disposing of a large inventory of old or excess chemicals from a laboratory, one sees the chemical containers placed into cardboard boxes. This may be appropriate. However, all too often, there are not enough boxes so chemicals are crammed together without regard to compatibility of chemicals. Sometimes containers with liquids are placed in boxes upside down, an action that creates a high likelihood of spills. Good quality boxes must be selected so that the bottoms and sides will not fail during handling.

Cardboard boxes do not provide good containment in the event of spills. As a result, polyethylene boxes are often a better choice. Even if polyethylene boxes are used, it is desirable to package the wastes to prevent container breakage. This can be done by packaging waste containers tightly to prevent jostling during transport or by using absorbent packaging material such as vermiculite or calcium bentonite.

3.3.3.4 Tracking of Wastes

Even if not required by regulations, most institutions will want to develop a scheme for tracking wastes from the laboratory to the storage facility to the off-site disposal facility. A tracking scheme does the following:

1. Provides information about wastes in the event of an emergency
2. Helps the institution in evaluating its hazardous waste management program
3. Provides information useful in redistribution or recycling programs
4. Facilitates reporting requirements to regulatory agencies

In the U.S., a multipart manifest is required for transportation of hazardous wastes along public roads. For on-site transport of wastes between the laboratory and the accumulation/storage area, the institution will probably wish to develop forms that provide more detailed information than the manifest.

CHAPTER 3.4

Off-Site Handling in the United States

Peter C. Ashbrook and Peter A. Reinhardt

TABLE OF CONTENTS

0-8493-682-5/94/$0.00 + $.50
© 1994 by CRC Press, Inc.

A comparison of off-site options for handling laboratory wastes is presented in Table 1.

3.4.1 INCINERATION

3.4.1.1 Incineration of Labpacks

The most common method used by universities for disposal of laboratory chemicals is incineration of labpacks. This method is expensive and the number of commercial facilities that are capable of incinerating labpacks is not very large — about a dozen in the U.S. Most of the incinerators are located in the Midwest or the South.

Incineration of labpacks is a good, general purpose disposal method for organics. For inorganics, it is not so good. Some metals, such as mercury and arsenic, must be carefully screened from wastes to be incinerated because they could be released with air emissions. Other metals must be limited to prevent them from concentrating in the ash. Most incinerators also have limits that prevent incineration of explosives, and several limit the quantities of reactive chemicals that can be burned per charge (drum). Wastes covered by these special limits can be extremely expensive to dispose of by this method.

3.4.1.2 Incineration of Liquids

Incinerators that can take labpacks can also take bulk liquids in drums. Bulk liquids are much easier to handle both in terms of labor and safety concerns. As a result, the unit costs for disposal of liquids by incineration is much lower than for labpacks. Many schools that have bulk liquids incinerated opt for fuel blending, which is about half the price of incineration. Those that pay the premium for incineration feel that the extra cost is justified by the very wide variety of chemicals in their wastes, the fewer handling steps compared with fuel blending, and because it is easier to track the ultimate disposal site for the waste.

3.4.1.3 Incineration of Bulk Solids

Since the handling of solids is much more difficult than for liquids, the cost to dispose of bulk solids by incineration is more expensive than for bulk

Table 1 Laboratory Waste Disposal Options in the U.S.

Disposal method	Suitable for	Typical costs	Comments
Incineration — labpacks	Organic materials, lab quantities	$20–30/kg	Small number of incinerators available
Incineration — bulk liquids	Solvents, combustible liquids	$1–2/kg	Good availability of disposal sites
Incineration — bulk solids	Organic materials	$2–4/kg	Limited number of available facilities
Fuel blending	Solvents	$0.5–1/kg	Good availability of disposal sites
Treatment — bulk liquids	Aqueous wastes	$1–4/kg	Good availability of facilities
Treatment — labpacks	Inorganic materials, lab quantities	$15–30/kg	Few facilities, availability improving
Returns	Pesticides, lecture bottles	Shipping costs	Not all companies accept returns
Reclamation	Liquid mercury	Credit	Typically shipped in 76-lb flasks
	Precious metals	Credit	May be hard to find interest in lab amounts
	Used oil	$0.05–0.10/kg	When possible, use rerefining in preference to burning
Normal trash	Nonhazardous solids, dry biohazardous wastes	0	Check with local landfill, label all containers "nonhazardous"

liquids. However, it is still cheaper to dispose of bulk solids than labpacks on a net weight basis. The number of facilities available to incinerate bulk solids is much more limited than for liquids. Most laboratories do not generate much waste in this category.

3.4.1.4 Fuel Blending

As noted above, there are a number of large industrial facilities that utilize flammable hazardous wastes for fuel. These include cement kilns, steel mills, and other large industrial facilities. When hazardous wastes can be used as a fuel supplement, the resource is being used more efficiently than if the waste is incinerated. Instead of the material being disposed of merely as a waste, some useful value (energy) is derived from the waste in the course of disposal. Because generators of hazardous waste are willing to pay for disposal of the material, the company burning the waste can be paid for the fuel rather than having to pay for fuel. While it is not quite that simple because of potential safety hazards, logistics, and the many regulations governing fuel blending, there is still an incentive for industrial facilities to accept such wastes for disposal. As a result of this incentive, generators who dispose of waste flammable solvents by this method find that their disposal costs may be only half of what are incurred if the waste were disposed of by incineration.

When this option is employed, wastes are typically first transported to a fuel blender. The fuel blender combines the wastes from many generators, characterizes the waste, and then works with interested facilities to arrange for deliveries in a timely fashion. One possible drawback to the generator for this method is that the fuel blender generally is paid by the generator when the waste is shipped from the generator's site. In some cases, when a fuel blender has had cash flow problems, the fuel blender will accept many loads of waste for disposal and get the money for the disposal, but then abandon the waste. When this happens, the generators will find that they are still responsible for disposal of their waste, as well as being responsible for cleaning up the site. Fuel blenders going out of business are not common, but it has happened. Generators can protect themselves by refusing to pay for services until evidence is presented that the waste has actually been disposed, usually in the form of a Certificate of Disposal. Insisting on shipment of waste directly to the disposal site is another possibility. As with any disposal site, generators should tour the facility before it is used and make their own judgment about whether it is a well-run facility.

3.4.2 TREATMENT FACILITIES

3.4.2.1 Treatment of Bulk Wastes

Treatment facilities for bulk wastes are best suited for treating wastes that are generated in tanker quantities. Most of these facilities will accept wastes in 55-gal drums, but not usually in smaller quantities. These facilities are designed to treat aqueous wastes with high levels of heavy metals, many of which are also corrosive. By far the most common treatment is precipitation of heavy metals followed by neutralization. The resulting sludge, which is considered to be stabilized, is dewatered and usually disposed of in a hazardous waste landfill. With the metals removed and the pH adjusted to neutral, the resulting wastewater can be disposed to the municipal sanitary sewer system. Depending on the types and amounts of metals in the original waste, it is sometimes feasible to recover metals from this treatment process instead of merely disposing of the sludge. There is good availability of facilities for treatment of bulk aqueous wastes. Most facilities will request a detailed analysis of these wastes for metals prior to acceptance of these wastes. If the contents of drums of such waste is highly variable, as is likely the case for wastes generated by laboratories, the required analysis can add significantly to disposal costs.

Some treatment facilities take a more basic approach to treatment of aqueous wastes. They may have a treatment pit into which they place lime, fly ash, or similar materials and add the contents of drums of wastes. The resulting mixture will stabilize the heavy metals from leaching. The mixture will often solidify, so that it can be disposed of as a stabilized waste in a hazardous waste landfill. When compared with the treatment process in the previous paragraph, there is a larger quantity of waste to dispose of by this latter method since all the water is bound in the waste that is finally disposed. Metals tend to be less concentrated, making the waste that is actually disposed slightly less hazardous. This latter method is usually used with wastes transported in drums or smaller containers since larger quantities overwhelm such treatment systems with their large volume.

3.4.2.2 Treatment of Labpacks

Facilities that treat labpack quantities of wastes have increased in number over the last few years. Because of the wide variety of laboratory wastes, the labor involved in individually treating a large number of containers can quickly become very large. Treatment facilities that handle labpacks operate by taking wastes from a number of sources, consolidating the wastes, and then treating wastes in batches that make more efficient use of labor.

3.4.3 RECLAMATION, RECYCLING, RETURNS

Facilities that can economically reclaim or recycle laboratory quantity wastes are not very common. As mentioned in the previous section, some general purpose facilities have recently been established that take laboratory wastes from a variety of sources. These facilities can then consolidate wastes into larger, bulk quantities for which there are more and cheaper treatment and disposal options.

Some facilities claim to accept partially used laboratory reagents with the intent of redistributing them to other facilities that may have a use for them. This concept is appealing; however, such facilities must be viewed with caution. Promises about success rates in recycling partially used reagents tend to be overly optimistic. Mineral acids are in high demand. Inorganic chemicals usually have reasonably long shelf lives. However, organic materials have shorter shelf lives. The management of redistribution facilities must be very good, both to accumulate a reasonably large inventory of used reagents and to distribute these materials to interested parties. From the point of view of an institution with used laboratory reagents to dispose, it is preferable to find interested parties directly rather than go through a third party.

In the U.S., a number of industrial materials exchange programs have been established. These exchanges have been established based on the concept that some companies may be able to use wastes from other companies as raw materials in production processes. The exchanges try to match companies producing wastes with companies interested in using the wastes. When matches are made, both companies benefit — one through lower disposal costs, the other through lower cost of materials. The public benefits as well because less material is disposed of as waste. Although waste exchanges are best suited to industries that produce waste of predictable quantity and quality, occasionally producers of laboratory wastes may find these exchanges beneficial to use.

With the exception of used oil, recycling or reclamation possibilities tend to be limited. Used oil can be accumulated in drums with little hazard. Because of the widespread use of oil, oil recyclers serve most communities. These recyclers serve small generators of used oil, such as automobile service stations, and therefore will usually be willing to collect used oil from laboratories. When working with an oil recycler, ask the person collecting the used oil how it will be recycled. If presented the choice, it is preferable to dispose of oil through companies that will rerefine the oil instead of companies that burn the oil for fuel.

Reclamation of other chemical wastes from laboratories is more difficult due to the small quantities generated. Some examples of laboratory wastes where successes are possible include liquid mercury, silver-containing wastes, and other precious metals. If you are interested in reclamation, be sure to get good cost figures from the company. Sometimes disposal costs may be greater for reclamation than for other options.

Some materials may be returned to manufacturers or redistributed. Types of wastes for which returns might be possible include lecture bottles of compressed gases, pesticides, and occasionally certain other specific chemicals. Most manufacturers of laboratory chemicals have not shown any interest in accepting excess or unused materials for returns.

3.4.4 MUNICIPAL SYSTEMS

In prior years, disposal of laboratory wastes was almost exclusively through the normal trash or via the sanitary sewer system. Over the past 20 years, this practice has become highly discouraged. Although not appropriate for hazardous wastes, many chemicals can be safely disposed of via municipal systems.

Putting waste chemicals down the drain to the sanitary sewer is acceptable when wastes are in compliance with the local sewer ordinance. Most sewer systems restrict wastes that are flammable, corrosive, or have levels of metals higher than specified limits. Laboratory workers frequently dispose of waste acid solutions down the drain believing that there will be adequate dilution of the waste by other wastes entering the sewer. In other cases, researchers are aware that all wastes go through acid neutralization basins prior to entering the sewer systems. Both practices should be discouraged: acid wastes should be neutralized before being poured down the drain.

The normal trash is suitable for nonhazardous solid wastes. Under no circumstances should liquids of any kind be placed in the normal trash. Modern municipal solid waste landfills are designed to keep liquids out, even the small quantities generated in labs. If the institution decides to allow disposal of nonhazardous chemicals in the trash, clear guidelines must be established. The guidelines should specify: (1) what chemicals may be disposed by this method, (2) how the chemicals should be packaged, and (3) how the chemicals should be labeled. Appropriate packaging is particularly important. If solid wastes escape their containers, they will form dusts when small trash receptacles are emptied into larger receptacles by the custodial staff. Since the custodial staff rarely has training in chemicals, the dusts can be a source of anxiety even when they are not hazardous. Proper packaging and labeling helps maintain good relations between laboratory workers and custodial staff.

3.4.5 OPTIONS FOR OFF-SITE HANDLING AND DISPOSAL OF BIOHAZARDOUS WASTE

Off-site treatment is usually necessary for isolation wastes from patients with highly communicable diseases, contaminated and uncontaminated sharps (e.g., needles, syringes, scalpel blades, glass slides and cover slips, pipettes and pipette tips), blood, body fluid, tissue, and animals and other materials contaminated with infectious agents, blood, or body fluids. Disposal options include the use of municipal systems for nonbiohazardous wastes and facilities specifically designed to destroy biohazardous waste.

3.4.5.1 Municipal Waste Landfill

Municipal solid-waste landfills (also called sanitary landfills; see Section 3.4.4) for normal trash, refuse, and garbage continue to be a disposal method for untreated biohazardous waste. According to the Agency for Toxic Substances and Disease Registry (ATSDR), "untreated medical waste can be disposed of in sanitary landfills, provided procedures to prevent worker contact with this waste during handling and disposal operations are strictly employed" (Public Health Implications of Medical Waste: A Report to Congress, U.S. Department of Health and Human Services, Public Health Service, Agency for Toxic Substances and Disease Registry, Atlanta, 1990). In many cases, the moist and warm landfill environment is hostile to pathological agents.

As noted by ATSDR, disposal of biohazardous waste in a municipal waste landfill requires that the waste be contained and undisturbed in transit and when placed in the landfill. A major disadvantage to this disposal method is that such careful handling is unusual. Normal trash is often subject to handling that disturbs the waste and any containment that may be present. Municipal solid waste is increasingly compacted, sorted, and examined for hazardous and recyclable materials. Intact and uncontained biohazardous waste can cause injury to refuse workers (e.g., from needles and other sharps) and aesthetic degradation of the environment if it escapes the landfill area. As a result, many municipal waste landfills are reluctant or refuse to accept biohazardous waste. In addition, landfilling of biohazardous waste is only appropriate for solid waste and absorbed liquids.

The principal advantage of landfilling is cost. Since biohazardous waste is disposed of with normal trash there is no marginal cost above the disposal of other institutional normal trash.

3.4.5.2 Municipal/Sanitary Sewage Treatment Plant

A modern and efficient sewage treatment plant is an appropriate disposal method for blood, other body fluids, and aqueous infectious waste. The environment of sewage treatment works are hostile to infectious agents. In addition, the great dilution of infectious agents in wastewater systems greatly reduce their potential for disease transmission. An important advantage of sanitary sewage disposal is its low cost. The municipal sewage treatment plant is likely to limit use of the sanitary sewer for wastes that exhibit a high biological oxygen demand, such as blood.

The primary disadvantage of this method is the potential occupational risks to those persons who discharge the liquid biohazardous waste into the sewer; they must be protected from direct contact with the liquids and any aerosols that often form when pouring and handling liquids. Engineering controls must be provided (e.g., containment, filtered ventilation, etc.), in addition to ample personal protection (e.g., appropriate gloves, apron, respiratory protection, and complete eye protection, etc.). Proper facilities and equipment can be expensive.

3.4.5.3 Municipal Waste Incinerator

Normal trash, refuse, and garbage are incinerated in some locales in a municipal waste incinerator. Disposal of biohazardous waste by incineration is advantageous because a properly operating incinerator will destroy infectious agents, sharps, and recognizable waste. As mentioned with municipal landfills, low cost is another advantage. There is no marginal cost above the disposal of other institutional normal trash.

As is true with use of other municipal systems, preventing worker contact with biohazardous waste is necessary but difficult when using a municipal incinerator for disposal of biohazardous waste. Another problem is that some incinerators cannot handle bulk blood and other liquids, or can do so only when liquids are absorbed or in very small volumes.

3.4.5.4 Regional Biohazardous Waste Incinerator

When municipal systems cannot or will not accept biohazardous waste, some biological and medical facilities have transported their waste to a regional biohazardous waste incinerator. These incinerators are owned either commercially or cooperatively (by their users). Because of the high cost of achieving compliance with today's transportation, occupational safety, and air pollution control laws, these larger facilities can keep costs low through the economy of scale. Liability for improper operation is also shared, which can be a benefit and a burden for users, depending on their relative risks. Cooperatively owned facilities have the advantage of directly controlling operating costs, while the cost of using privately owned facilities is subject to the waste disposal market. Privately owned facilities that are operated by large waste disposal firms may have better resources to deal with increasing regulatory scrutiny.

3.4.5.5 Other Off-Site Disposal Methods

An alternative to a regional biohazardous waste incinerator includes a retort or autoclave for steam sterilization. Although not used widely, these facilities deserve more attention. Other technologies are offered or are approaching the market, but none have the long successful track record of steam sterilization. Be sure to carefully examine both the technology (for effectiveness) and the site (for environmental risks and impacts) before using any off-site facility.

3.4.6 HAZARDOUS WASTE LANDFILL

In the U.S., direct disposal of all hazardous wastes, including those from laboratories, has been banned, except in cases where specific exemptions have been granted. Direct landfill disposal was completely banned with the final implementation of the land disposal restrictions by the U.S. Environmental Protection Agency (EPA) in 1990. These restrictions have not made landfills obsolete. The restrictions have prevented virtually all hazardous wastes from being disposed of in landfills without some kind of pretreatment. The U.S. EPA has specified treatment standards for every waste listed as hazardous. Laboratories are unlikely to ship wastes directly to hazardous waste landfills.

On-Site Treatment System of Laboratory Waste from Research and Educational Activities

CHAPTER 4.1

Responsibility for Selection of an On-Site Treatment System

Takashi Korenaga

TABLE OF CONTENTS

0-8493-682-5/94/$0.00 + $.50

4.1.1 INTRODUCTION

The global environmental protection has now become a problem of society worldwide. It is therefore necessary to have universities which are in harmony with their environment. Research staff and students who can integrate environmental awareness in their daily campus life will play an important role not only in reducing environmental degradation, but also in gaining economic advantages for their university.[1] University clean-up technologies with a goal of expediting research and educational laboratory experiments while polluting less are a real challenge, but are mandatory for a university, because campus life can be characterized as a system into which certain elements are input and from which different elements are output.[2-5] To simplify this idea, we can list several input elements:

1. Raw chemicals, paper, foods, pesticides, and herbicides which are necessary for campus activities
2. Water and air as solvents, cleaning, cooling agents, and for food preparation
3. Manpower
4. Capital and time
5. Energy use of different forms

Some output elements are

1. Experience gained from academic training and education
2. Publications
3. Know-how and patents
4. Graduate degrees (Bachelor, Master, and Doctorate)
5. Wastes (gas, liquid, and solid)

In years past, someone would just collect these wastes and take them directly to the local dump, where they might receive some sort of special attention. But this usually involved just standing by to see if anything unusual would happen. Sometimes a hazardous material might diffuse due to the dissolution of metallic compounds caused by acid rain phenomenon. Often, fires would occur as a result of this type of disposal. Even universities located in rural areas may not have disposed of waste chemicals by open burning 20 or more years ago. By the end of the 1970s, some universities in the U.S. and Japan were packaging their chemical wastes in steel drums for long-term storage, during which some of the contents would diffuse directly into the soil.[2-5] Most of these management techniques involved little in the way of actual out-of-pocket costs. The labor required for the collection of hazardous wastes and monitoring of the university effluent was usually considered just an added administrative responsibility for existing official staff. Total disposal cost was unlikely to exceed a couple of tens of thousands of dollars.

University laboratory waste management has some dangerous aspects if waste is stored at least six months, so certain waste management procedures are necessary:

1. Creating of a laboratory waste management system
2. Advance planning
3. Reducing waste as much as possible
4. Segregating waste and keeping it clearly labeled
5. Disposing of waste soon after collection
6. Means of treatment when chemical waste is spilled at a laboratory bench or elsewhere on campus

Waste reduction items are

1. Buying only the quantity required for each experiment
2. Working on a micro or semi-micro scale
3. Trying to recycle or reuse waste
4. Concentrating waste aqueous solution
5. Neutralizing/disposing of acids and bases
6. Precipitating metals from solutions
7. Utilizing/disposing of small quantities using specific methods

Recyclable waste items are

1. Silver and noble metal wastes
2. Nickel and valuable transition metal wastes
3. Organic solvent wastes

Various methods of on-site and off-site treatment of laboratory waste with multiple, random, and minimal discharge characteristics, and the responsible systems within a university administration, are discussed for different examples of Japan,[3] the U.S.,[6] and Korea.[7]

4.1.2 SELECTION OF APPROPRIATE FORM OF WASTE TREATMENT IN JAPANESE UNIVERSITIES

Many universities in Japan first recognized that their laboratory waste represented a special disposal problem in the 1970s. Prior to that time, most chemical waste disposal was made in the sewer or trash, and some was directly disposed of in the rivers, lakes, and sea. Problems with those disposal routes included fires in dumpsters, toxic odors coming from floor drains, and corrosion of plumbing systems. The industrial environmental concerns resulting in the 1970 Basic Law for Environmental Pollution Control (BLEPC) and the Water Pollution Law (WPL) certainly caused university personnel to examine their own laboratory waste disposal practices. Hazardous chemical wastes and wastewater from university laboratories then became a legal concern with the passage of the improved WPL at the end of 1974.

Universities came up with a variety of solutions to the management of hazardous waste during the 1970s. Chemistry departments would often take the first steps since they were the ones producing the most waste. Most of the initial efforts involved some sort of simple collection and use of off-site treatment process, for instance, a laboratory or campus clean-up once or twice a year.[3]

4.1.2.1 Establishment of the Japanese Association for Laboratory Waste Treatment Facilities of Universities (JALWTFU)

In July 1981, systematic improvement of WPL resulted in total load regulations being imposed to control water pollution. A total load value of chemical oxygen demand (COD) was mandated in areas with a concentration of industries and population near semiclosed bodies of water; their protection from organic pollutants was particularly necessary to reduce the number of red tide incidents. From the late 1970s to the early 1980s, almost all national universities had serious problems of hazardous laboratory waste. It was for this reason that a temporary Association of Liquid Waste Treatment Facilities at Japanese National Universities was established in November 1979 through the cooperation of the Environmental Centers of the University of Tokyo and Kyoto University and with the encouragement by the Ministry of Education, Science and Culture of Japan. In November 1983, the Association was made permanent and renamed the Japanese Association for Laboratory Waste Treatment Facilities of Universities (JALWTFU). National universities cooperating were Okayama, Nagoya, Hiroshima, Tokyo Institute of Technology, and others.

Hazardous and nonhazardous wastes from universities and their systems of treatment are defined on the basis of extensive investigations funded by the Nissan Science Foundation of Japan.[3] Among the characteristics deemed hazardous are: ignitable, corrosive, reactive, and toxic. A waste is considered toxic if it contains an unacceptably high level of any one of six heavy metals or cyanide, or of seven chlorinated solvents and pesticides. Several hundred chemicals used in Japanese university laboratories were grouped and listed according to delivery informations on each department purchased. The regulations imposed were largely the same as companies producing hazardous waste totaling more than 50 m³ of their effluent volume per day. A manifest system was established in the late 1980s to track wastes from the generator to the disposal site. Perhaps most significantly, hazardous wastes could only be disposed of at permitted treatment, storage, or disposal sites. These regulations made it necessary for all universities using hazardous chemicals to hire a full-time individual to manage the ensuring waste. Disposal costs for large universities now typically range from $80,000 (U.S. $) to $800,000 in Japan. After lengthy administrative discussions, all the universities decided not to send their waste chemicals packed in steel drums to a landfill for disposition.

4.1.2.2 Substitution of Nonhazardous Alternatives

Alternatives for acidic solutions with heavy metals are quite straightforward. The first option is for researchers to neutralize their own corrosive waste if it has no other hazardous characteristics. Separation of heavy metals from acidic waste will increase the amount of the latter which can be neutralized. If heavy metals are present, they can be precipitated out by raising the pH. We decided to look at mercury waste, not because it composes a large portion of laboratory waste, but becasue almost every laboratory uses mercury. Substitution of another material is the best way to minimize mercury waste. Good alternatives are available, with thermometers, although for certain organomercury reagents in common use, good alternatives are often difficult to find. We also selected chromic acid because it is commonly used for cleaning glassware, and we were aware that many alternatives were available. We found that in most cases chromic acid was being used from force of habit; often detergents would do the same job. In other cases, a stronger agent such as potassium hydroxide and ethanol may be required. If a strong oxidizing, acidic cleaner is needed, there are agents available that do not use chromium(VI). Neutralization is necessary, however, for alkaline or acidic detergent solutions when discarded.

4.1.2.3 Downsizing to Microscale Laboratories

Two waste management concepts could be applied by universities to prevent predictable pollution and minimize various wastes. Economic pressure caused by high disposal costs and safety concerns have had a significant impact on the production of chemical waste from teaching laboratories. The advisability of selecting experiments compatible with the global environment has recently been recognized, and "microscale laboratories" is now a key phase stirring much interest in junior, senior, and graduate student courses involving research. At Okayama University, some of the introductory teaching courses are usable videotapes and a sightseeing tour of the Center for Environmental Science and Technology (CEST), which handles liquid waste from laboratories and encourages downsizing the volume of chemical waste. Over 1000 junior, senior, and graduate students experience such courses every year. A current investigation by CEST on details of the acceptance of microscale laboratory concepts is holding the waste generated by these laboratories to only several percent of the total generated at Okayama University.

Research laboratories are much more significant sources of chemical waste. Because of changing research methods and projects, efforts to reduce the production of waste from research have apparently not been as successful as in teaching laboratories, although there have been many successful examples of reduction. We recognize both the importance of waste minimization and the difficulties in implementing microscale practices in a university setting. Therefore, CEST first designed microscale laboratories based on downsized experiments, such as the application of flow-injection analysis (FIA) to environmental water analysis in 1980.[8] By introducing FIA methodology to all laboratories on campus which produced hazardous wastes, minimization would be attained to some extent.[9-11] The purposes of the 1990 microscale trial were to better understand how chemicals were used in the laboratory and to identify practices of waste minimization being followed.

4.1.2.4 Selection of an On-Site Treatment System in Japan

At Okayama University, we have found that approximately 20% of laboratory hazardous waste comes from nonlaboratory sources, such as medical and dental practices[12] and photo and art processing. Most of the latter consist of materials such as pigment, paint, paint thinner, degreasing solvent, used oil, and complexed/mixed wastewater. Other sources include batteries, copier fluids, glass bottles, certain cleaning materials, light sources, and electrical equipment that sometimes contain polychlorinated biphenyls (PCBs). Common dry batteries and fluorescent light tubes usually contain mercury compounds. Many universities make the mistake of focusing only on laboratory waste; however, regulatory agencies are very aware of these other sources, and nonlaboratory hazardous wastes account for a large share of the regulatory violations.

Therefore, each Japanese university should have as part of its overall laboratory waste management program the responsibility of deciding about on-site treatment, so that direct, fast, and flexible administrative judgements can be made and the diffusion of hazardous waste avoided. At over 100 national and private universities/colleges in Japan, an on-site waste treatment is also effective in making students, teachers, and official staff recognize the importance of sustainable environmental education. Details are described in the following sections of this chapter.

4.1.3 WASTE TREATMENT SYSTEMS IN THE U.S.

The head of the Waste Management Section at the University of Illinois at Urbana-Champaign made the following points in his invited lecture at the First Asian Symposium on Academic Activity for Waste Treatment (AAWT) in Tokyo in August 1992.[6] Hazardous waste management at universities in the U.S. had undergone many changes during the past 20 years. This period of change was continuing today with the current emphasis on pollution prevention and waste minimization. All large universities were finding the need to develop a central hazardous waste management staff. However, the most successful of these programs would be those finding ways to work with researchers in the huge number of university laboratories to minimize waste generation at the source, in overall consideration of protecting the global environment.

As started, changes in the manner of handling hazardous waste began in the 1970s with the recognition of problems in this area. Many formal management programs began as a result of the first regulations of the Resource Conservation and Recovery Act (RCRA), which became effective in the U.S. in November 1980. At the end of the 1980s, the land disposal restrictions caused further changes in university waste chemical management programs. Now, although not yet mandated by government regulation, the current emphasis is on waste minimization and pollution prevention, and hazardous wastes are defined as falling into the four categories listed earlier (ignitable, corrosive, reactive, or toxic). Several hundred chemicals are also considered hazardous when discarded. A laboratory waste is considered toxic if it contains an unacceptably high level of one of eight heavy metals or one of six chlorinated pesticides based on RCRA.

4.1.3.1 Recognition of Problems with University Waste

In the late 1980s, a manifest system was established to track waste from the cradle (the generator) to the grave (the disposal site). The regulations were directed primarily at companies that produced more than 1000 kg of hazardous waste per month. Less stringent regulations applied to those generating more than 100 kg but less than 1000 kg per month. Perhaps most significantly, hazardous wastes could only be disposed of at approved treatment, storage, or disposal sites. These regulations made it necessary for many of the large universities to hire full-time personnel to manage their hazardous wastes. Disposal costs for the largest universities typically ranged from $10,000 to $50,000 in the U.S.

After Congress passed the Hazardous and Solid Waste Amendments (HSWA) in 1984, direct disposal of hazardous wastes in a landfill was prohibited unless the wastes had been treated to make them as nonhazardous as possible. Provisions in the HSWA act were called "land disposal restrictions". At the direction of Congress, the U.S. Environmental Protection Agency (EPA) developed treatment standards for every material considered hazardous waste.

The last of the land disposal restrictions became effective in 1990, and many universities in the U.S. experienced difficulty when these were initially put into effect. The main reason for the delay was that most universities shipped off a large portion of their waste from the university laboratories for direct landfill disposal. When the new regulations became effective, alternatives to direct landfill were somewhat limited and were changed to incineration or suitable treatment. Recognizing a business opportunity, several hazardous waste facilities have recently developed new programs to handle these laboratory chemicals. Most universities can now find several companies willing and able to meet their disposal needs. Costs took another jump since alternatives to landfill disposal were usually more expensive, and many large universities have annual hazardous waste disposal costs of $100,000 or more.

4.1.3.2 Hazardous Waste Management and Overall Waste Minimization

In the U.S. most large universities recognize that hazardous waste management requires a continuing commitment of both staff and money. There is really no way to prevent the improper use of the drain and the trash can because they are so convenient. However, there is strong evidence that laboratory workers are acting responsibly. For example, at the University of Illinois at Urbana-Champaign, waste chemical generation rose from 20,000 kg in 1982 to 80,000 kg in 1992. The use of chemicals did not increase by a factor of four, rather, the proportion of waste chemicals disposed of through proper channels accounted for most of the increase.

Both the importance of waste minimization and the difficulties in implementing waste minimization practices had to be recognized in the university setting. We undertook a two-year survey of laboratory waste minimization of the University of Illinois campus which began in the fall of 1990 using three approaches:

1. A questionnaire was prepared and given to all 300 laboratories on campus which produced hazardous waste in 1990. The purposes of this survey were to better understand how chemicals were used in the laboratories and to identify existing waste minimization practices.
2. Part of the study took a look at several broad groups of waste chemicals to evaluate minimization methods.
3. Six different laboratories were chosen for detailed study and to present some of the new findings. Of the 300 questionnaires distributed, about 270 were returned.

One finding that encouraged the study staff was that researchers showed a high degree of interest in minimizing waste.

1. One would expect that teaching and research would be the primary concerns of laboratory workers. So it was a pleasant surprise to find that these individuals showed a deep interest in waste minimization. Not only were the researchers interested, they requested information about alternative waste practices. Also, as expected, we identified many waste minimization procedures that researchers had established on their own initiative.

2. The study focused on the following general wastestreams: solvents, acidic liquid-waste containing heavy metals, mercury, and chromic acid in the second phase. One of the first approaches we looked into was a redistribution program for "pre-owned" chemicals. About 10% of the chemicals being held as waste were in perfectly good condition, many of them even in the original unopened container. Rather than disposing of them, it was better to find another researcher who could use these items. Redistribution means that the second researcher saves money on purchase costs, and the university saves on disposal costs. Redistribution was not the sole answer to minimizing waste, however, and, in fact, only reduced waste disposal by a few percent. While there were clear economic benefits to redistribution, the real value of the program was that it raised the awareness of waste minimization among researchers.

3. The final part of the survey involved a detailed examination of six laboratories in U.S. universities. Each laboratory was given the detailed questionnaire, and from these, we developed information on the experimental procedures actually being used. We also attempted to arrive at an overall balance by looking at the chemical purchases and waste disposal records. The single most effective waste minimization strategy was good housekeeping; this was hardly a radical finding, but the best solutions are sometimes overlooked because the laboratories are so unconcerned in management of their waste.

4.1.3.3 Selection of an Off-Site Treatment System

The detailed survey of the six laboratories was an attempt to study the overall balance. We traced all purchases, uses, and disposals of chemicals in these laboratories over a one-year period. Inventories of waste chemicals were conducted every three months over the same year. We found that the typical laboratory had 300 to 600 different chemicals in storage or use at a given time. One large organic chemistry research laboratory housed over 2700 different chemicals in various quantities. To simplify, we looked only at those in a quantity of at least 100 g. This overall balance study provided interesting information about how chemicals were used and disposed of. We were able to account for only half of the solvents bought by the organic chemistry laboratory; presumably the remaining half had either evaporated or gone down the drain. In a histology laboratory, in contrast, we found that purchases and disposal amounts of xylene were almost equal.

Options are available to minimize solvent waste. The scale of an experiment can be reduced. Less hazardous solvents can be substituted for those more hazardous. Distillation must be considered in minimizing solvent waste; however, from the standpoint of cost reduction, there are alternatives that are more effective than distillation. Solvents account for two-thirds of the waste generated at most universities so that any method such as distillation that can significantly reduce the amount of waste requiring disposal should receive close attention. We discovered that in the chemistry laboratories, about 50% of the solvents bought ends up as waste. If distillation were readily available, however, it is believed 65% of the solvents purchased might be recovered. Many researchers routinely redistill the solvents they buy to assure themselves of high purity; therefore, the stills should be readily available. Development of a distillation program for organic solvent wastes ought to be straightforward.

In the U.S., there is some interest evinced in requiring facilities to track all hazardous materials from purchase to disposal. For a single laboratory, or perhaps even a small number of laboratories, such tracking might be possible, but, for a large university it is almost inconceivable; the costs involved would far exceed the benefits. Each laboratory should maintain a good inventory system of the chemicals they use, but that is different from having a university-wide inventory.

There are no simple, low-cost options that will meet all hazardous waste management needs at universities. On the other hand, the solutions available are not technically complex. Each university is unique and must take the responsibility for developing a program to meet its unique requirements. Such a program should clarify the responsibilities of not only laboratory workers but also students, as well as the duties of the central hazardous waste management staff. All large universities, therefore, need to develop a central hazardous waste management staff and to select a reliable off-site treatment system for their laboratory waste. The most successful program will find a way to reduce the waste being generated at each university laboratory source.

4.1.4 WASTE TREATMENT SYSTEM IN KOREAN UNIVERSITIES

The Research Director of the Institute of Environmental Science and Engineering at Seoul National University, Korea also presented an invited lecture at the First Asian Symposium on AAWT in Tokyo in August 1992.[7] Since the treatment plant of laboratory waste was constructed and put into operation at Seoul National University in 1982, only four other national universities have practiced waste treatment which meets wastewater discharge requirements. Similar facilities are now under construction in several other universities. Laboratory waste, other than radioisotopes, is classified into six groups (two of organics and four of inorganics) and collected in different colored plastic bottles. Most universities are facing the following problems:

1. Sometimes the classification catalogue gives an indefinite answer to the type of waste generator to be used, even when the proper bottle is chosen.

2. There are no compulsory measures preventing hazardous laboratory waste from being discharged directly into the sewer.
3. In Environment Centers of Korean universities, the number of full-time professors, assistants, and operators is insufficient to carry out both research and management of laboratory waste.
4. There is no compulsory administrative edict mandating that the generators of waste become familiar with the capabilities and limitations of treatment plants.

4.1.4.1 Clean-Up Technology in Laboratory Waste Management

The so-called "end of pipe technology" was the first solution to treating generated waste so that it is not detrimental to the environment, thus fighting pollution. This technology, however, was utilized only by the lower two terms in the university waste management hierarchy and had two areas of inconvenience:

1. In many cases there was an additional cost (for investment and operation) for something which did not pay.
2. There was a decoupling of waste generation and waste treatment, thus creating a kind of dilution of responsibility.

The clean-up technologies did, however, offer a different way of thinking by introducing two concepts:

1. Reduction of waste at the source by a more rational use of raw materials and energy.
2. Enhancement of the value of raw materials, products, and energy available in the waste generated.

The idea underlying clean-up technology was that it made far more sense for a generator inside the campus not to produce waste rather than to develop extensive treatment schemes to insure that the wastesteam posed no threat to the quality of the environment.

4.1.4.2 Waste Minimization and an Acceptable On-Site Treatment System

Relatively small quantities of a large variety of chemicals are generated from university laboratories. However, the treatment and disposal of these hazardous wastes are both difficult and expensive. Many approaches are therefore being made to introduce clean-up technologies as a waste management strategy in Seoul National University (SNU) and other of the country's universities. Some research and educational programs are currently being done or are anticipated in the near future.

Waste could be minimized at the source by scrutinizing procedures to determine if their modification would result in waste reduction, by recovering and/or recycling waste where possible. For example, the second or third rinse water in a laboratory should be stored and reused for the first rinsing of a subsequent need.

Laboratory teachers are recommended to reorganize their laboratory courses to reduce the scale of experiments or substitute less hazardous chemicals for hazardous ones, as long as students are able to gain both a practical understanding of experiments and experience in experimental techniques. This would surely substantially reduce waste volume and toxicity.

Waste exchanges have been organized with considerable success by locating someone other than the generator who can use the waste. Used ultrapure water or solvent in one laboratory, for example, can be reused as fresh in another laboratory.

To convert ferrite residue to valuable material, research on the catalytic activity of chromium-substituted ferrite is being done by dehydrogenating ethylbenzene to styrene.

Water consumption increased at SNU by 70% between 1981 and 1991. Thus, wastewater from the cafeteria and wash basin was being considered for treatment and reuse in washing floors and flushing toilets, at least at the School of Engineering, which will be relocated within 5 years.

4.1.5 RESPONSIBILITY FOR ON-SITE TREATMENT SYSTEMS

Economic pressures due to high disposal costs and safety concerns have had a significant impact on the production of chemical wastes from teaching laboratories. Microscale laboratories are a topic in U.S., Canada, Japan, and elsewhere that has received much interest in recent years. At the University of Illinois, and several Japanese universities,[13] some introductory teaching courses are opened using videodisks to eliminate the need for wet chemistry laboratories, and videotapes illustrate both an on-site waste treatment system and microscale laboratories for waste minimization. Overall, however, teaching laboratories accounted for only a small percent of the waste generated in the U.S. universities.

The waste management concept for a university laboratory is different from an industrial one because the latter system is fundamentally different in its waste production. The concepts applied to a university are pollution prevention and waste minimization, meaning that there are attempts to prevent the generation of waste. If complete prevention is not possible, then at least the amount or toxicity of the waste that is generated should be reduced.

Almost all Japanese national universities possess an on-site treatment and have established a system of administrative responsibility for total waste management. These systems also have several new viewpoints and advantages in university research and education projects. These are useful so that university students, teachers, and official staff can recognize the importance of other related problems, such as technology transfer, which advances global environmental protection to save our only earth, and the need for safety management education to assure sustainable development.

REFERENCES

1. Armour, M.-A., Crerar, J. A., Klemm, R., McKenzie, P. A., Renecker, D M., and Spytkowski, G., Disposal methods for hazardous/carcinogenic inorganic and organic chemicals, Abstracts/Manuscripts of 1st Asian Symp. Academic Activity for Waste Treatment, Tokyo, July/August, 1992, 62.
2. Committee on Hazardous Substances in the Laboratory, Prudent Practices for Disposal of Chemicals from Laboratories, National Academy Press, Washington, D.C., 1983.
3. Japanese Association of Waste Treatment Facilities of Universities, Methodology and Technology for On-Site Treatment at Universities, Okayama, Japan, Nishi-Nippon Houki Shuppan, 1989 (in Japanese).
4. Kaufman, J.A., *Waste Disposal in Academic Institutions,* Lewis Publishers, Chelsea, MI, 1990.
5. Ashbrook, P.C. and Renfrew, M.M., *Safe Laboratories: Principles and Practices for Design and Remodeling,* Lewis Publishers, Chelsea, MI, 1991).
6. Ashbrook, P.C., Hazardous waste management at U.S. universities, Abstracts/Manuscripts of 1st Asian Symp. Academic Activity for Waste Treatment, Tokyo, July/August, 1992, 73.
7. Chung-Hak, Lee, Research activities of Seoul National University and other Korean universities in laboratory waste treatment, Abstracts/Manuscripts of 1st Asian Symp. Academic Activity for Waste Treatment, Tokyo, July/August, 1992, 53.
8. Korenaga, T. and Ikatsu, H., Flow injection analysis of chemical oxygen demand in waste waters from laboratories, *Bunseki Kagaku,* 29, 497, 1980 (in Japanese).
9. Korenaga, T. and Stewart, K.K., Seeking high-sensitivity flow injection analysis, *Anal. Chim. Acta,* 214, 87, 1988.
10. Korenaga, T., Izawa, M., Fujiwara, T., Takahashi, T., Muraki, H., and Sanuki, S., A double plunger micropumps for small flow rate down to several μl/min in flow injection analysis, *Anal. Sci.,* 7, 515, 1991.
11. Korenaga, T., Zhou, X., Izawa, M., Takahashi, T., Moriwake, T., and Shinoda, S., High-sensitivity flow method for the determination of proteins with microflow plunger pumps, *Anal. Chim. Acta,* 261, 67, 1992.
12. Reinhardt, P.A. and Gordon, J.G., *Infectious and Medical Waste Management,* Lewis Publishers, Chelsea, MI, 1991.
13. Tamaura, Y., Nakanishi, J., Takatsuki, H., Kitamura, M., Nakamura, I., Shinoda, S., Korenaga, T., Shoto, E., Hasegawa, N., and Yoshida, M., Environmental education and laboratory waste treatment in the Japanese universities, Abstracts/Manuscripts of 1st Asian Symp. Academic Activity for Waste Treatment, Tokyo, July/August, 1992, 105.

CHAPTER 4.2

Educational Program on Control of Hazardous Waste in the Laboratory

Hiroshi Tsukube

TABLE OF CONTENTS

4.2.1 INTRODUCTION

Who is the person responsible for control of hazardous chemicals in your laboratory? "Everyone," would be the ideal response but can everyone answer the following questions?

Is there a list of the hazardous chemicals used in your laboratory?
Do you have a full inventory of them?
Do you know the quantities of these chemicals consumed last year?
Can you justify the methods of their storage and disposal?
Do you know the legal regulations for handling and disposal of your hazardous waste?

All laboratory members should know how to do things correctly. This is the primary reason for conducting a hazardous waste education program.

Universities and companies often encounter many difficulties in establishing and carrying out educational programs, though these can ensure the proper and economical control of hazardous waste. Students and other members having broad knowledge and experience with chemicals routinely handle many kinds of hazardous materials. To improve laboratory safety and to assure compliance with legal regulations, these individuals must be taught the approved procedures for the handling, storage, transport, and disposal of these chemicals. It is the responsibility of the manager in a university or a company to develop an information program appropriate for his organization in which every member in the laboratory should be encouraged to participate. Planning a good educational program is not so easy, but putting it into practice and maintaining the momentum are much more difficult.

Most chapters in this volume are concerned with the control of hazardous chemicals and wastes, but this section is organized from a somewhat different standpoint. Several approaches are provided for those responsible for designing the needed information programs. Although offerings for university personnel and students may sometimes differ from those for industrial employees, successful programs will have many points in common. Described here are prototypes of educational programs which have been designed, tested, and improved in Okayama University. Included are a preliminary textbook on environmental science, written directions for laboratory procedures, information obtained from visiting a waste treatment facility, and comments from expert staff and laboratory supervisors. Depending on the size and character of his/her organization, the reader can modify this program to fit his need. Legal and technical specialists in waste management and other references should be consulted as necessary.

4.2.2 OVERALL STRATEGY FOR CONTROL OF HAZARDOUS WASTE

Before planning an information program, the overall management of hazardous waste should be carefully addressed. Since an effective educational program requires an understanding of all levels of the organization, overall management must be designed to support the educational program as well as to ensure compliance with the legal requirements of handling hazardous waste. Overall management decisions should be made with the cooperation and commitment of as many departments and individuals as possible. A statement of intent/purpose from the university or company president will set the initial tone and should assign a responsible person to plan and carry out the program. Drawing up a plan for a waste management strategy typically involves:

1. Review of management policies
2. Organization for hazardous waste control
3. Selection of waste treatment system
4. Safe and legal storage
5. Inventory control
6. Recycling and reuse of spent materials
7. Safe treatment and legal disposal
8. Minimization of hazardous waste
9. Information program for all levels involved

Figure 1 illustrates the organization of Okayama University's hazardous waste control operation. Okayama University is one of the major, multidisciplinary universities in Japan, a coeducational institution consisting of more than ten colleges as well as several graduate schools and hospitals. The university has on-site treatment facilities for various hazardous wastes which are maintained by the Center for Environmental Science and Technology. The president of the university appoints the director of the Center and organizes a Management Committee and an Operation Committee composed of several faculty members of related departments who develop the policies and plans for control of hazardous waste. The Center has facilities for treatment and monitoring of all wastewater created by the university and also runs several educational programs for students and faculty members. Despite the numerous colleges and departments, the members studying, investigating, or working in the laboratories have a direct line of communication to the Center. This type of system allows educational programs to be easily and effectively offered to the various groups.

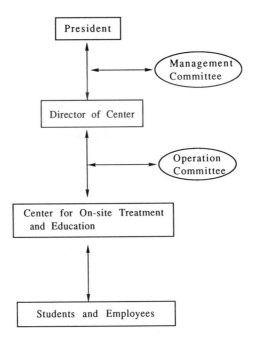

Figure 1. Flow chart of responsibility for hazardous waste control.

4.2.3 DESIGN OF EDUCATIONAL PROGRAM

Educational programs in a university or company are aimed at a broad audience and must therefore have a broad spectrum:

1. Undergraduate students
2. Researchers and graduate students
3. New and established employees
4. Line workers
5. Office workers
6. Foreign visitors and employees
7. Others

Of greatest importance among these groups are individuals investigating or working in the laboratory. Information in the program offered must convey "how and why" of hazardous waste management as well as the approved procedures to be followed in the laboratory. Information given by instructors in the initial orientation and follow-up guidance should include:

1. Legal regulations on wastewater
2. Overall management system for hazardous waste
3. Nature and toxicity of the chemicals in their laboratories

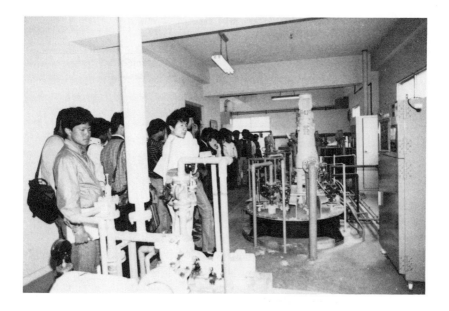

Figure 2. On-site visit to waste treatment facility.

4. Safe procedures to avoid chemical hazards
5. Correct storage and disposal procedures
6. Emergency action and responsible persons to notify
7. Reference manuals and books
8. Waste minimization

In addition, shorter courses for students majoring in human/social sciences and for office workers are valuable. They can benefit by a chance to recognize the unique atmosphere surrounding studies in environmental science and to realize a representative part in society.

A program may use any of several pedagogical approaches in providing instruction to a group: a pure lecture form, a written format, audiovisual materials, and an on-site visit to a waste treatment facility. Audiovisual materials are particularly effective when the training program must be replicated for groups of students or employees, and visiting a treatment facility leaves a deep impression which is long remembered (Figure 2).

This section will first outline an example of an educational program for new people coming into a laboratory. The program was designed for first- and second-year students of Okayama University and was intended to reduce risk by teaching identification and control of hazardous waste and also to assure compliance with legal regulations. The target audience is primarily individuals who are for the first time handling, storing, transporting, or treating hazardous chemicals in any quantity. The emphasis is on instructing students in safe and correct procedures for dealing with these chemicals

Table 1 Composition and Size of Chemistry Lab-Work Course

No.	Major (year)	Students per Class
1	Mechanical engineering (1st)	30
2	Electronics (1st)	40
3	Medicine (2nd)	25
4	Dentistry (2nd)	20
5	Pharmacology (1st)	25
6	Agriculture (2nd)	40
7	Computer engineering (1st)	20
8	Chemistry (1st)	35
9	Biotechnology (1st)	30

because they have, as yet, little knowledge of chemicals. They are exempt from legislation but obliged as socially responsible citizens. As described below, the program can be modified to fit other audiences, such as undergraduate and graduate students and employees in various types of organizations. The training may provide specific information on potential hazards in their particular laboratory, of either a chemical or physical nature, as well as general information on legal regulations surrounding the control of waste.

4.2.4 PROGRAM FOR JUNIOR STUDENTS IN THE LABORATORY

The College of Liberal Arts and Science of Okayama University holds a "Chemistry Lab-Work Course" (CLC) for first- and second-year students. CLC consists of four series of preliminary chemistry experiments: (1) test-tube analysis of metal cations; (2) test-tube analysis of inorganic anions; (3) acid/base and oxidation/reduction titrimetry; and (4) colorimetric determination. Students have had no prior training in a chemistry laboratory and for the first time handle, react, store, treat, and dispose "unfamiliar" chemicals. Their major fields of study are surprisingly diverse: chemistry, physics, biology, medicine, dentistry, pharmacology, mechanical engineering, electronics, agriculture, biotechnology, and other natural sciences. The composition and sizes of the CLC are summarized in Table 1. Although most have little knowledge of hazardous chemicals, students are eager to begin the experiments and often require most of the course time to complete it.

First impressions are the most important in every case. In the CLC orientation, the minimum requirements of waste control are presented in such a way that the students recognize they can protect the campus and meet legal standards by doing things the right way. These requirements are presented orally, are printed in the laboratory manual, and are distributed as a handout.

**Table 2 List of Selected Chemicals
Employed in Chemistry Lab-Work Course**

Inorganic acids	Inorganic bases
HCl	NH_3
HNO_3	NaOH
H_2SO_4	$Ca(OH)_2$
Hazardous metal salts	Organics
$K_3[Fe(CN)_6]$	CH_3OH
$KMnO_4$	CCl_4
$Pb(CH_3CO_2)_2$	CH_3CO_2H
$Cu(NO_3)_2$	
$MnCl_2$	Others
	H_2O_2

They are briefly but clearly worded and strictly enforced. Table 2 lists the selected chemical reagents that are employed in the experiments. They are common acid/base and inorganic/organic reagents in the laboratory but should be correctly controlled to adhere to legal regulations. Although CLC has been well organized and simplified to maintain safety in the laboratory, from their first exposure to these hazardous reagents, students learn "do things right".

The Center for Environmental Science and Technology of Okayama University began cooperative research on the educational programs in 1984 with the Department of Chemistry in the College of Liberal Arts and Science and prepared several documents giving the minimum requirements for hazardous waste control:

1. Information Sheet for Waste Disposal
2. Question and Answer Book on Environmental Science
3. Guide to the Center for Environmental Science and Technology
4. Videotape Entitled "Waste Control on Campus"

These materials are useful not only in the initial orientation but also in the continuing follow-up guidance. Instructors can use some of them to regain a student's attention, considering his/her background and progress in CLC. Subsequent instructions are designed to persuade students to comply with guidelines using rational arguments. As the term proceeds, students experience various situations in the laboratory and note that these practices may be troublesome but are rewarded. The form and substance of these materials are briefly described below.

The Information Sheet for Waste Disposal was prepared for students who handle, store, transport, or treat chemical waste in the laboratory. It contains information on the campus identification of hazardous wastes and the correct procedures for their storage and disposal, as well as legal regulations on

Table 3 Okayama University Campus Standards for Storage of Hazardous Waste

Container	Content
Inorganic liquid waste	
White 20 liter (l) polyethylene safety cans	Liquid waste containing toxic metal ions (Cu, Zn, Fe, Mn, Cr, Cd, Pb, etc.); 1st and 2nd washings of vessels.
Red 20 l polyethylene safety cans	Wastewater containing mercury salts; 1st, 2nd, 3rd, and 4th washings of vessels.
Blue 20 l polyethylene safety cans	Wastewater containing cyanide compounds retaining alkaline (above pH 12).
Organic liquid waste	
White 10 l polyethylene safety cans marked "combustible waste"	Self-combustible; also mixes easily with kerosene (benzene, toluene, xylene, hexane, ethyl acetate, ether, oil)
White 10 l polyethylene safety cans marked "noncombustible waste"	Formalin, etc.

wastewater (Tables 3 and 4). A copy is kept in each laboratory, and students consult it when they deal with any chemical hazards.

The Question and Answer Book on Environmental Science is a preliminary textbook for various study groups in natural sciences, engineering, geology, art, certain aspects of social science, public health, medical, and environmental sciences. Twenty topics were selected as key items from the various fields of environmental science, and typical ones are listed in Table 5. Students are thus given an opportunity to open the door to previously unknown but interesting fields.

The Guide to the Center for Environmental Science and Technology is used for visitors to the Center. More than 300 undergraduate and graduate students visited in 1991 and inspected line processes and/or learned about the treatment procedurs for hazardous wastes. This guide also describes the campus overall waste management system and has a rough sketch of the treatment facilities (Figures 3 to 6).

The videotape offers a visual explanation of the waste control and treatment on the campus. It is particularly effective when a training program must be replicated for several groups of students and also allows easy review by individuals or small groups. Some groups are shown this before their visit to the Center. Such audiovisual material is also helpful for students to overview the campus waste management, even if they have no time to visit the Center. Since an educational program for a junior student is a comprehensive "cradle to the grave" process, an instructor must combine several methodologies to teach his students.

Table 4 Japan's Legal Regulations for Disposal of Wastewater

Toxic Substances	
Items	**Allowable Limit**
Cadmium and its compounds	Cadmium, 0.1 mg/l
Cyanide compounds	Cyanide, 1 mg/l
Organic phosphates (parathion, methyl demeton, etc.)	1 mg/l
Lead and its compounds	Lead, 1 mg/l
Hexavalent chromium compounds	Hexavalent chromium, 0.5 mg/l
Arsenic and its compounds	Arsenic, 0.5 mg/l
Mercury and its compounds	Mercury, 0.005 mg/l
Alkylmercury compounds	Not detected
PCB	0.003 mg/l
General Standard	
pH	5.8–8.6
BOD	120 mg/l (average/d)
COD	120 mg/l (average/d)
Suspended solids	150 mg/l (average/d)
Normal hexane extract	
Mineral oil	5 mg/l
Fatty oil	30 mg/l
Phenolic compounds	5 mg/l
Copper	3 mg/l
Zinc	5 mg/l
Soluble iron	10 mg/l
Soluble manganese	10 mg/l
Chromium	2 mg/l
Fluorine	15 mg/l
Number of coliform group	3000 cells/cm^3 (average/d)

Table 5 Selected Topics in Question and Answer Book

• Environmental Pollution: What and Why?
• Typical Examples of Environmental Pollution
• Legal Regulations for Hazardous Material Control
• Responsibility for Hazardous Material Control
• Campus Standards for Hazardous Waste Control
• Definition and Identification of Laboratory Waste
• Correct Procedures in the Laboratory

4.2.5 PROGRAM FOR SENIOR STUDENTS IN THE LABORATORY

The significance of this type of program as part of the undergraduate and graduate course curriculum is increasing in chemistry and related natural

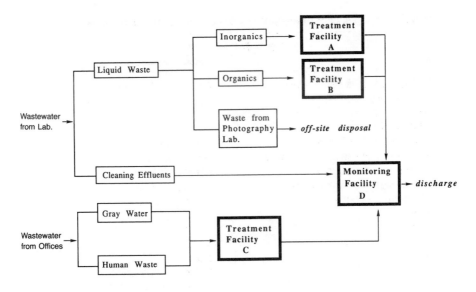

Figure 3. Wastewater treatment and disposal systems.

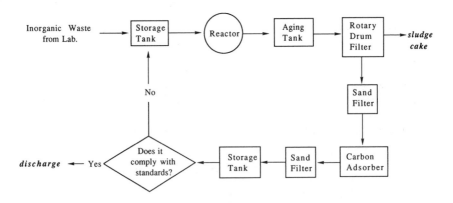

Figure 4. Clean-up process of inorganic liquid wastes (Facility A).

sciences. Educational programs offered senior students should include lecture by the research supervisor, printed manuals, and group seminars. The supervisor plays a particularly important role, with both moral and legal responsibilities, and greatly influences the attitude of the student. Before beginning a new laboratory experiment the supervisor should provide students with various information about the physiological properties of the chemicals to be used and the proper way of handling hazardous chemicals. This is an excellent opportunity for a student to seek information in the literature about the dangerous and toxic properties of these chemicals. Students thus are trained

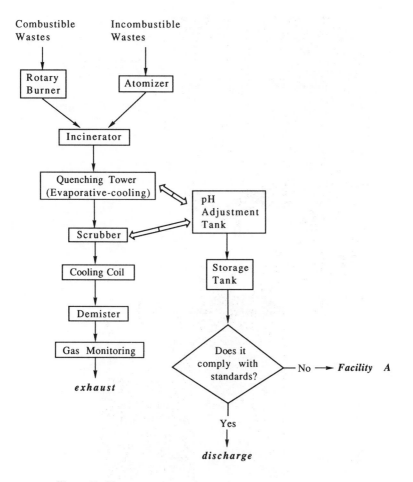

Figure 5. Clean-up process of organic liquid wastes (Facility B).

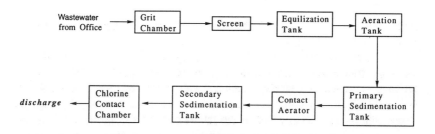

Figure 6. Clean-up process of wastewater from offices (Facility C).

Figure 7. Hands-on experience at a treatment facility.

to recognize potential hazards in their own research and to take the appropriate actions.

A valuable addition to the laboratory program is practical experience in the actual transporting and treatment of hazardous wastes. By hands-on experience at a treatment facility, a student can better understand some of the difficulties in satisfying the legal requirements and can recognize significance of how the material is handled in the laboratory. This program for a senior student should be continuous and organized by one or more knowledgeable and enthusiastic individuals.

4.2.6 PROGRAM FOR UNIVERSITY OR COMPANY EMPLOYEES

An educational program developed for an employee of a university or company has similar criteria for students. It essentially involves an initial orientation for new employees and follow-up seminars for experienced persons. The responsible individual should provide a program commensurate with the background and career experience of an employee.

As stated earlier, the initial orientation is the most important. A new employee must understand the basic rules of the laboratory and his particular responsibility in hazardous waste control. These should be clearly explained verbally by his direct supervisor and he/she should be given a manual for later reference. The orientation should include details of handling, storage, transport, treatment, and disposal of hazardous chemicals. When a number of new employees join the organization at the same time, this information may be presented to them in a series of lectures, demonstrations, and discussions. The minimum requirements should be determined by each laboratory and should be reviewed and modified as needed after an appropriate period.

Depending on the type of job and the responsibility of an individual, a follow-up program should be scheduled. An employee should be encouraged to take advantage of programs available to him within the company and/or in an outside teaching institution. A seminar of one day or less is generally effective in rekindling attention to detail and can be conducted by a guest speaker. Such an event underlies the commitment of management to make the laboratory a safer place and to enforce the legal regulations. The described programs for university students are also adaptable to nonstudent laboratory workers.

4.2.7 CLOSING REMARKS

Programs defining good practices in a university laboratory have two main purposes: (1) the reduction of accidents involving hazardous waste and the enforcement of prescribed regulations and (2) teaching students the practical meaning of environmental science. Management of hazardous materials is an integral part of becoming a chemist, a physician, a pharmacist, an engineer, or any related profession. At the same time, the protection and maintenance of a good and healthy environment is something for which everyone shares responsibility. A well-organized and systematized program can offer good examples of ways in which natural sciences and social sciences operate complementarily. An educational program prepared for company personnel has an added advantage in that effective waste control and minimization result in greater economic benefits. The control of hazardous chemicals is a necessary and essential factor to be addressed/recognized today and in the future.

ACKNOWLEDGMENTS

The author thanks his colleagues, Profs. Naomi Hayama, Tadashi Iwachido, Tatsuo Higashiyama, Kentaro Takagi, and Hiroyuki Ishida of Okayama University for their cooperation. Financial support for promoting these educational programs from Okayama University (1984 and 1986) is also acknowledged. He especially thanks Mrs. Yuri Migaki and Jane Clarkin for their patient help in typing, editing, and proofreading the manuscript.

REFERENCES

All materials prepared by Okayama University can be obtained at cost from the following address: Center for Environmental Science and Technology, Okayama University, Okayama 700, Japan.

Comprehensive Guidebooks

Young, J.A., Ed., *Improving Safety in the Chemical Laboratory: A Practical Guide,* John Wiley & Sons, New York, 1987.
Dawson, G.W., Mercer, B.W., *Hazardous Waste Management,* John Wiley & Sons, New York, 1986.

ACS Safety Books

The following volumes contain columns reprinted from monthly issues of *J. Chem. Educ.* They offer many useful hints in designing an educational program.

Steere, N.V., Ed., *Safety in the Chemical Laboratory,* Vols. 1–3, ACS, Easton, 1967, 1971, 1974.
Renfrew, M.M., Ed., *Safety in the Chemical Laboratory,* Vol. 4, ACS, Easton, 1981.

Journal Issues

Much useful information is also contained in continuing issues of the *J. Chem. Educ.*

Armour, M.A., Browne, L.M., and Weir, G.L., Tested disposal methods for chemical wastes from academic laboratories, *J. Chem. Educ.,* 62, A93, 1985.
McKusick, B.C., Classification of unlabeled laboratory waste for disposal, *J. Chem. Educ.,* 63, A128, 1986.
Lang, P., Disposal, labeling and storage in a high school chemistry laboratory, *J. Chem. Educ.,* 63, 887, 1986.

Hall, S.K., Labeling in hazard communication, *J. Chem. Educ.*, 63, 225, 1986.

Walton, W.A., Chemical wastes in academic labs, *J. Chem. Educ.*, 64, A69, 1987.

Houck, C. and Hart, C., Hazards in a photography lab: a cyanide incident case study, *J. Chem. Educ.*, 64, A234, 1987.

Fischer, K.E., Certifications for professional hazardous materials and waste management, *J. Chem. Educ.*, 65, A282, 1988.

Armour, M.A., Chemical waste management and disposal, *J. Chem. Educ.*, 65, A64, 1988.

Zwaard, A.W., Vermeeren, H.P.W., and Gelder, R., Safety education for chemistry students: hazard control starting at the source, *J. Chem. Educ.*, 66, A112, 1989.

Gerlovich, J.A. and Miller, J., Safe disposal of unwanted school chemicals: a proven plan, *J. Chem. Educ.*, 66, 433, 1989.

Gannaway, S.P., Chemical handling and waste disposal issues at liberal art colleges, *J. Chem. Educ.*, 67, A183, 1990.

Bretherick, L., Chemical laboratory safety: the academic anomaly, *J. Chem. Educ.*, 67, A12, 1990.

Carpenter, S.R., Kolodny, R.A., and Harris, H.E., A novel approach to chemical safety instruction, *J. Chem. Educ.*, 68, 498, 1991.

Peterson, D.N. and Thompson, F., The Minnesota technical assistance program offers intern experience in industrial waste reduction, *J. Chem. Educ.*, 68, 499, 1991.

CHAPTER 4.3

Fabrication of a Large-Scale Treatment Facility at a Japanese University

Kazuaki Isomura

TABLE OF CONTENTS

0-8493-682-5/94/$0.00 + $.50

© 1994 by CRC Press, Inc.

4.3.1 INTRODUCTION

There are many kinds of waste generated in the course of the research and educational activities of an academic institution. Even though the total quantity of waste is not large, some portion may be strongly hazardous materials. As described in other chapters of this book, universities are being forced to consider waste management, as environmental regulations are becoming more stringent and generators of smaller amounts of waste are now also subject to the regulatory system of the government and the region.

One option for the handling of waste is off-site treatment by some contracting company, but this imposes a high cost on universities. An alternate option is on-site treatment on university grounds, where the waste was generated. A convenient monograph, *Waste Disposal in Academic Institutions* edited by J. A. Kaufman (Lewis Publishers) was published on this matter in 1990, detailing the regulations on waste and options for its managements. One example of a treatment facility established within a university is also described with a comparison of the costs between that and commercial off-site disposal by a contractor.

Many academic institutions in Japan established waste treatment facilities in the early 1970s, when the Japanese archipelago was suffering from severe pollution. As the treatment of waste was an urgent problem at that time, facilities were constructed with regularly little consideration given to how they should be managed. It took a long time to overcome the many problems that resulted as the facilities continued to operate. Now, many universities have plans to rebuild because of the deterioration of the plant and increasing amount of wastes requiring treatment.

In this chapter, factors to be discussed and resolved in establishing a waste treatment facility in an academic institution are described in the earlier sections, and the case history of waste treatment in one university in Japan is shown in later sections.

Although emphasis is placed on the installation of the waste treatment facility, the importance is shown of establishing a consensus among all members of the institution to cooperate in preserving the environment in order for treatment to be conducted safely and surely.

4.3.2 REQUIRED CONSENSUS

With a facility installed on-site, wastes, especially liquid wastes which cannot be disposed of as they are generated, can be treated and the volume reduced, making the residue acceptable to a waste dump site. The cost for disposal is thus lowered. For safe and economical treatment fulfilling the regulated quality of effluent, flue gas, and solid residue, to be discharged or disposed of, the wastes must satisfy the qualities required for the processes to be conducted by the plant. Researchers and students who produce wastes are asked to segregate them into categories depending on the content. The collecting wastes must be safely stored until they are transported to the waste treatment facility. If the university is large, it is impossible for each waste generator to take his waste to the facility so certain personnel in each department are charged with collecting and storing them. This requires that a manual covering the handling and segregation of waste be prepared stating the method and rules for systematic collection as set by the university administration.

In the facility, the wastes must be treated in a manner to avoid any secondary pollution of the surrounding environment. Effluent and flue gas must be monitored continuously, and the residue produced in the course of the treatment must be disposed of according to environmental laws.

To conduct these tasks properly, the treatment facility needs staff having expertise in the treatment of waste and knowledge of the regulations in place. Such individuals can be hired from an engineering company, but if university members are assigned to such post, the future of the facility should be considered at the administrative level. Two or three years after the facility begins operation, the treatment would be routine for the staff, despite their difficult and dangerous nature. One of the best perspectives would be the establishment of a research center on the environment on the campus which would also be knowledgeable about environmental education.

4.3.3 PLANNING

Once consensus among university personnel has been reached, a project team should be organized to oversee the installation of the facility. This team should be composed of representatives from all faculties or departments concerned.

4.3.3.1 Fundamental Requirements for the Installation

In installing a waste treatment facility, there are three fundamental principles which should be met:

1. Sure and economically reasonable treatment of various wastes
2. Establishment of a safe environment in the facility
3. Prevention of secondary pollution

These principles actually apply to all aspects of waste management. The following sections describe inspection of the wastes generated and treated in a university, a feasibility study on the potential processes which can be used, actual designing and installation of the plant, and other matters.

4.3.3.2 Inspection of Wastes Generated

The first task for the project team is inspection of the quality and quantity of wastes generated in the university. Quantities of hazardous materials present in the wastes might be estimated by the amount of chemicals used in an academic year. In the actual treatment, knowing the concentration of hazardous materials is important. However, even more important in planning the facility is knowledge of how the chemicals are mixed in the wastes. In some cases, special additional procedures are required before the waste is subjected to the usual facility process.

In the case of the treatment of inorganic waste solutions, the plant is usually designed to treat waste containing only inorganic materials, and there are instances where no consideration has been given to the organic substances which are present in "inorganic wastes". Trace amounts of an organic chelating reagent, such as ethylenediamine tetraacetic acid (EDTA), will solubilize metallic ions by chelation even under high pH conditions, interfering with the treatment of heavy metals by neutralizing sedimentation. Even organic material having no chelating ability poses problems to a facility. Researchers and students might consider that 1% of organic material is too minimal to cause any problems in treatment. But reduction of 10,000 ppm (1%) of an organic substance to several tens of ppm or below, as enforced by regulations, is very difficult. In such cases, the contaminating organic substances should be separated or decomposed before treatment by an auxiliary process, or decomposed by incineration, followed by treatment of the water used to scrub the exhaust gas from the incinerator.

In the case of the organic wastes, a variety of wastes can be generated, some wastes are flammable and some inflammable. For incineration treatment, both types must be fed to the furnace in a proportion assuring steady incineration. Therefore, the content and the quantity of each waste sent to the facility must be known as precisely as possible.

It should be noted that not all organic waste can be treated by incineration. For example, polychlorinated biphenyls (PCBs) can be decomposed; however, even more hazardous polychlorinated dibenzodioxines or polychlorinated dibenzofuranes are produced if the incineration temperature is low. Incineration of PCBs is forbidden in Japan.

Another important matter to be considered in this phase of the project is the management of wastes containing radioactive materials and those containing microbial creatures which might infect facility staff.

4.3.3.3 Inspection of Regional Conditions

Operation of the facility causes inevitable emission of flue gas and aqueous effluent as by-products of the treatment, and the legal limits for such emission must be checked. Lows governing the environment are very complicated, however, and it is recommended that the proper governmental or regional office of the environmental agency be consulted for clear understanding.

There may be cases in which negotiations are required with people living in the area surrounding the location of the planned facility to reach an understanding about the flue gas and effluent.

4.3.3.4 Technical and Economical Evaluation

There are several processes which can be applied to the treatment of wastes generated in academic institutions. In determining what type of plant is optimal, the following points must be discussed for each of the candidate plants:

1. The kind of waste that can be treated by each process
2. Construction and operating costs
3. Method of treatment of the residue produced

There is no single plant which can treat all of the wastes produced in a university. Therefore, inspection on the treatability of all expected wastes should be made by each candidate plant being considered by the criteria shown below.

1. Waste that can be treated as it is produced.
2. Waste that requires pretreatment before the routine process of the plant.
3. Waste that is untreatable by the plant.

The most desirable plant is one which is capable of treating the majority of wastes generated with required pretreatments which are simple and easy, although the costs of investing and operating the plant must be taken into account. Construction cost depends on the degree of sophistication of the plant, especially on the degree of automation. The more automated, the fewer are the workers required to operate it. Installation of auxiliary processes also has a bearing on the cost. Instruments to monitor environmental quality of the working area and to prevent secondary pollution must be installed for smooth operation.

4.3.3.5 Management Planning for Operation of the Facility

Another important task of the project team is to decide on how the facility is to be managed.

1. The extent researchers and students should cooperate in the actual treatment of waste
2. The number of staff and workers required for the facility
3. The interval of waste collection throughout the institution

Without the cooperation of waste generators in minimizing waste production and separating by the content into the desirable categories, the time required to treat the wastes will be longer and more workers will be needed. At the very least, the researchers should segregate their wastes into designated categories; when they are transported to the facility, a document showing the content should be delivered to the facility staff. Lacking this information, treatment cannot be performed safely and thoroughly.

Classification should be determined based on the treatment processes which can be conducted by the plant installed in the facility, although, as stated, there may be some wastes which cannot be handled without some pretreatment. If pretreatment can be done easily and safely by the generator, no additional instrument is required in the facility. Nowadays, however, even in departments where chemicals were not previously used, chemicals are being employed. For example, in electronics research or with sophisticated technologies, chemicals are used to etch the surface or for other purposes. For researchers in these laboratories, even simple pretreatment of waste would not be easy. Although it is difficult to draw a line, in respect to pretreatment, between the responsibility of the facility and the waste generator, this line must be established. Naturally, installation of the pretreatment apparatus in the facility means a higher investment cost and the need for more manpower for the treatment, while the waste generator is relieved of this task. The decision on this issue requires thorough discussion and agreement.

4.3.3.6 Selection of Optional Instruments and Measures

The fact that experimental wastes contain many hazardous materials and, in some cases, microbial creatures means that facility personnel can be exposed to infection in their handling of wastes. If the waste has been produced in a medical or other laboratory investigating biorelated science, the vessel containing the wastes may also be contaminated. The possibility of direct contact with such waste must, therefore, be avoided to the greatest extent possible. This can be achieved by using automation, thus reducing manual handling; this can also reduce the manpower required.

From waste storage tanks, reaction vessels for treatment of the inorganic waste and in the working area where organic wastes are fed into the incinerator, bad smelling gas and organic solvent vapor are inevitably produced. These vessels and reaction tanks, thus, must be closed structures, and the gas evolving must be led to a scrubbing apparatus or deodorizing instruments in order to maintain an atmosphere in the facility which is safe for the personnel. That area of the facility where poisonous gas might be evolved must be well-ventilated by a system equipped to absorb vapor to prevent pollution of the outdoor atmosphere. Furthermore, gas sensors to detect poisonous gases such as HCl, HF, HCN, Cl_2 etc. should be placed in appropriate locations. The number of sensors and the kinds of gases requiring detection depend not only on the process; the possibility of noxious gases which might result from errors in segregation or inadequate pretreatment by the generator must also be taken into account, as shown later.

4.3.3.7 Countermeasure for the Secondary Pollution

There are many and various wastes and reagents used in the treatment in a facility. Countermeasures to cope with waste leakage must be considered. Contamination of groundwater by waste is against the law. Seeping water from the sealed part of any of the pumps in the facility is usually collected in underground tanks. To avoid groundwater contamination, these tanks should be placed in concrete basins.

4.3.3.8 Final Designing and Installation

Once the type of plant has been determined, designing is begun. Installation of additional instruments after the facility has been constructed is difficult, so the layout design is important. The design must be drawn up on the basis of the fundamental principles described earlier. The apparatuses for treatment, as well as the instruments for monitoring the atmosphere in the facility and the equipment and measures to assure against secondary pollution, must be taken into consideration in designing the building.

The instruments must be arranged so as to ease maintenance of the apparatuses, as well as the routine tasks.

4.3.4 CASE HISTORY OF WASTE TREATMENT IN ONE UNIVERSITY

4.3.4.1 Installation of the First Facility

About 20 years ago when the Japanese archipelago suffered from severe pollution problems, waste management became an urgent problem to be addressed by each university. At Kyushu University, a committee was formed by representatives from the faculties of engineering, science, agriculture, medicine, pharmaceutical science, and other affiliated institutes involved in scientific research. The biggest problem was how to treat experimental waste solutions containing mercury, heavy metals, or cyanide ion since the majority of solid wastes could be disposed as generated at an appropriate disposal site. Initially, the waste solutions were handled by a contractor, but the university then began to consider on-site treatment on the campus. However, at that time there was no plant suitable for installation on the university campus. The committee therefore had to design the facility. After many meetings, a coagulative sedimentation method was adopted in which heavy metal ions were precipitated by neutralization after reduction of chromate ion into chromium ion. If mercury ion was present, sulfide ion was added to reduce the concentration of mercury. Cyanide ion was decomposed by oxidation using calcium hypochlorite.

The plant consisted of several storage tanks of wastewater, a reaction vessel equipped with a mechanical agitator, a thickener, filter press, sand filter, activated carbon packed columns, mercury absorber packed with chelating resin, and tanks for sulfuric acid and sodium hydroxide used to adjust pH. The wastewater, which was segregated into that containing mercury, that containing heavy metal, and that containing cyanide, was collected from all over the university once a month.

4.3.4.2 Operation of the First Facility and the Problems Experienced

In the first several years after the plant began operating in 1972, the treatments were conducted smoothly. However, organic material contaminating the wastewaters gradually increased, probably because of the change in chemicals used in the research being conducted (Figure 1). In 1980, 35 m^3 of wastewater was treated. Annual load of organic contaminant, as measured by chemical oxygen demand (COD), was 661 kg, whereas that of inorganic substances (Fe, Cr, Cu, Zn, Hg, cyanide, and others) was 129 kg. In 1981, operators of the facility asked waste generators to reduce the organic contaminants in wastewater; however, the COD load was still more than three times that of inorganic substances. As the amount of organic contaminants grew, coagulative sedimentation of metallic ions became much more difficult,

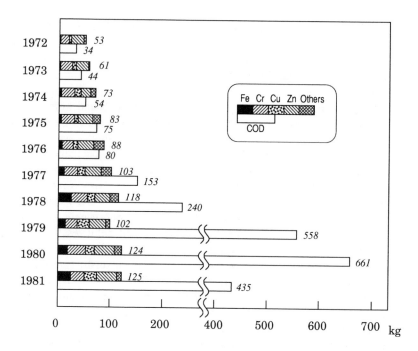

Figure 1. Annual load of organic contaminants, as measured by chemical oxygen demand (COD), and inorganic substances (Fe, Cr, Cu, Zn, and others) at the plant facility of the Kyushu University from the year 1972 to 1981.

and two-step coagulation was required. A large part of the metal ions were precipitated by the neutralization and by the action of sulfide ion in the first sedimentation procedure, but their concentrations were nonetheless usually larger than the regulation specified. This means contaminating organic materials interfered with the sedimentation. Actually, COD of the filtrate of this treatment was very large. Then, in the second step, the filtrate was treated with calcium hypochlorite to decompose the organic substances. This two-step sedimentation requires thorough washing of the reaction vessel, thickener, filter-press, and pipelines connecting these apparatuses, because even a small amount of residue from the first step could liberate chromate ion by oxidation with chlorine. Furthermore, using a great deal of hypochlorite sometimes caused the release of chlorine gas in the facility, a gas very toxic to humans and corrosive to the machines. To make the situation worse, the initial consensus within the university to cooperate in preserving the environment seemed to weaken year by year, as shown by the examples below.

Example 1. The rule governing cyanide solutions was settled as follows: they must be kept in a blue-colored container and the water should be kept alkaline. In one case, however, the blue container used was that for waste sulfuric acid, and it was transported to the facility. If the facility personnel had assumed the vessel contained cyanide solution and had mixed it with other cyanide solutions, a large amount of poisonous hydrogen cyanide gas would have evolved. In this instance, fortunately, they noticed that the wastewater was much heavier than usual cyanide solution, and examination revealed it was waste sulfuric acid.

Example 2. Methanol was transported to the facility in a yellow vessel which was usually used for wastewaters containing heavy metals. This made the concentration of organic substances intractably high, if it had been mixed with other wastewaters for treatment.

Example 3. Hydrofluoric acid in a yellow container was sent to the facility. If the waste had been mixed with other heavy metal wastewaters, which were usually strongly acidic, a large amount of HF gas would have been released.

Example 4. An even worst instance was the presence of chlorinated solvents such as chloroform or methylene chloride in the wastewaters. The great density of these solvents interferes with detection with the eyes. They can easily dissolve the bottom of a wastewater storage tank, as these tanks are usually made of plastic, and allow the wastewater to flood the facility.

These types of dangerous situations in the facility are ascribable to many factors. One is that there was no information as to the content and nature of the waste in a container. The only information the facility staff had to go by was the color of the container. The rule was mainly aimed at preventing the researchers and students from flushing the waste down the drain, which was occasionally checked by the regional environmental agency. This was very easy for the waste generators so that they may have thought segregation of wastes was not very important. In some cases, glassware and disposable plastic tips were even found in the waste ''solution''.

The other factor was the absence of a steering committee for the facility, although there was such a committee to discuss environmental problems of the university. Problems within the facility, however, was not reported to this committee except in extreme cases. As these cases continued to occur and no effective measures were taken by the committee, it seemed to be useless to report any longer.

4.3.4.3 Project to Construct New Facility

After ten years of operation of the first facility, because of deterioration and an increasing volume of waste, the university was forced to construct a new plant. At this time, again a committee to decide the most appropriate treatment plant was formed, and reconsideration of the rules governing wastes was deemed necessary. Selection of the new plants to be settled in the new facility was discussed simultaneously with the amendment of the former relatively rough rules on wastes.

It was in the course of these discussions that the three fundamental principles described earlier were adopted by the committee members. It decided to install two units in the new facility, one to treat wastewater containing only a small amount of organic contaminants and one to decompose organic components in inorganic wastewaters by incineration (Figures 2 and 3). By constructing an incineration plant, treatment of waste organic solvents would also be possible.

4.3.4.4 Selection of the Plants

The main processes, which can be conducted by the plant selected to handle the wastewaters, were sedimentation and electrolytic flotation. Metal ions, as is known, can be removed by neutralizing sedimentation. However, in the case of cadmium, neutralization is effective only under very high pH conditions, where lead or zinc ions are redissolved. In the electrolytic flotation process, the cadmium ion can be removed at the same pH range as sedimentation in the electrolysis vessel where aluminum alloy is used as the anode; this releases aluminum into the water collecting the cadmium ion. The wastewater treated by this process is further successively passed through a sand filter, two packed columns of activated carbon, and a column of mercury absorbing chelating resins as the final treatment.

The incineration process selected was the emulsion incineration, by which wastewaters containing a large amount of organic substances were mixed with waste organic solvents and emulsified by adding surfactant; the resulting oil-in-water type emulsion is fed to the pyrolyzing furnace. The exhaust gas from the furnace is quenched by an alkaline water solution and washed by passing through a Venturi scrubber. The flue gas is passed through a column packed with a mercury absorbing resin, as shown in Figure 3.

The capacities of these units, based on the annual increase of the waste during the life of the first facility, were determined adequate for 10 years (Tables 1 and 2).

The following measures were decided to fulfill the second and third principles of maintaining a safe working environment and preventing secondary pollution.

4.3.4.4.1 Reduction of Direct Contact with Wastes

The waste solutions would be transferred to the stock vessels using pumps. Many treatment procedures were automated.

4.3.4.4.2 Gas Sensors at Proper Locations

Gas sensors to detect chlorine, hydrogen fluoride, and hydrogen cyanide would be placed where these gases might be released.

4.3.4.4.3 Scrubber and Deodorizing Column

The wastewater reservoirs and reaction vessels would all be closed type and connected to a scrubber system and deodorizing column packed with

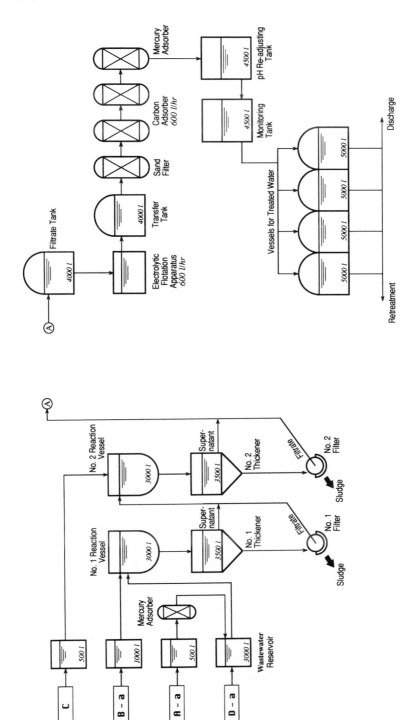

Figure 2. Treatment plant unit to treat wastewater containing only a small amount of organic contaminants. See Table 3 for treatment codes.

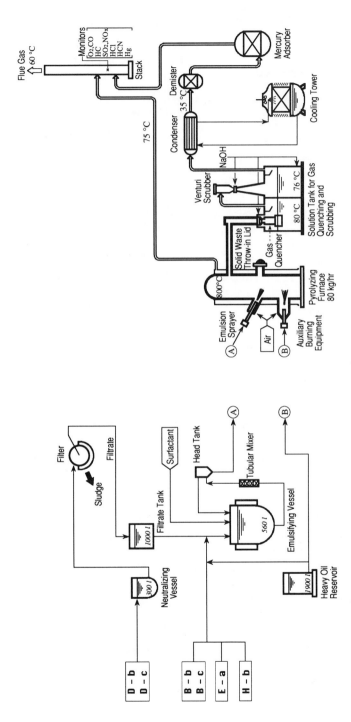

Figure 3. Treatment plant unit to decompose organic components in inorganic wastewaters by incineration. See Table 3 for treatment codes.

Table 1 Capacity of the Plant for Inorganic Waste Solutions

Waste	Concentration Limit (mg/l)
Mercury	500
Cyanide and arsenic	1000
Fluorine	4000
Heavy metal	1000

Note: Treatments are performed by batch reaction (2600 l/batch).

Table 2 Capacity of the Plant for Incineration

Waste	Feed Rate
Organic	40 kg/h
Aqueous	40 kg/h
Flammable solid	10 kg/d

activated carbon. The poisonous gases evolved in these vessels would be led to the scrubbing system by an induced fan and exhausted to the outer atmosphere.

4.3.4.4.4 *Strong Ventilation System*

The air of the treatment room would be well ventilated by two large fans, one for bringing fresh air into the room and the other for venting the air of the room out of the building. A deodorizing filter would be built into the air vent to avoid pollution of the outside air. In the area of the incineration unit the emulsion room was insulated from other parts of the facility, as a large amount of organic solvent vapor could be released there. This room would be the most strongly ventilated area, and the air vent from this room would be connected to a large deodorizing column packed with activated carbon.

4.3.4.4.5 *Prevention of Secondary Pollution*

As is obvious from the description of ventilation of the facility, all of the vessels and instruments except the vessels for treated water would be housed in a building in order to prevent bad odors or poisonous gas from contaminating the outdoor atmosphere. These instruments can both keep the atmosphere in the facility clean and also prevent harmful or unpleasant substances from escaping.

The other important factor in secondary pollution is the contamination of groundwater by a spill from the facility. In this university, the groundwater is pumped up and used as tap water. To prevent its contamination, even in a contingency such as breakage of a waste storage tank, a large tank was placed in an underground concrete pit coated with acid-proof paints to intercept any leaked wastewater and prevent its seepage into groundwater. Suspected leakage from this underground tank to the pit can be checked at any time.

Before construction of the second facility, the direction of the flow of groundwater around the construction site was checked and the water analyzed. After 12 years of operation no evidence of the contamination of groundwater was detected by repeated analyses.

For incineration to be operated smoothly, a large blower and compressor are required which would emit loud noises to the surrounding area. These machines would, therefore, be set in a soundproof room.

4.3.4.4.6 Analytical Instruments

The content of a bottle or a container found in the corner of a laboratory is often not known. Without knowing the content no treatment can be performed. Therefore, instruments to examine the component of an unknown substance would be needed in the facility, along with analytical instruments to check the water after treatment.

4.3.4.5 Establishment of New Rules

Recognizing that the reasons for the dangerous situations in the first facility might have been the absence of a steering committee, a committee to oversee the second facility was established by the concerned faculty and university-affiliated institute representatives. The director of the facility chairs the committee, and all matters of daily operations and management of the waste control system are discussed.

The director is also a member of the Committee for the Preservation of the Environment, composed of the deans of faculties and institutes and presided over by the president of the university. In this top committee, university environmental matters are discussed, and rules concerning environmental preservation, including those involving wastes, are mandated.

The rules for the handling of wastes are also set forth based on the three fundamental requirements. Every person engaged in scientific research in the University is required to know the nature of the chemicals they are using, and the waste must be segregated according to the categories listed in Table 3. Every time they pour waste into a vessel, they are requested to log it. When the vessel is transported to the treatment facility, it must be accompanied by a multicopy card showing the pH value and contents of the waste solution it holds.

Table 3 Classification and Treatment of Wastes

Waste	Code	Treatment[a]	pH[b]	Contents
Wastewater containing mercury	A-a	L	<2	Inorganic mercury compound
	A-b	E and L[c]	—	Organic mercury compound, e.g., methyl or phenyl mercury
	A-c	E and L[c]	—	Inorganic mercury compound containing more than 0.2% of organic substance
Wastewater containing cyanide and arsenic	B-a	W	>10.5	Inorganic cyanide and arsenic compound
	B-b	W	>10.5	Nondissociative cyanide and complex
	B-c	W	>10.5	Cyanide and arsenic compound containing more than 0.4% of organic substance
Wastewater containing fluoride	C	W	>6	Inorganic fluoride
Wastewater containing heavy metal	D-a	W	—	Inorganic heavy metal
	D-b	W	—	Inorganic heavy metal containing more than 10 ppm of chelating reagent
	D-c	W	—	Inorganic heavy metal containing more than 0.4% of organic substance
Photography wastewater	E-a	W	—	Photography developing fluid
	E-b	F[d]	—	Fixing solution
Wasteacid and alkali	F	R[e]	—	Acid or base containing no toxic substance

Waste	Code	Treatment[a]	Description
Wastewater containing organic phosphorous	G	R[f]	Parathion, methylparathion, ethyldimethone, and EPN
Organic liquid waste	H-a	F	Organic waste containing halogen
	H-b	W	Organic waste containing hydrocarbon and organic oxygen compound with flush point >70°C
	H-c	R[g]	Organic waste mixed with toxic substance
	H-d	F	Other organic waste
Solid waste containing mercury	I-a	F	Inflammable solid waste containing mercury
	I-b	F or W[h]	Flammable solid waste containing mercury
Solid waste containing toxic material	J-a	F	Inflammable solid waste containing toxic material
	J-b	F or W[h]	Flammable solid waste containing toxic material

[a] W, treated in the waste treatment facility; F, collected in the department and treated in the waste treatment facility; R, treated by a contractor; R, treated by the researcher.

[b] pH range requiring adjustment.

[c] Converted to the waste of code A-a by oxidation and treated in the waste treatment facility.

[d] Collected in the department and sold to a company recovering silver.

[e] Neutralized by the researcher and flushed down the drain.

[f] Decomposed under strongly alkaline conditions and flushed into drain after neutralization.

[g] Separated into toxic substance and organic waste to fit the waste of another code by the researcher.

[h] Mercury or toxic materials are washed out to give waste of another code and residual flammable waste is treated in the waste treatment facility.

An associate professor is also assigned to the facility and is in charge of supervising and advising on the treatment administered. Even more important is education on environmental protection. Although the importance of environmental protection is generally accepted nowadays, in a busy daily schedule of research and education one may tend to forget it. Continuous education is, therefore, necessary for older members of the university, as well as students who undertake experiments for the first time. The associate professor may be requested to talk about the importance of environmental protection, methods to treat wastes generated in experiments, and to show the facility for those who will generate waste.

4.3.5 CONCLUSION

Problems about wastes generated in academic institutions are also recognized in the U.S., Korea, and other eastern Asian countries. But, there are few waste treatment facilities in universities in these countries.

The problems involved with the construction of such a facility and its operation have been described here using a university in Japan as an example. Following its construction of the first facility of this kind in Japan, many national and private universities in the country built a facility to handle waste; this was when we were suffering from severe pollution in many areas of Japan. After the facilities were established and treatment was being conducted smoothly, however, the importance and the struggles of the staff of the facility gradually tended to dim in the minds of members of the university. Although a generator of waste would rarely be injured, a mistake in the segregation of his or her wastes could cause a severe accident in the treatment facility. This is the primary reason why a facility should be developed to the level of a research center which can also function as a center for environmental education. For a facility to function well and without interruption, the importance of the environment and the role waste management plays in it must be recognized from the administrative level all the way to the student level of a university.

Section
5

Monitoring and Analysis of Effluents and Wastes

CHAPTER 5.1

Analysis of Hazardous Complexed Wastewaters

Etsu Yamada

TABLE OF CONTENTS

0-8493-682-5/94/$0.00 + $.50
© 1994 by CRC Press, Inc.

5.1.1 INTRODUCTION

Small quantities of a wide variety of wastewaters of hazardous chemicals from universities, research institutes, and hospitals present a difficult problem to analyze. Treatment of inorganic liquid wastes containing mercury, heavy metals, cyanide, fluoride, and so on is usually done in universities using the ferrite process or the iron hydroxide coprecipitation method. In a waste treatment process, the quality of wastewaters and treated waters must be determined according to legal regulations. Analysis of hazardous chemicals should be done accurately and rapidly to assure safety and so that no pollution will result. However, it has been found that even if hazardous chemicals are present in these complexed wastewaters, sometimes they cannot be detected by conventional method even after pretreatment of digestion, evaporation, solvent extraction, or ionic exchange; this is due to the interference of coexisting inorganic or organic substances. Additionally, wastewaters from an organic waste incinerator containing mercury or fluoride ions present difficulties in both treatment and analysis. In an attempt to solve these problems, methods of analyzing mercury, heavy metals, fluoride, total cyanide, and so on to eliminate the interference of coexisting substances were investigated. In this chapter, the analytical methods developed are presented.

5.1.2 MERCURY

Analysis of mercury and its compounds in environmental samples is very important because these are highly toxic. The most widely used method for the determination of mercury is the cold-vapor atomic absorption (CVAA) analysis after the reduction of mercury with tin(II) chloride in acidic solution because of its high sensitivity and simplicity; it is cited as a Japanese Industrial Standard (JIS).[1] However, iodide and its related compounds interfere with the determination of trace mercury contained in complexed wastewaters even after digestion. In such iodide-containing samples, mercury cannot bedetected even if a considerable amount is present. Umezaki and Iwamoto,[2] Omang,[3] and Taguchi et al.[4] reported the negative interference of iodide on mercury determination, but considered that the existing means of its determination by CVAA analysis is applicable to environmental sample solutions with low iodide concentration such as river water, lake water, and sewage. Wastewaters discharged from universities, research institutes, hospitals, and factories, however, often contain a large amount of iodide and its related compounds.

The effects of potassium iodide, potassium iodate, potassium periodate, and iodine were examined (Figure 1). Even after digestion with acidic permanganate, more than 1.0 ppm of iodide or its related compounds interfered with the reduction of mercury at 0.75 M sulfuric acid. Interference of iodide became more evident with an increase in the concentration of sulfuric acid.

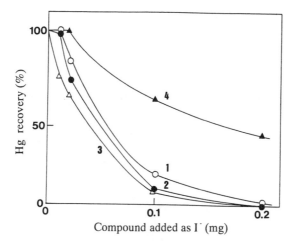

Figure 1. Interference of iodide and its related compounds on reduction of mercury by the JIS method. 1, KI; 2, KIO$_3$; 3, KIO$_4$, 4, I$_2$.

In wet digestion by acidic permanganate at 95°C, iodide is oxidized to iodine (I$_2$), which is easily released from a sample solution. However, iodine is further oxidized to iodate ion or an ion-pair and I$_3^-$ is formed at a high concentration of iodide. Hence, mercury complexes with iodide ion such as HgI$_4^{2-}$ would be formed because iodide ion in a sample solution could not be released completely in the digestion procedure.

An improved method with alkaline tin(II) reduction was devleoped by Korenaga and Yamada et al.[5] The determination procedure is as follows: an aliquot (less than 100 ml) of sample solution is taken into a reaction vessel after digestion with potassium peroxodisulfate in a diluted sulfuric acid solution heated at 95°C for 1 h. To the solution, 10 ml of 5 *M* sodium hydroxide, 2 ml of 1000 mg/l Cu^{2+} solution, 10 ml of 5% potassium zinc cyanide solution, and 2 ml of 10% tin(II) chloride solution are added, and the evolved mercury is measured with an atomic absorption spectrometer at 253.7 nm. The reducing power of tin(II) in alkaline solution is stronger than that in acidic solution because the standard redox potential for tin(II) is −0.93 V vs. normal hydrogen electrode (NHE) at pH 14 (0.15 V vs. NHE in acid). Potassium zinc cyanide is added as a masking agent for silver(I) ion. The detection limit and precision of the alkaline method are 0.5 μg/l and 3%, respectively. However, it is preferable to determine mercury in an acidic solution because sample treatment for mercury is usually done in sulfuric acid solution. The use of potassium zinc cyanide is not as safe as once thought, nor is it easy to dispose. Yamada et al.[6] then developed yet another method of determining mercury which eliminated the interference of iodide by reducing mercury with a mixture of sodium borohydride (NaBH$_4$) and tin(II) chloride in acidic solution. The pretreament of samples is done by the usual method. After transferring

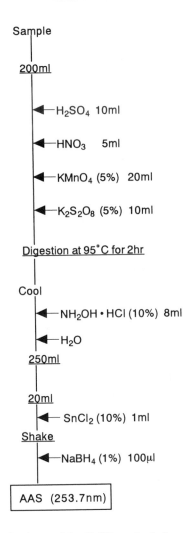

Figure 2. Analytical scheme of the NaBH₄ method of mercury determination.

20 ml of pretreated sample solution containing mercury (<1.0 μg) into a reaction vessel, 1 ml of 10% tin(II) chloride solution is added and the mixture is shaken; then 100 μl of 1% NaBH₄ solution is added to the sample solution, followed by immediate air-bubbling through the reaction vessel. The concentration of mercury is measured at 253.7 nm by CVAA spectrometer. The analytical scheme is shown in Figure 2. Using this method, the concentration of sulfuric acid in the sample solution must be more than $0.5\,M$ before addition of tin(II) chloride. Iodide up to 20 mg (1000 ppm) does not interfere with the determination of mercury, and even the addition of 100 mg of iodide (5000 ppm) allows 85% recovery of mercury. The calibration curve by the

Table 1 Comparison of Analytical Results of Mercury in Sewage from the Kyoto Institute of Technology

Sample	Hg Content (µg/l)		
	JIS Method	Alkaline Method	NaBH$_4$ Method[a]
1	0.44	—	0.49
2	0.61	—	0.59
3	0.33	—	0.35
4	6.74	—	6.58
5	1.69	—	1.58 (1.66)
6	0.73	—	0.60 (0.72)
7	0.65	0.62	0.66 (0.65)
8	0.68	0.74	0.60 (0.63)

[a] The values in parentheses were obtained by the standard addition method.

Table 2 Analytical Results of Mercury in Wastewaters from a University Using the JIS, Alkaline, and NaBH$_4$ Methods

Sample	Hg content (µg/l)			I$^-$ (mg/l)
	JIS Method	Alkaline Method	NaBH$_4$ Method[a]	
A	N.D.[b]	4.58	4.49 (4.50)	85
B	N.D.	1.39	1.20 (1.22)	50
C	N.D.	7.23	6.93 (6.93)	14
D	N.D.	14.85	14.30 (14.56)	127
E	N.D.	22.40	19.80 (19.90)	98
F	N.D.	9.08	8.54 (8.60)	89
G	N.D.	9.82	11.63 (11.75)	120
H	536.9	—	546.9	N.D.

[a] The values in parentheses were obtained by the standard addition method.
[b] N.D., not detected.

recommended method was linear from 0.004 to 1.0 µg (per 20 ml). The relative standard deviation (n = 10) of the wastewater sample (Hg: 0.04 µg, I$^-$: 1.6 mg) was 1.67%. This technique was applied to the determinations of mercury in complexed samples such as sewage, wastewaters, and treated waters. The results obtained are listed in Tables 1 and 2 and compared with those obtained by the conventional method (JIS) and the alkaline method. In sewage samples containing no iodide, the results by this method were in good

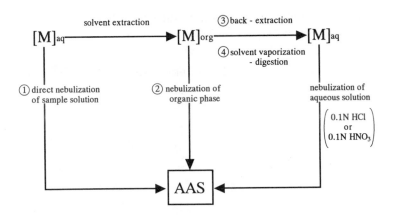

Figure 3. Pretreatment procedures for the determination of heavy metals by atomic absorption spectrometry (AAS). [M], metal; aq, aqueous phase; org, organic phase.

agreement with those obtained by the conventional and the alkaline methods. In wastewaters and treated waters containing iodide, however, the conventional method could not detect mercury and showed 0% recovery of mercury when known amounts of mercury had been added. On the other hand, mercury content obtained by the $NaBH_4$ method using both calibration curve and standard addition methods showed good agreement. The recovery of mercury was ~94 to 102% by the $NaBH_4$ method after an addition of known amounts of mercury, in contrast to ~84 to 112% by the alkaline method. The $NaBH_4$ method has been successfully applied to the determination of mercury in the process monitoring of wastewater treatment.

5.1.3 HEAVY METALS

Hazardous heavy metals such as cadmium, lead, chromium, copper, zinc, and manganese in effluents, wastewaters, and treated waters must be determined following legal regulations. Atomic absorption spectrometry (AAS), colorimetry, and inductively coupled plasma atomic emission spectrometry (ICP-AES) are generally used for these determinations.

Flame atomic absorption spectrometry is particularly widely used because of its selectivity and simplicity. However, pretreatments by solvent extraction, ionic exchange, and coprecipitation are necessary for the determination of heavy metals in complexed wastewaters because of the interference of co-existing substances or insufficient sensitivity. In Figure 3, the procedures for prior extraction of metals to be determined by AAS are shown. Diphenyl-thiocarbazone (dithizone), diethyldithiocarbamate (DDTC), or ammonium pyrrolidine dithiocarbamate (ADPC) is commonly used as a chelating reagent. Extraction provides a useful means of separating heavy metals from solutions

Table 3 Analytical Results of Heavy Metals in Treated Waters by ICP-AES and AAS

Sample	Method	Cr	Cu	Fe	Mn	Pb	Zn	Salt (w/v%)
1	ICP	0.31	0.08	0.45	0.77	0.15	0.09	3.6
	AAS	0.31	0.08	0.46		0.2	0.1	
2	ICP	0.11	0.06	0.59	1.96	0.09	N.D.	1.8
	AAS	0.12	0.06	0.59	1.98	0.1	N.D.	
3	ICP	0.12	0.03	0.17	0.11	N.D.	0.01	1.6
	AAS	0.12	0.05					
4	ICP	0.40	N.D.[a]	0.41	0.20	N.D.	0.01	3.6
	AAS	0.41		0.41	0.18		N.D.	
5	ICP	0.22	N.D.	0.27	0.10	N.D.	N.D.	1.8
	AAS	0.25		0.25	0.08			
6	ICP	0.70	0.10	0.66	1.33	N.D.	0.44	4.4
	AAS	0.66	0.10	0.66	1.36		0.5	

Note: N.D., not detected; data for Cr, Cu, Fe, Mn, Pb and Zn is mg/l.

that contain heavy loadings of foreign elements, such as Ca, K, Mg, and Na, the presence of which can cause errors due to nonspecific background absorption. Furthermore, cadmium, for example, cannot be extracted into organic phase with dithizone in the presence of a large amount of iron, so cadmium must be separated from iron by ionic exchange or another method prior to solvent extraction. Many AAS have built-in background correction systems which can compensate for scatter or a great deal of molecular absorption. The simplest system passes light from a continuum source, commonly a hollow cathode lamp (HCL) containing deuterium (D_2), through the flame. The D_2 lamp provides almost an exclusive measure of the contribution to the absorbance signal of molecular absorbance or scatter. This signal can therefore be subtracted from the signal obtained with the determinant lamp. The Smith-Heifiji system or the Zeeman system is also used.

Inductively coupled plasma atomic emission spectrometry, which has the characteristic of a very wide concentration range for linear calibration and multielement determination, is also now used. The high temperature of the argon ICP compared to that of the nitrous oxide-acetylene flame means that the extent of chemical interference resulting from stable compound formation is greatly reduced. On the other hand, spectral interference becomes a potentially very serious problem. An analytical wavelength for each element should therefore be carefully selected without spectral interference, and an adequate background correction system should be used. An internal standard such as Y or Co is often used to compensate for instrumental drift. The analytical results of treated waters obtained by ICP-AES were in good agreement with those obtained by flame AAS (Table 3). The sample solutions were directly nebulized by both methods, and the torch for high salt concentration was used by ICP-AES because salt concentration of treated waters was very high. Pretreatments by solvent extraction and ionic exchange should be performed for the determination of heavy metals in wastewaters that contain a large amount of foreign elements.

Comparing these methods, X-ray fluorescence spectrometry (XRF) is still rarely used to determine heavy metals in water and wastewaters. Direct energy-dispersive X-ray spectrometer (EDX) analysis is possible for the determination of metals at a milligram per liter level in samples such as wastewaters from university laboratories. A few milliliters of whole or filtered water samples placed in a cup with a thin Mylar or polyethylene bottom is measured in a nonevacuated instrument. Smits and Van Grieken[7] and Murata et al.[8,9] proposed to spot sample solution (~1 to 1.5 ml) on a specially designed filter paper in order to reduce the spreading solution in the filter and to evaporate the water by dried air. The detection limits were found to be about 100 μg/l for most elements. Accuracy and precision became better when an internal standard was added to the solution prior to the spotting. Further, various preconcentration methods of XRF for trace metals were investigated.[10] A suitable routine analysis of heavy metals contained in effluents from universities is as follows:[11] the essential procedure consists of the preconcentration of trace elements as DDTC complexes on a Millipore® filter and direct multielement determination by EDX. The following process is recommended for the preconcentration procedure of the elements. For the determination of total metals, samples are digested with nitric acid and perchloric acid. Then, 10 ml of 1% DDTC is added to the solution after neutralization by ammonia or acetic acid. After standing for 1 h, the precipitate of DDTC complexes is collected on a 0.45 μm Millipore filter and dried prior to the XRF determination. To determine total Cr the reduction of Cr(VI) to Cr(III) using ethanol in hydrochloric acid solution is necessary before the addition of ferric ion as a carrier. Then, 1% DDTC is added to the solution after adjustment of pH to 8.5. The detection limits for Cu, Fe, Mn, Ni, Pb, and Zn are 1, 2, 6, 4, 7, and 2 μg/l, respectively. These preconcentrated samples can be measured by a wavelength dispersive X-ray spectrometer, the same as EDX.

5.1.4 FLUORIDE

Fluorine-containing organic compounds have recently been widely used in research projects in the fields of organic synthesis, medical science, and agriculture because of their interesting characteristics. As a result, wastewaters from exhausted gas washers of organic waste incinerators contain a large amount of fluoride ions because many kinds of fluoride compounds are present in the organic liquid wastes discharged from universities and institutes.[12] Urban disposal of wastes, including many Teflon products, presents a similar problem. Further, borofluoride compounds, now often used in the field of electronics, present difficulty problems of analysis and treatment.

Table 4 Effect of Metal Ions on Determination of Fluoride Ion[a]

Metal Ion Added	Amount Added (mg)	Buffer[b] (TISAB)	Recovery of F⁻ (%)		
			Calib.	KA[c]	GP[d]
Ca(II)	50	1	85	101	100
		2	86	101	100
Mg(II)		1	82	99	102
		2	85	102	101
Pb(II)	10	1	102	102	99
		2	103	103	101
Fe(III)	10	1	94	99	99
		2	98	101	101
Al(III)	0.1	1	70	96	97
		2	96	101	101
	1	1	2.4	119	99
		2	71	99	98
	2	2	53	97	97

[a] 100 μg fluoride ion and 10 ml TISAB were present in a sample solution of 100 ml.
[b] Total ionic strength adjustment buffer (TISAB) contains sodium citrate (1) and CyDTA (2).
[c] KA, Known addition.
[d] GP, Gran's plot.

A spectrophotometric method using alizaline complexone is reliable and sensitive but requires preliminary separation of fluoride. An ion selective electrode (ISE) for fluoride is often used because of its simplicity and selectivity. The results, however, are frequently subject to interference, and large errors may occur unless the measurement system is carefully calibrated. Fluoride concentrations would be underestimated in the presence of metal ions (such as Al^{3+}, Pb^{2+}, Ca^{2+}, and Mg^{2+}) and boric acid that form stable fluorocomplexes, or at a sufficiently low pH when an appreciable amount of HF forms. Ion activity is depressed at high ionic strength, so variations in the ionic strength between samples and standards can cause inaccuracy. These interferences are removed by a total ionic strength buffer (TISAB) with the use of a known addition of potentiometry or Gran's plot potentiometry (Table 4).[13-15] TISAB consists of:[16] (1) sodium chloride to give a relatively high ionic strength that minimizes intersample variations in ionic strength; (2) glacial acetic acid/sodium hydroxide to provide a pH buffer at 5.2 ± 0.1 which minimizes OH^- and HF influences; (3) 1,2-cyclohexanediaminetetraacetic acid (CyDTA) which complexes metal ions, thereby leaving free F^- in the solution. The electrode response should be calibrated by the method of known addition or Gran's plot instead of the usual direct calibration procedures with standard solutions. By the known addition method, the potential, E_1, is first measured for a sample solution of volume V_0 and unknown fluoride concentration C_A. Then, known volume V_S of a standard solution with an F^- concentration C_S is added and the potential, E_2, recorded. The change in potential, ΔE, is due to the standard aliquot added, and the unknown F^- concentration is calculated from the expression

$$C_A = C_S(V_S/V_0)(10^{E/S} - 1)^{-1} \qquad \text{(ue1)}$$

where S is the Nernstian slope of the electrode. This can be determined by calibrating with the two standards. In the case of Gran's plot, by plotting $(V_S/V_0) \cdot 10^{E/S}$ vs. V_S a straight line is obtained which intercepts the abscissa for a V_S^* value. The unknown fluoride concentration is obtained from $C_A = -C_S V_S^*/V_0$.

The determination procedures are as follows:[12] the Nernstian slope of electrode S is determined by calibrating with 100 ml of fluoride standard solutions of 1 and 10 ppm containing 10 ml of TISAB solutions. The potential of 100 ml sample solution containing fluoride (F^-: ~0.02 to 1.0 mg) with addition of 10 ml of TISAB solution is measured. Standard solution (50 μl) with 1000 ppm fluoride concentration is added, and the potential is measured. Similar procedures are carried out four times. The unknown fluoride concentration is calculated by known addition or Gran's plot method as described previously. If borofluoride compounds are contained in wastewaters, the following rapid and simple pretreatment to decompose the borofluoride should be done prior to ISE measurement. An aliquot of sample solution (BF_4^- as $F^- \leq 0.5$ mg) is taken into a reaction vessel and diluted to 50 ml with distilled water after adding 2.5 ml of $AlCl_3$ solution (1000 mg/l as Al^{3+}) and adjusting pH 2 with HCl or NaOH, and then is heated for 90 s in a microwave oven. Measurements of fluoride concentration between ~0.1 to 100 mg/l are straightforward, with a relative standard deviation of about 2%. The results of wastewater samples obtained by the present method are shown in Table 5 compared with the results by spectrophotometric and ISE methods with distillation.

5.1.5 TOTAL CYANIDE

Analysis of total cyanide is important because of the high toxicity of the substance. Total cyanide in water and wastewater is usually determined by colorimetry with pyridine-pyrazolone (JIS), 4-pyridinecarboxylic acid-pyrazolone (JIS) or pyridine-barubituric acid (ISO, ANSI, EPA),[17] and cyanide-selective electrodes after distillation. However, it has been reported that hydrogen cyanide (HCN) is formed by the reaction of organic compounds with

Table 5 Determination of Fluoride Ion in Wastewater by Various Methods

Sample[a] (TISAB)		Colorimetry[b]	ISE	ISE[c] Calib.	ISE[c] KA[d]	ISE[c] GP[e]
A-1	(1)	114		110	110	108
	(2)		111	106	110	115
A-2	(1)	131		130	133	131
	(2)		128	133	134	134
A-3	(1)	298		280	284	276
	(2)		275	285	289	287
B-1	(1)	9.2		8.3	10.7	10.2
	(2)		9.7	8.5	10.7	10.3
B-2	(1)	15.7		13.8	15.6	15.2
	(2)		15.3	14.5	16.0	16.0
B-3	(1)	14.3		12.7	14.4	14.1
	(2)		14.5	13.2	14.3	14.2
B-4	(1)	13.0		12.0	13.1	13.0
	(2)		13.6	11.9	13.2	13.4
C-1	(1)	4.3		3.8	4.0	4.0
	(2)		4.3	4.0	4.1	4.1
C-2	(1)	2.0		1.6	1.7	1.5
	(2)		1.6	1.7	1.5	2.0

[a] A, wastewater from the organic incinerator; B, wastewater after removing fluoride ion by $CaCl_2$ from wastewater A; C, wastewater after inorganic material treatments; (1) contains sodium citrate; (2) contains CyDTA.
[b] ISE with distillation; mg/l.
[c] ISE without preliminary distillation; F^- mg/l.
[d] KA, known addition.
[e] GP, Gran's plot.

nitrogen-containing compounds in the waste solutions during distillation and is detected even if cyanide compounds are not initially present in the solutions.[18-20] Since various kinds of organic compounds are involved in the formation of hydrogen cyanide, it is difficult to eliminate their effects in wastewater. Nonomura[21,22] reported that although nowadays toxic cyanide compounds are not incorporated in photographic processing solutions, hydrogen cyanide is found in photographic waste solutions, formed by a reaction between hydroxylammonium salts and organic compounds, such as aminocarboxylic acids and aromatic amines (the color-developing agent), during distillation. Several studies have been made on the decomposition of cyanide during distillation and the interference of coexisting compounds on the determination of cyanide by colorimetry or ISE.[23,24]

Hydrogen cyanide was formed during distillation when EDTA (JIS) was used in the presence of sodium nitrite, hydroxylammonium hydrochloride, or

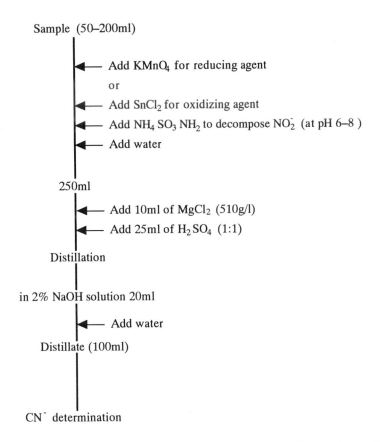

Sample (50–200ml)

← Add KMnO₄ for reducing agent

or

← Add SnCl₂ for oxidizing agent

← Add NH₄ SO₃ NH₂ to decompose NO₂⁻ (at pH 6–8)

← Add water

250ml

← Add 10ml of MgCl₂ (510g/l)

← Add 25ml of H₂SO₄ (1:1)

Distillation

in 2% NaOH solution 20ml

← Add water

Distillate (100ml)

CN⁻ determination

Figure 4. Recommended procedures of distillation for the determination of total cyanide.

hydrazonium dihydrochloride.[21] Thus, it is preferable to use magnesium chloride and sulfuric acid instead of EDTA and orthophosphoric acid as additives and to use tin(II) chloride instead of L-ascorbic acid, sodium arsenite, and sodium thiosulfate as reducing agents in determining total cyanide. Nonomura recommended the following distillation procedure (Figure 4).[22] In wastewaters containing sulfide, zinc acetate should be added and filtered before distillation. It was found that cyanide compounds such as NH₄Fe[Fe(CN)₆], not easily decomposed, can be analyzed quantitatively by a threefold increase in EDTA addition at distillation.[25] After distillation, titrimetry, colorimetry, ion-selective electrode analysis, or ion chromatography is useful for the determination of total cyanide in wastewater. In ion chromatography, cyanide ion is oxidized by sodium hypochlorite to cyanate ion and measured indirectly by a conductivity detector.[26] Chemiluminescence detection is used to determine trace amounts of cyanide ion with the uranine-sodium hydroxide-didodecyldimethylammonium bromide (surfactant) system.[27,28] The emission induced by cyanide ion is efficiently sensitized by uranine in an organized surfactant aggregate solution.

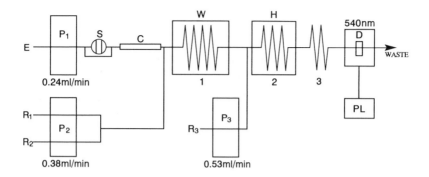

Figure 5. Flow diagram of the apparatus for measuring carboxylic acids. W, water bath (40°C); H, heater (80°C); P_1, P_2, P_3, pumps; S, sample injector; C, column; R_1, 0.15 M EDC; R_2, 0.01 M ONPH; R_3, 1.3 N NaOH; E, 0.02 M KH_2PO_4 + 0.05 M H_3PO_4; 1, reaction coil (0.5 mm × 10 m); 2, reaction coil (0.5 mm × 5 m); 3, coil (0.5 mm × 2 m).

5.1.6 CARBOXYLIC ACIDS

It is difficult to treat heavy metal ions in wastewaters containing carboxylic acids such as citric acid, tartaric acid, and others because water-soluble metal complexes may be formed with carboxylic acids. As chromium(III) complexes with carboxylic acids are inert in exchange reactions, the treatment of chromium is particularly difficult. Therefore, carboxylic acids in wastewaters should be determined before hazardous metal ions are treated. Special color reaction of carboxylic acids with 2-nitrophenylhydrazine hydrochloride (ONPH) was applied to a detection system for reversed-phase high pressure liquid chromatography (HPLC) determination of carboxylic acids in complexed wastewaters.[29]

1-Ethyl-3-(3-dimethylaminopropyl)carbodiimide hydrochloride (EDC) was used as a condensation reagent. A flow diagram of the apparatus for measuring carboxylic acids is shown in Figure 5. A column (Cosmosil $5C_{18}$-AR, 4.6 I.D. × 250 mm; Nakarai Tesque) was used. The absorbance was measured at 540 nm. Almost all carboxylic acids and β-amino acids reacted with ONPH to develop colors. Saccharides, esters, amides, sulfonic acids, phosphate ion, and sulfate ion gave no color and most α-amino acids showed weak coloration. Tolerable amounts of foreign ions and salt (NaCl) in the determination of carboxylic acids are listed in Table 6. From these results, it was apparent that the present reaction is highly selective to carboxylic acids. A chromatogram of seven carboxylic acids is shown in Figure 6. Calibration curves of various carboxylic acids were linear between $10^{-5} M$ and $5 \times 10^{-3} M$. The coefficient of variation (n = 10) of $10^{-4} M$ malic acid was 0.6%. The present method was applied to the determination of carboxylic acids in wastewaters and treated waters with satisfactory results.

Table 6 Tolerable Amounts of Foreign Ions and Compound in the Determination of Carboxylic Acids

Ion and Compound	Acetic Acid (ppm)	Malic Acid (ppm)	Citric Acid
Cu^{2+}	1	1	1 ppm
Ni^{2+}	1	1	1 ppm
Pb^{2+}	10	10	10 ppm
Zn^{2+}	10	10	10 ppm
Mn^{2+}	10	10	10 ppm
Cr^{3+}	10	10	10 ppm
Fe^{3+}	50	20	1 ppm
S^{2-}	—	—	100 ppm
I^-	—	—	500 ppm
NaCl	—	—	5%

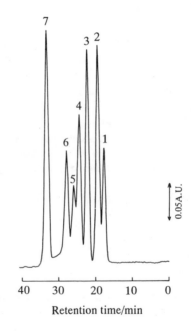

Retention time/min

Figure 6. Chromatogram of carboxylic acids; 1, glutamic acid; 2, tartaric acid; 3, malic acid; 4, malonic acid; 5, lactic acid; 6, acetic acid; 7, citric acid.

REFERENCES

1. Japan Industrial Standard (JIS), Testing Methods for Industrial Wastewater, K0102, 1986.
2. Umezaki, Y. and Iwamoto, K., Determination of submicrogram amounts of mercury in water by flameless atomic absorption spectrophotometry, *Bunseki Kagaku*, 20, 173, 1971.

3. Omang, S.H., Trace determination of mercury in biological materials by flameless atomic absorption spectrometry, *Anal. Chim. Acta,* 63, 247, 1973.

4. Taguchi, T., Yasuda, K., Hashimoto, M., and Toda, S., Some improvements for mercury determination in marine organisms by atomic absorption spectrometry, *Bunseki Kagaku,* 28, T33, 1979.

5. Korenaga, T., Yamada, E., Hara, Y., Sakamoto, H., Chohji, T., Nakagawa, C., Ikatsu, H., Izawa, M., and Goto, M., Elimination of interference by iodide in determination of mercury by cold-vapor AAS, *Bunseki Kagaku,* 36, 194, 1987.

6. Yamada, E., Yamada, T., and Sato, M., Determination of trace mercury in the environmental waters containing iodide by cold-vapor atomic absorption spectrometry, *Anal. Sci.,* 8, 863, 1992.

7. Smits, J. and Van Grieken, R., Optimization of a simple spotting procedure for X-ray fluorescence analysis of waters, *Anal. Chim. Acta,* 88, 97, 1977.

8. Murata, M. and Murokado, K., A formed filter paper medium for microdroplet analyses of liquid samples by X-ray fluorescence spectrometry, *X-Ray Spectrom.,* 11, 159, 1982.

9. Omatsu, M., Mushimoto, S., and Murata, M., A concentration apparatus for microdroplets X-ray fluorescence spectrometry, *Adv. X-Ray Chem. Anal.,* 17, 113, 1985.

10. Van Grieken, R., Preconcentration methods for the analysis of water by X-ray spectrometric techniques, *Anal. Chim. Acta,* 143, 3, 1983.

11. Yamada, E. and Sato, M., Multielement determination of metal ions in wastewater by energy dispersive X-ray fluorescence spectrometry, *Bunseki Kagaku,* 32, 654, 1983.

12. Yamada, T., Yamada, E., and Sato, M., Simplified determination of fluoride ion in wastewater by an ion selective electrode, *Bunseki Kagaku,* 37, T61, 1988.

13. Moody, G.J. and Thomas, J.D.R., Ion-selective ion sensitive electrodes, Merrow, England, 1971.

14. Manscini, M., *Ion-Selective Electrode Reviews,* Vol. 2, Thomas, J. D. R., Ed., Pergamon Press, Oxford, 1981, 17.

15. Gran, G., Determination of the equivalence point in potentiometric titrations, *Analyst (London),* 77, 661, 1952.

16. Frant, M.S. and Ross, J.W., Use of a total ionic strength adjustment buffer for electrode determination of fluoride in water supplies, *Anal. Chem.,* 40, 1169, 1968.

17. *Standard Methods for the Examination of Water and Wastewater,* 16th ed.,: APHA, AWWA, WPCF, American Public Health Association, Washington, D.C., 1985.

18. Koshimizu, T., Takamatsu, K., Kaneko, M., Fukui, S., and Kanno, S., Formation of cyanide ion at the distillation of the reaction mixture, *Eisei Kagaku,* 20, 332, 1974.

19. Koshimizu, T., Takamatus, K., Kaneko, M., Fukui, S., and Kanno, S., Formation of cyanide ion by the reaction of aromatic amines and nitrite ion, *Eisei Kagaku,* 21, 1, 1975.

20. Koshimizu, T., Takamatsu, K., Kaneko, M., Fukui, S., and Kanno, S., Formation of cyanide ion by the reaction of glycine and nitrite ion at various pH and room temperature, *Eisei Kagaku,* 21, 326, 1975.

21. Nonomura, M., Endogenous formation of hydrogen cyanide during distillation for the determination of total cyanide, *Toxicol. Environ. Chem.,* 17, 47, 1988.

22. Nonomura, M., Pretreatments for the determination of total cyanide-interferences, effects of reducing agents and additives, *Int. J. Environ. Anal. Chem.,* 35, 253, 1989.

23. Higuchi, Y., Aida, T., Morita, M., and Mimura, S., Detailed examination on the interference during the determination of cyanide ion in water, *Ann. Rep. Tokyo Metr. Res. Lab. P. H.,* 25, 417, 1974.

24. Hirata, H., Automated determination of total cyanide in waters using cyanide ion-selective electrode, *Kankyo Gijutsu,* 12, 687, 1983.

25. Yamada, E., Nagaoka, K., Takeuchi, T., Ikatsu, H., Nakagawa, C., Tanaka, M., Fujiwara, I., Hara, Y., Kojima, H., Sakamoto, H., Goto, M., and Korenaga, T., A round robin test for the total cyanide determination of complicated wastewater samples, *Bunseki Kagaku,* 41, T103, 1992.

26. Nonomura, M., Indirect determination of cyanide compounds by ion chromatography with conductivity measurement, *Anal. Chem.,* 59, 2073, 1987.

27. Ishii, M., Yamada, M., and Suzuki, S., Didodecyldimethylammonium bromide bilayer vesicle-catalyzed and uranine-sensitized chemiluminescence for determination of free cyanide at picogram levels by flow injection method, *Anal. Lett.,* 19, 1591, 1986.

28. Ishii, M., Yamada, M., and Suzuki, S., Determination of ultratraces of cyanide ion by FIA with surfactant bilayer vesicle-catalyzed and uranine-sensitized chemiluminescence, *Bunseki Kagaku,* 35, 955, 1986.

29. Yamada, E., Taguchi, H., and Sato, M., Separation analysis of carboxylic acids in hazardous complexed wastewaters, Abstr. 1st Asian Symp. Academic Activity for Waste Treatment (AAWT), 1P19, 1992.

CHAPTER 5.2

Monitoring of University Effluents

Edison Munaf and Toyohide Takeuchi

TABLE OF CONTENTS

0-8493-682-5/94/$0.00 + $.50

© 1994 by CRC Press, Inc.

5.2.1 INTRODUCTION

Analytical chemistry and environmental chemistry play significant roles in process control and environmental protection. Improvements in industrial process control often proceed from developments in process analytical chemistry, the goal of which is to supply quantitative and qualitative information about the chemical processes.

Waste problems are due to accelerated human activities. Research activities in university and industry laboratories can be candidates. Wastewaters from such laboratories are, in nature, composed of various materials. In order to assess whether the wastewater is hazardous or not, rapid, sensitive, reliable methods for determination of substances in wastewaters are essential for environmental conservation and for rationalization of related industrial processes.

Process analyzers are assembled based on either a conventional batch mode or an automated continuous flow mode. The term "flow analysis method" refers to analytical procedures with which the analyte concentration is measured under a continuous stream of gas or liquid.

Three types of automated flow analysis methods have been developed for this purpose, as illustrated in Figure 1, namely, segmented flow analysis (SFA), flow injection analysis (FIA), and continuous flow analysis (CFA).[1] A continuous segmented flow analyzer has been commercialized by Technicon as the AutoAnalyzer system. The classical version was first described by Skeggs[2] in 1957, in which the samples were aspirated sequentially, and air bubbles separated (segmented) the sample stream.

Unsegmented flow analysis methods such as FIA and CFA do not use air bubbles for the separation of successive sample streams. The absence of a gas phase in these methods has several advantages, such as better signal reproducibility and less complex instrumentation.[3] Furthermore, these unsegmented flow techniques offer a wide range of possible methodologies. FIA in its conventional (or normal) form (nFIA, defined as a technique with injection of sample into a continuous flow of reagents) allows analysis for pollutants with minimum consumption of sample solution. On the contrary, reverse FIA (rFIA) is defined as one of the FIA techniques in which the sample solution is continuously supplied and the reagents are periodically injected for the determination. rFIA is favored in the case of monitoring of wastewater, where the reagents determine the major cost for the measurements. On the other hand, when continuous monitoring of sample composition is necessitated, CFA is ideal, in which both sample and reagents are continuously supplied.

The potential of flow analysis has recently gained much attention in continuous monitoring of process relevant parameters or contaminants such as chemical oxygen demand (COD), nitrogen, phosphorus, cyanide, and trace heavy metals. Flow analysis has been the subject of several scientific con-

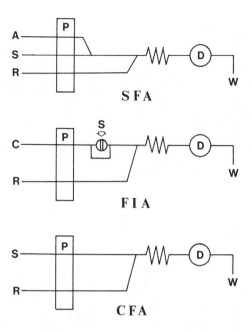

Figure 1. Schematic comparison of three automatic flow analysis methods, SFA, FIA, and CFA. Manifold specifications: A, air; S, sample; R, reagent; C, carrier; D, detector; P, pump; W, waste.

tributions as described hereinafter. This section deals with flow analysis of contaminants in wastewater discharged from university laboratories.

5.2.2 FLOW ANALYSIS

Automated flow analysis methods for the monitoring of industry and university wastewater have been described by several groups[4-13] using batch or automatic flow analyzers. Disadvantages of the batch analyzers lie in the fact that the instruments required are mechanically complex due to many moving parts and that a single measurement usually requires as much as 1 h. On the other hand, the flow analyzers based on SFA and FIA have advantages in that they are suitable for the analysis of a large number of samples and that the instruments employed are less complex. In SFA and nFIA, reagent flow rates are around 1 ml/min levels, resulting in the large consumption of the reagents for continuous measurements. Such disadvantages are not encountered in rFIA, or they can be overcome by reducing the reagent flow rates.

In CFA and rFIA, the sample is continuously supplied as a flowing stream, while SFA and nFIA have discontinuous sample flow. In Table 1 the features of SFA, FIA, and CFA for monitoring of industry and university effluents are compared.

Table 1 Comparison Between Three Types of Continuous Flow Analysis Methods

Parameter	SFA	FIA	CFA
Kind of flow	Segmented	Unsegmented	Unsegmented
Sample introduction	Aspiration	Injection	Aspiration
Reagent flow rate	1 ml/min	1 ml/min	0.1 ml/min
Response time	2–30 min	3–60 s	1–2 min
Sample throughput	<80 s/h	<300 s/h	Continuous
Tubing bore	2 mm	0.5–0.7 mm	0.5 mm

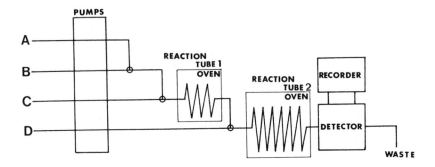

Figure 2. Schematic diagram of the flow system for continuous monitoring.

The key perspective to the monitoring of wastewater is perhaps immediate response besides the accuracy, reproducibility, sensitivity, selectivity, and economy. Reduction of time lag between the sampling and the analysis has been accomplished by speeding up the sample transport or reducing the distance between the plant (or source of contamination) and the analyzer (in-site laboratory). The ideal situation, however, would be direct sampling from the process stream or effluent. In some cases, wastewater sample can be continuously supplied through a pipe to the analytical laboratory.

5.2.3 CONTINUOUS MONITORING

The diagram of the flow system for continuous monitoring is illustrated in Figure 2. A single or multiple peristaltic pump is usually employed as the pump for supplying reagents and sample solution in flow analysis. Reaction tubes are placed in ovens where the reaction temperature is controlled. A to D are the flow lines; for FIA, a sample injection valve is inserted in flow line C. The manifold depends on the type and the number of reaction processes.

5.2.3.1 Chemical Oxygen Demand (COD)

Discharge of organic matters to water-course is strictly controlled by water authorities. When an ecological balance is maintained, natural waters are purified by biochemical oxidation by microorganisms, in which the polluting substances are utilized as a source of carbon while consuming dissolved oxygen for respiration.

In many countries, a widely used index of wastewater quality is COD which relates to the oxygen required for complete oxidation of the constituents. In the common methods, COD has usually been determined by manual titration using potassium permanganate[14,15] or potassium dichromate solutions.[16,17] Despite its importance, the manual methods have several disadvantages as follows: these common methods are complicated and time-consuming, and they require toxic reagents. Furthermore, the precision strongly depends on the operator skill, and about 3 h are required to obtain any data. Such a time lag becomes a problem for industry and university laboratories which continuously produce a large volume of wastewaters. As a result of these shortcomings, alternatives to the traditional COD measurement method has been requested to be developed.

Adelman[18] has described the determination of COD by SFA, while Ikatsu and Korenaga,[19,20] as well as Appleton et al.,[21] have described the automated methods based on the FIA technique. In the latter methods, either permanganate[19] or dichromate solution[20,21] is used. The detection limits achieved by these methods are much better than those achieved by the conventional manual methods. Goto et al.[22] have developed a CFA system for COD determination in which the system illustrated in Figure 2 is used. The system has 4 manifolds carrying each stream at a flow rate of around 50 μl/min by peristaltic pumps. The sample (A in Figure 2) is introduced continuously, merged first with an acid reagent stream (B in Figure 2, a mixture of 40% phosphoric acid and 10% sulfuric acid), then mixed with a catalytic reagent stream (C in Figure 2, 0.05% silver nitrate), and finally with an oxidizing reagent stream (D in Figure 2, 1.6 mM potassium permanganate). The final mixture is then introduced into a (poly)tetrafluoroethylene (PTFE) oxidation reactor (reaction tube 2, 0.6 mm bore and 18 m length) placed in a boiling water bath, and the solution is subjected to heating for 30 min under continuous flow. In this system, the reaction tube 1 is not necessary. After the reaction, the solution passes through a thin layer electrolytic flow cell operated at a constant potential and controlled by a potentiostat (0.5 mV vs. Ag/AgCl). The reduction current of the residual permanganate is continuously recorded, and the COD value of the sample is estimated from the decreased current. Figure 3 shows continuous monitoring of COD for university wastewater.[22]

5.2.3.2 Cyanide

The toxicity of cyanide made it indispensable as a determinant for the content in a great variety of samples. Rios et al.[23] developed unsegmented

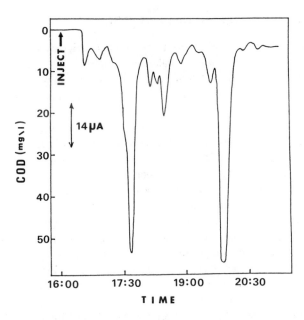

Figure 3. Continuous monitoring result of COD in sewage discharged from chemical laboratories. (Reproduced from Goto, M., *Trends Anal. Chem.*, 2, 92, 1983 with permission from the copyright holder.)

flow methods for the determination of cyanide involving nFIA, rFIA, and CFA. In their method, all of the solutions, (namely, buffer solution, chloramine-T, sample, and pyridine-barbituric acid) are pumped at the same flow rate (300 μl/min) from the lines A, B, C, and D in Figure 2, respectively. The buffer solution (pH 6.3) is mixed first with the chloramine-T reagent and then with the sample solution. The reaction takes place in a 25 cm × 0.35 mm i.d. tube (reaction tube 1 in Figure 2). The subsequent confluence of cyanogen chloride (CNCl) is mixed with the pyridine-barbituric acid solution and introduced into a 5.25 m × 0.35 mm i.d. tube (reaction tube 2 in Figure 2) kept at 35°C. The violet product is introduced into a flow cell (inner volume 18 μl) of the UV-visible absorbance detector, and the absorbance at 578 nm, corresponding to the total cyanide concentration, is continuously recorded as shown in Figure 4.[23] The cyanide comes from a 2-l container which initially contained 500 ml of distilled water. Another vessel containing cyanide solution is fixed to this tank. Periodically, different volumes of cyanide or distilled water are added to the container in such a way that cyanide concentration is changed.

5.2.3.3 Phosphorus

The continuous monitoring of phosphate in wastewater can be carried out by using the manifold shown in Figure 2.[24] The flow line A is not required

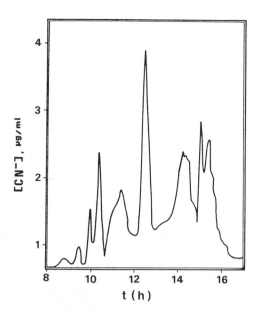

Figure 4. Simulation of continuous monitoring of cyanide concentrations in wastewater during a working day. (Reproduced from Rios, A., Luque de Castro, M.D., and Varcárcel, M., *Talanta,* 31, 673, 1984, with permission from the copyright holder.)

in this case. The measurement is made based on the molybdenum blue method. Sample solution, an oxidizing reagent, and a color-forming reagent are continuously supplied with a peristaltic pump at flow rates of 200, 100, and 100 μl/min from the lines B, C, and D in Figure 2, respectively. The sample is first mixed with the oxidizing reagent, 2% potassium peroxodisulfate, and the mixture is then introduced into an 8 m × 0.5 mm i.d. oxidation reaction tube (reaction tube 1 in Figure 2), placed in an aluminum block bath heated at 140°C, where phosphorus compounds are converted to phosphate ion. A platinum wire with a diameter of 0.2 mm is inserted into the reaction tube to serve as a catalyst for the oxidation reaction. The stream from the reaction tube is then mixed with a stream of the color-forming reagent, which is composed of a mixture of aqueous solutions of molybdic acid and ascorbic acid. The molybdenum blue formed in the mixed stream is introduced into a flow cell of a photometer, and its color intensity is continuously measured at 880 nm to monitor total phosphorus in a wastewater sample, as demonstrated in Figure 5.[24]

5.2.3.4 Silicon

The manifold for the continuous monitoring of trace silicon is basically the same as that for phosphorus. The determination of silicon is also based on the molybdenum blue absorption spectrophotometry.[25] Sample solution, a

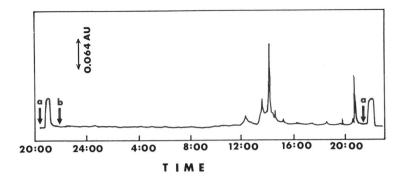

Figure 5. Continuous monitoring of total phosphorus concentrations in wastewater discharged from chemical laboratories; a, 200 μg/l standard; b, wastewater. (Reproduced from Goto, M., Hu, W., Ishii, D., *Environ. Sci.,* 2, 41, 1989, with permission from the copyright holder.)

molybdic acid solution, and a reducing reagent (a mixture of oxalic acid and 4-amino-3-hydroxy-1-naphthalenesulfonic acid) are continuously pumped with a peristaltic pump at the same flow rate of 75 μl/min from the lines B, C, and D in Figure 2, respectively. The sample is mixed with the molybdic acid reagent and then introduced into an 8 m × 0.5 mm i.d. PTFE tube (reaction tube 1 in Figure 2) left at room temperature (23°C), in which the molybdosilicic acid is formed. A platinum wire with a diameter of 0.2 mm is inserted into the reaction tube. It does not work as the catalyst, but just assists in the mixing of sample with the reagent stream. The stream is then mixed with a reducing reagent to form the molybdenum blue which is finally introduced into a spectrophotometric flow cell (10 mm light pass length and 8 μl volume). The absorbance at 815 nm is continuously recorded to monitor the total concentration of silicon in the sample.

Figure 6 demonstrates a typical monitoring result of silicon in industrial wastewater.[25] Under the above conditions, oxalic acid employed as the reducing agent also works as the masking agent for phosphate ion.[25] The interference with 20 mg/l phosphate is 1.5% for the determination of 0.2 mg/l silicon.

5.2.3.5 Mercury

Analysis and monitoring of total mercury in environmental samples are extremely important because of the high toxicity of mercury and its compounds. Therefore, checking the total content of mercury in wastewater is essential for environmental water control. Continuous monitoring systems of mercury based on CFA have been reported.[26-28]

Figure 7 shows a schematic diagram for CFA of total mercury.[26] Sample solution, an oxidizing reagent (2% potassium peroxodisulfate solution

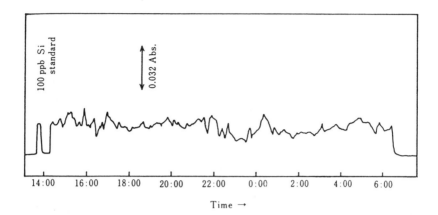

Figure 6. Continuous monitoring of trace silicon in wastewater. Sample: wastewater from chemical laboratories diluted to 161 times with pure water. (Reproduced from Goto, M., Hu, W., and Ishii, D., *Bunseki Kagaku,* 38, 419, 1989, with permission from the copyright holder.)

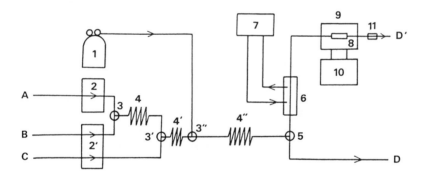

Figure 7. Schematic diagram of continuous monitoring apparatus for total mercury. 1 = Argon gas; 2,2' = peristaltic pumps; 3,3',3'' = mixing joints; 4,4',4'' = reaction tubes; 5 = gas-liquid separator; 6 = condenser; 7 = refrigerated circulating bath; 8 = flow cell; 9 = UV detector; 10 = recorder; 11 = mercury vapor absorbent; A = sample; B = oxidizing and catalytic reagent; O = reducing reagent; D,D' = wastes. (Reproduced from Goto, M., Munaf, E., and Ishii, D., *Fresenius Z. Anal. Chem.,* 332, 745, 1988, with permission from the copyright holder.)

containing copper(II) as the catalyst), and a reducing reagent (2% tin chloride in 3.4 *M* sodium hydroxide) are continuously pumped at the flow rates of 200, 100, and 100 μl/min, respectively. The sample stream is joined first with the oxidizing reagent and introduced into a 27 cm × 0.5 mm i.d. PTFE tube left at room temperature. The stream is then mixed with the reducing reagent and introduced into a 4 cm × 0.5 mm i.d., followed by mixing with argon gas flowing at 9.0 ml/min. The stream passes through a 60 cm ×

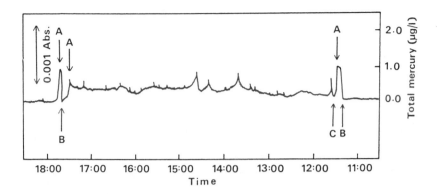

Figure 8. Continuous monitoring of total mercury in wastewater discharged from chemical laboratories. (Reproduced from Goto, M., Munaf, E., and Ishii, D., *Fresenius Z. Anal. Chem.*, 332, 745, 1988, with permission from the copyright holder.)

0.5 mm i.d. PTFE tube and a laboratory-made gas-liquid separator, where mercury vapor is extracted into the gas phase. The gas phase is finally swept into the photometric flow cell (40 mm light pass length and 1.13 ml volume), and the absorbance at 253.7 nm is continuously recorded to monitor total content of mercury.

Figure 8 demonstrates continuous monitoring of total mercury present in environmental waters.[26] Under the conditions in Figure 8, the response time for the determination of total mercury is about 4 min, and the detection limit is 0.1 μg/l. The effect of various matrices, (e.g., cations, anions, and organic compounds) on the signal response for the mercury determination is found to be much smaller than that in conventional automatic methods.[29]

It is well known that inorganic mercury compounds are converted into more toxic methylmercury by the aquatic organism. Since organomercury(II) compounds (such as alkylmercury) are much more toxic to humans than inorganic mercury compounds, speciation of the chemical forms of mercury compounds contained in environmental samples is required. Coupling of the flow system described above with a microcolumn liquid chromatographic system allows the speciation of mercury compounds. The experimental procedure is reported by Munaf et al.[30]

5.2.3.6 Chromium

The continuous monitoring of chromium has been reported by Ruz et al.[31] using the diagram illustrated in Figure 2.[31] A loop injector is inserted in flow line C, and flow lines A and B are not necessary in their system. The sample solution containing chromium(VI) and (III) is continuously pumped at a flow rate of 1.2 ml/min from line C.[31] When an oxidant solution (500 mg/l cerium(IV)) is introduced with the loop injector, the oxidation of chromium(III)

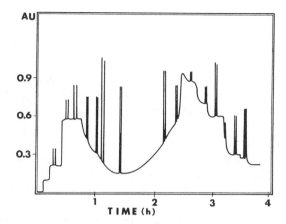

Figure 9. Simulation of the speciation analysis of chromium(VI) and chromium(III) over a working morning. (Reproduced from Ruz, J., Rios, A., Luque de Castro, M.D., and Varcárcel, M., *Fresenius A. Anal. Chem.*, 322, 499, 1985, with permission from the copyright holder.

to chromium(VI) takes place in a 6 m × 0.5 mm i.d. reactor (reaction tube 1 in Figure 2), otherwise chromium(III) is not detected. The solution is merged with the indicator, which is composed of 0.17% 1,5-diphenylcarbazide dissolved in 40% ethanol aqueous solution and pumped at a flow rate of 1.4 ml/min from line D, and then introduced into a 60 cm × 0.5 mm i.d. reactor (reaction tube 2 in Figure 2), followed by monitoring at 540 nm. In this manner, chromium(VI) is monitored, and the transient signal is observed when the oxidant is injected, which is related to the chromium(III) concentration in the sample.

Since the presence of chromium(III) does not affect the chromium(VI) absorbance signal, the concentration of the latter could be directly obtained. This type of configuration is suitable for the continuous monitoring of chromium(VI) as well as periodical measurement of chromium(III) in wastewater.

Figure 9 shows the recording obtained for hypothetical wastewater.[31] For this purpose, a 2-l reservoir (furnished with a magnetic stirrer) initially containing 200 ml of water, has been used. From this, the water is continuously aspirated and analyzed. There are also three bottles containing solutions of chromium(VI), chromium(III), and water, respectively, which pour their contents randomly into the reservoir.

5.2.3.7 Selenium

Selenium has been recognized as an element which has both essential and toxic activities to humans. The importance of its determination can be seen in a number of analytical methods that have been developed.[32-35]

Spectrophotometry was the most often used technique in the past and is still widely used in many laboratories.

For the determination, Hwang et al.[36] used selenium as the catalyst on the reduction of picrate ion in the presence of sodium sulfide. The catalytic-spectrophotometry determination of selenium has been automated by Linares et al.,[37] Shiundu and Wade,[38] and Munaf (unpublished). The methods involve a redox reaction in which selenium acts as the catalyst. The intermediate product coupled with the color-forming reagent and the colored product is detected by a photometric detector.

Linares et al.[37] described the simultaneous determination of selenium(IV) and selenium(VI) by FIA in which selenium(VI) must be reduced to selenium(IV) prior to the determination, although the procedure described was rather complicated.

Shiundu and Wade[38] developed a catalytic method for selenium by nFIA, stopped flow, and rFIA. These methods are capable of conducting only periodical determinations of selenium in wastewater.

Munaf et al.[38] described a CFA method based on catalytic-spectrophotometry for the determination of selenium, and the determination is conducted by the system shown in Figure 2.[32] Sample solution, oxidizing reagent (2% potassium peroxodisulfate solution), and combination of 0.01% hydrazine and 0.01% *o*-tolidine in 1.0 *M* hydrochloric acid are transferred to the system by using a peristaltic pump at flow rates of 50, 15, and 15 μl/min from the lines B, C, and D in Figure 2, respectively. For the determination of selenium, the sample is first mixed with 2% potassium peroxodisulfate and passed through the PTFE tube (reaction tube 1 in Figure 2) and then mixed with the solution of hydrazinium sulfate/*o*-tolidine. The reaction occurs in reaction tube 2 (0.5 mm i.d. and 2 m length) placed in an aluminum block bath, where selenium catalyzes the redox reaction between potassium peroxodisulfate and hydrazinium sulfate at a temperature of 65°C. The intermediate product of the redox step reacts with *o*-tolidine, and the yellow-colored product is monitored at 440 nm. The sample is pumped for about 4 min because the response reaches a steady state in about 3 min after the injection. The sample is pumped after the response for the preceding sample is returned to the base line.

5.2.4 CONCLUSIONS

During the last decade, parallel with rapidly developing technology and increasing research activity in universities, we have been witnessing alarming phenomena on the pollution problems. As we come to consider the waste problem, continuous monitoring of waste effluents from universities is necessary, and for this purpose, the monitoring methods based on flow analysis seem to be ideal.

REFERENCES

1. Varcacel, M. and Luque de Castro, M.D., *Flow Injection Analysis, Principles and Application*, Ellis Horwood, Chichester, England, 1987.
2. Skeggs, L.T., Jr., Automatic method for colorimetric analysis, *Am. J. Clin. Pathol.*, 28, 311, 1957.
3. Luque de Castro, M.D., Continuous monitoring by unsegmented flow techniques: state of the art and perspectives, *Talanta*, 36, 591, 1989.
4. Bailey, B.W. and Lo, F.C., Automated method for determination of mercury, *Anal. Chem.*, 43, 1525, 1971.
5. El-Awady, A.A., Miller, R.B., and Carter, M.J., Automated method for the determination of total and inorganic mercury in water and wastewater samples, *Anal. Chem.*, 48, 110, 1976.
6. Oda, C.E. and Ingle, J.D., Jr., Continuous flow cold vapor atomic absorption determination of mercury, *Anal. Chem.*, 53, 2030, 1981.
7. Fang, Z., Xu, S., Wang, X., and Zhang, S., Combination of flow-injection techniques with atomic spectrometry in agricultural and environmental analysis, *Anal. Chim. Acta*, 179, 325, 1986.
8. Thomson, J., Jhonson, K.S., and Petty, R.L., Determination of reactive silicate in seawater by flow injection analysis, *Anal. Chem.*, 55, 2378, 1983.
9. Hansen, E.H., Ghose, A.K., and Růžička, J., Flow injection analysis of environmental samples for nitrate using an ion-selective electrode, *Analyst (London)*, 102, 705, 1977.
10. Korenaga, T., Apparatus for measuring chemical oxygen demand based on flow injection analysis, *Bunseki Kagaku*, 29, 222, 1980.
11. Kuroda, R. and Mochizuki, T., Continuous spectrophotometric determination of copper, nickel and zinc in copper-base alloys by flow-injection analysis, *Talanta*, 28, 389, 1981.
12. Hirai, Y., Yoza, N., and Ohashi, S., Flow injection analysis of phosphates in environmental waters, *Bunseki Kagaku*, 30, 465, 1981.
13. Kamidate, T., Yamaguchi, K., Segawa, T., and Watanabe, H., Lophine chemiluminescence for determination of chromium(VI) by continuous flow method, *Anal. Sci.*, 5, 429, 1989.
14. Japan Industrial Standard (JIS) Handbook, *Testing Methods for Industrial Water*, K0102, Shin-nihonhoki Shuppan, Tokyo, 1325, 1986.
15. Japan Industrial Standard (JIS) Handbook, *Testing Methods for Industrial Wastewater*, K0101, Shin-nihonhoki Shuppan, Tokyo, 1120, 1986.
16. Japan Industrial Standard (JIS) Handbook, *Testing Methods for Industrial Water*, K0102, Shin-nihonhoki Shuppan, Tokyo, 1331, 1986.
17. Kehoe, T.J., Determining TOC in waters, *Environ. Sci. Technol.*, 11, 37, 1977.
18. Adelman, M.H., *Automation in Analytical Chemistry*, Vol. 1, Technicon Symposia 1966, Medead, NY, 1967.
19. Korenaga, T. and Ikatsu, H., Continuous-flow injection analysis of aqueous environmental samples for chemical oxygen demand, *Analyst London*, 106, 653, 1981.
20. Korenaga, T. and Ikatsu, H., The determination of chemical oxygen demand in wastewaters with dichromate by flow injection analysis, *Anal. Chim. Acta*, 141, 301, 1982.
21. Appleton, J.M.H., Tyson, J.F., and Mounce, R.P., The rapid determination of chemical oxygen demand in waste waters and effluents by flow injection analysis, *Anal. Chim. Acta*, 179, 269, 1986.

22. Goto, M., Monitoring of environmental water using continuous flow, *Trends Anal. Chem.*, 2, 92, 1983.
23. Rios, A., Luque de Castro, M.D., and Varcárcel, M., Spectrophotometric determination of cyanide by unsegmented flow methods, *Talanta*, 31, 673, 1984.
24. Goto, M., Hu, W., and Ishii, D., continuation of total phosphorus monitoring in wastewater, *Environ. Sci.*, 2, 41, 1989.
25. Goto, M., Hu, W., and Ishii, D., Continuous monitoring method of trace silicon in industrial water using continuous micro flow analysis, *Bunseki Kagaku*, 38, 419, 1989.
26. Goto, M., Munaf, E., and Ishii, D., Continuous micro flow monitoring method for total mercury at sub-ppb level in wastewater and other waters using cold vapor atomic absorption spectrometry, *Fresenius Z. Anal. Chem.*, 332, 745, 1988.
27. Munaf, E., Haraguchi, H., Ishii, D., Takeuchi, T., and Goto, M., Determination of total mercury concentration in wastewater by continuous microflow analysis with cold vapor atomic absorption spectrometry, *Sci. Total Environ.*, 99, 205, 1990.
28. Munaf, E., Takeuchi, T., Goto, M., Haraguchi, H., and Ishii, D., Digestion method for the determination of mercury in continuous microflow analysis, *Anal. Sci.*, 6, 313, 1990.
29. Munaf, E., Goto, M., and Ishii, D., Matrix effects in the determination of mercury by continuous micro flow-cold vapor atomic absorption spectrometry in alkaline medium, *Fresenius Z. Anal. Chem.*, 334, 115, 1989.
30. Munaf, E., Haraguchi, H., Ishii, D., Takeuchi, T., and Goto, M., Speciation of mercury compounds in waste water by microcolumn liquid chromatography using a preconcentration column with cold-vapor atomic absorption spectrometric detection, *Anal. Chim. Acta*, 235, 399, 1990.
31. Ruz, J., Rios, A., Luque de Castro, M.D. and Varcárcel, M., Simultaneous and sequential determination of chromium(VI) and chromium(III) by unsegmented flow analysis, *Fresenius Z. Anal. Chem.*, 322, 499, 1985.
32. Verlinden, M., Deelstra, H., and Adriaensens, E., The determination of selenium by atomic-absorption spectrometry: a review, *Talanta*, 28, 637, 1981.
33. Cutter, G.A., Species determination of selenium in natural waters, *Anal. Chim. Acta*, 98, 59, 1978.
34. Chau, Y.K., Wong, P.T.S., and Goulden, P.D., Gas chromatography-atomic absorption method for the determination of dimethyl selenide and dimethyl diselenide, *Anal. Chem.*, 47, 2279, 1975.
35. Osburn, R.L., Shendrikar, A.D., and West, P.W., New spectrophotometric method for determination of submicrogram quantities of selenium, *Anal. Chem.*, 43, 594, 1971.
36. Hwang, J.M., Wei, T.S., and Chen, Y.M., Catalytic photometric determination of selenium by flow injection analysis, *J. Chin. Chem. Soc. Taipei*, 33, 109, 1986.
37. Linares, P., Luque de Castro, M.D., and Valcárcel, M., Spectrophotometric determination of selenium(IV) and selenium(VI) with flow injection, *Analyst London*, 111, 1405, 1986.
38. Shiundu, P.M. and Wade, A.P., Development of catalytic photometric flow injection methods for the determination of selenium, *Anal. Chem.*, 63, 692, 1991.

CHAPTER 5.3

Process Analytical Chemistry in a Waste Treatment Facility

Takashi Korenaga and Xiaojing Zhou

TABLE OF CONTENTS

0-8493-682-5/94/$0.00 + $.50
© 1994 by CRC Press, Inc.

5.3.1 INTRODUCTION

Chemical process analysis for process control is a research priority in both industry and academic institutions. The reason is that the application of useful chemical process analysis techniques results in products of better quality with less waste; these techniques also minimize the possibility of industrial pollution. Furthermore, environmental pollution offers special challenges to process analytical chemistry, and the development of new analytical methods or the adaptation of existing methods for the transport of pollutants is highly desirable in the environmental process and for process control of wastewater treatment. In view of the growing demand, the process analytical techniques already developed should be automated and, in many instances, operated on an automatic basis as batchwise and flow systems.[1-3]

Batch methods are usually mechanized versions of manual procedures. In a batch system, each discrete sample is assigned a container, within which it is held through all the steps necessary to perform the process analysis. The advantages of a batch method are that cross-contamination between samples is not possible; the samples preserve their identity and cannot be mismatched since each container carries an identification label. However, the greatest drawbacks of the batch method are the mechanical complexity and high cost of operation. Expansions in this technique have, however, promoted the development of many aspects of chemical process analysis. The continuous flow method is the most flexible way to perform the number of operations necessary in process control for wastewater treatment. There are two general types of continuous flow methods: segmented and unsegmented. In a segmented flow system, the samples are aspirated sequentially, and air bubbles separate the flow; these air bubbles are usually removed before they reach the detector cell. An unsegmented flow method differs in that the flow is not segmented by air bubbles; the sample is individually injected or continuously delivered, and neither physical nor chemical equilibrium has been attained by the time the signal is recorded.

As described in detail in the previous section, sensing systems based on a flow method are serious candidates as a new generation of chemical probes for process control of wastewater treatment, because there is a growing interest in process analytical techniques that combine versatility with the possibility of achieving rapid and simple analyses.[4-7] Some of the striking features of such a sensing system can be mentioned to emphasize the properties which make it a reliable technique for supporting process control of wastewater treatment: continuous process analyses are performed in either an on-line or in-line mode; maximum sampling frequencies can be attained when a sample is continuously delivered as an analytical stream; and immediate response can be obtained. This chapter deals with the application of a novel sensor and sensing system based on flow method for the process analysis/control of wastewater treatment.

A recent advance in the field of water quality and process analysis of wastewater treatment is the application of intelligent laboratory robotics. Robotics have often been applied to certain pharmaceutical analyses[8-10] since the commercial introduction of the first laboratory robots by Zymark Corporation in the U.S. in 1982. An individual making the analysis, together with a robotics system, is particularly advantageous when conditions make the work hazardous or impossible for a human, or when it is too tedious for a human to perform with sustained accuracy and efficiency. A commercially available robot manipulator has been examined to develop an automated procedure for analysis of chemical oxygen demand (COD) and to evaluate the wastewater treatment process.[11]

5.3.2 ANALYTICAL TECHNIQUES FOR CONTROL OF A BIOLOGICAL TREATMENT PROCESS

A biological means involving aerobic and anaerobic treatment is usually selected to handle various wastewater which contains biodegradable organic materials. To realize high degradation efficiency and ensure effluent quality, most efforts have been put into the development of new analysis techniques to control the biological treatment process. Reports relating to this research and its practical application have become more numerous in recent years.

5.3.2.1 Rapid Determination of BOD by a Microbial Sensor

Biochemical oxygen demand (BOD) is a widely used parameter to determine biodegradable organic compounds in wastewater. The conventional BOD measurement, however, requires a 5-d incubation period[12] and is thus unsuitable for process control of wastewater treatment. A more rapid estimation of BOD is possible by using a microbial sensor containing whole cells immobilized on an oxygen electrode. Such a BOD sensor was first reported by Karube et al.,[13] and in 1990, an analytical technique for rapid determination of BOD with a microbial sensor was incorporated in the Japanese Industrial Standards.[14]

Microorganisms for the BOD sensor should have a wide substrate spectrum, but it is difficult to prepare a sensor with activated sludge containing mixed populations that will give reproducible results.[15] As has been demonstrated, a sensor using pure cultures of microorganisms is more suitable.[15-17] Most BOD sensors use the stationary state measurement and show a response time of 15 to 20 min, but with a microbial sensor it is possible to use the dynamic or kinetic mode (differential current time-curve method) for more quick determination of BOD. These latter sensors are based on the measurement of the acceleration of respiration rate by the change in the electric current and allow a response time of less than 1 min.[17,18] More recently, extensive work

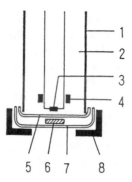

Figure 1. Scheme of BOD biosensor. (1) Isolator; (2) electrolyte; (3) anode; (4) cathode; (5) polyethylene membrane; (6) *Trichosporon cutaneum* PVA membrane; (7) dialysis membrane; and (8) ring.

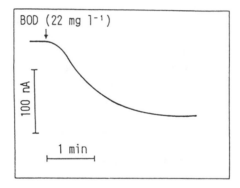

Figure 2. Response curve of BOD biosensor.

has been done on the determination of BOD with a microbial sensor by Riedel et al.[19] who used *Trichosporon cutaneum* cells immobilized in polyvinyl alcohol. The scheme of this biosensor is shown in Figure 1. A modified oxygen electrode is coated with a polyethylene membrane; the immobilized microorganisms are placed on this membrane and covered with a dialysis membrane. One of the most important features of this sensor is that it allows BOD measurement within a very short response time (30 sec). A typical response curve of the BOD sensor (shown in Figure 2) was found when a solution containing glucose (15 mg/l) and glutamic acid (15 mg/l) with a BOD value of 22 mg/l was employed as a standard solution and injected into the measuring system. A commercial instrument for rapid BOD measurement has been developed using such a sensor (Figure 3).[20]

In contrast to the conventional method which requires 5 days and reflects the various metabolic reactions of a mixed population, the determination of BOD with a microbial sensor is a test with a selected microbial species. The

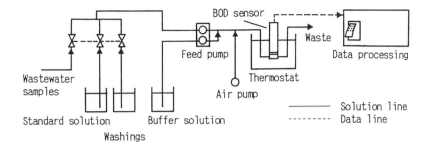

Figure 3. Commercial instrument using BOD biosensor.

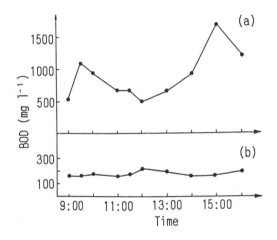

Figure 4. Day profile of municipal wastewater. (a) Inflow to the activated sludge tank; (b) sewage works effluent.

correlation of biosensor BOD value with the 5-d BOD value of wastewater is very important for evaluation of its reliability and availability. In general, the microbial sensor shows lower BOD values than those determined by the conventional method; however, an increase in BOD sensor sensitivity can be obtained by pretreatment of the sensor with a desired substrate. Good comparative results were obtained between BOD estimated by the microbial sensor and the conventional method when this sensor was incubated with the substrate determined. After incubation with wastewater, the microbial sensor can be used to measure the daily cycle of the organic load of a wastewater plant (Figure 4) and to determine the progress of aerobic/anaerobic sludge treatment (Figure 5). The short measuring time offers the possibility of obtaining a detailed picture of the time course of organic load and of the succession of peak loads in different steps of a wastewater plant and their degradation by the biochemical process.

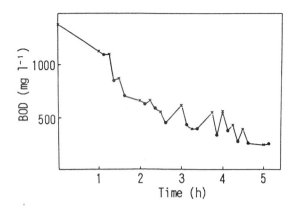

Figure 5. BOD alteration during aerobic/anaerobic treatment of sludge. (●) Measurement after aerobic phase; (×) measurement after anaerobic phase.

The other advantage of this BOD sensor is its remarkable stability over a 48-d period. Because of its short response time and long-term stability, it is possible to obtain an accurate and reliable measurement of the release and assimilation of organic substances at any time during sludge teratment as a result of rapidly changing aerobic and anaerobic conditions.

5.3.2.2 Application of a Robotics System to Automatic COD Measurement

COD is also a major index for the determination of organic pollution in water and for control of organic loading during wastewater treatment. There are several standard methods published by official organizations; these are generally manual and based on the titration of the aqueous sample with potassium permanganate[21] or potassium dichromate.[12] These manual methods are complicated and time-consuming, and their precision depends heavily on an operator's skill. The development of an automated method would improve the efficiency of pollution control and thus the operation of effluent treatment plants. An AutoAnalyzer procedure to determine COD was investigated by Adelman in 1966,[22] and Korenaga et al.[23,24] reported extensively on the determination of COD by flow-injection analysis (FIA). Very recently, an automatic instrument for measuring COD was developed by applying a robotics system;[11] use of this system assures high precision and reproducibility, as well as reliability. The apparatus was constructed from commercially available parts, which include a robot manipulator, a personal computer for process control and data processing, and analytical instruments (an automatic conductometric detector, an automatic oxidation-reduction potential or ORP titrator, etc.) (Figure 6). Figure 7 illustrates an experimental procedure for COD measurement using the robot manipulator, which is based on official manual methods incorporated in the Japanese Industrial Standards.[12]

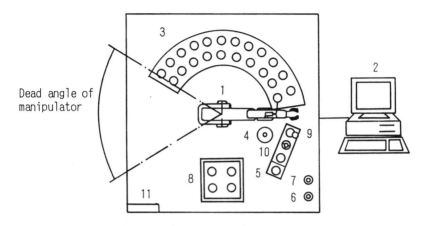

Figure 6. Basic composition and layout of the intelligent robot manipulator system for COD measurement. (1) Robot manipulator; (2) personal computer; (3) sample table; (4) automatic ORP titrator with an auto-buret for 0.025 *N* potassium permanganate; (5) automatic conductometric detector; (6) auto-buret for sulfuric acid; (7) auto-buret for 0.025 *N* sodium oxalate; (8) water bath; (9) silver sulfate powder hopper; (10) magnetic stirrer; and (11) compressed air control unit.

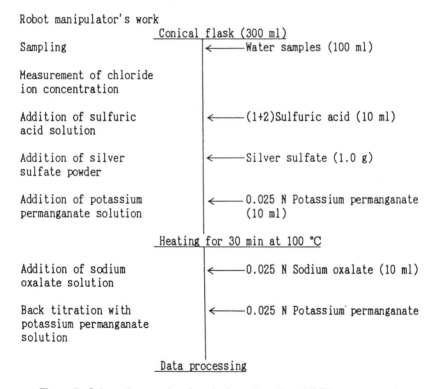

Figure 7. Schematic procedure for robotics automation of COD measurement.

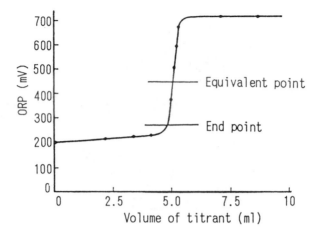

Figure 8. Representative ORP titration curve obtained by using robotics system.

The addition of silver sulfate was important both for masking the chloride ion and for catalyzing the degradation of organic substances. In the present analytical system, a conductometric detector was adopted for the rapid determination of chloride ion in water samples. The advantage of this method is its ease of operation, free maintenance property, and immediate response. The step of weighing silver sulfate power is thus accomplished using this automatic conductometric detector and a direct weighing system.

An ORP titration method for end-point detection was developed to resolve a major problem of turbid components in sample solutions resulting from earlier use of a spectrophotometric method based on the fading of permanganate. Figure 8 shows a representative ORP titration curve obtained when the excess sodium oxalate solution is back-titrated with 0.025 N potassium permanganate solution. The back titration step was carried out using artificial intelligence which was supported by a flexible common expert system.

Application of this robotics system to COD measurement for various wastewaters containing organic pollutants resulted in a coefficient of variation of about 0.3%, which is more satisfactory than the official manual method (i.e., 3%). Furthermore, the analytical results were in good agreement with those obtained by the official manual method. For 23 COD measurements, the regression equation and correlation coefficient were Y = 0.97X + 0.47 and 0.993, when X and Y denoted the COD value obtained with the system and the official manual method, respectively.

5.3.2.3 Toxicity Test Based on Flow Analysis

As biological treatment is used increasingly for complex wastewater that may contain high concentrations of toxic compounds, there is a need for

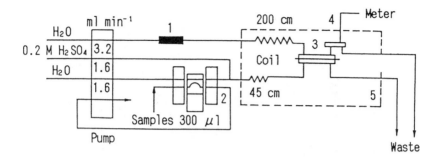

Figure 9. Flow-injection manifold for CO_2 measurement: (1) Mixed-bed ion-exchange resin; (2) sample introduction valve; (3) diffusion cell; (4) conductance II; and (5) water bath.

analytical techniques to evaluate and characterize possible inhibitory effects of these compounds on their own degradability, removal, and degradation.

Anaerobic toxicity assays were developed initially by Hungate[25] and can be used to screen wastewater for potential toxicity. This method is a batch procedure intended to provide reproducible assay conditions where toxicant concentrations are the only parameter varied. In this test, an active methanogenic culture is fed to some easily degradable compounds and various concentrations of toxicant. A decrease in the rate of methane production with increasing concentration of the test compound is indicative of toxicity. However, this batch procedure is time-consuming for the obtaining of toxicity data. Many studies with respect to the evaluation of toxicity have been accomplished using microbes as the test organism.[26,27] Such a bioassay provides rapid and reliable data with which to evaluate the toxicity of chemicals and to overcome disturbances in treatment performance caused by the presence of inhibitory or toxic substances.

The inhibition of some vital microbial functions is usually easier to measure. Respiration, for instance, is a reliable parameter to monitor in an acute toxicity test. The net rate of CO_2 production by microorganisms reflects a complete series of biochemical reactions which constitute the respiration process of the organism. Inhibition of this process alters the amount of CO_2 produced and trapped in the aqueous culture media. An *Escherichia coli* electrode was developed by Dorward and Barisa[28] and applied to assess the acute toxicity of metals, anions, gases, and organic compounds. More recently, short-term toxicity tests using *E. coli* were carried out for different stressing agents such as heavy metal ions.[27] Inhibition of the microbial respiration was monitored using FIA with a conductometric detector. The FIA manifold is shown in Figure 9. In the procedure, 300 μl of the sample containing the carbonate species was injected into a carrier stream (deionized water) which meets a diluted acidic solution (0.2 *M* H₂SO₄). The carbon dioxide formed in the acidic stream is allowed to diffuse through a

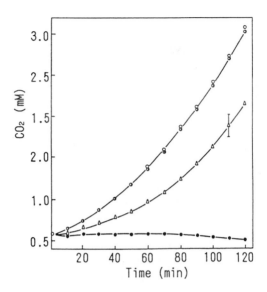

Figure 10. Bacterial response to Hg(II) ions used at concentrations of (◑) 10, (△) 50, and (●) 100 ppb, compared to the control (○) that received no metal.

poly(tetrafluoroethylene) (PTFE) membrane and collect in a stream of deionized water continuously flowing through the conductometric detector. The change in conductance is proportional to the total CO_2 concentration in the sample.

Bacterial responses to various concentrations of Hg(II), Cd(II), and Cu(II) ions are shown in Figures 10 and 11, respectively. As can be seen, a severe inhibition in bacterial respiration is obtained at 50 ppb of mercury and 5 ppm of copper and cadmium; for a mercury concentration of 100 ppb alone, the metabolic process is totally suppressed. This determination of CO_2 made by combining FIA with an electrochemical sensor can be directly utilized in acute toxicity tests and is feasible for evaluating the toxicity of chemicals to microorganisms which are required for biological treatment.

5.3.2.4 Monitoring of a Biodegraded Intermediate/Product

Anaerobic degradation of complex wastewater involves a series of reactions carried out by a consortium of bacterial species. Monitoring of biodegraded intermediates and/or products might eventually play a valuable role in the control of industrial scale anaerobic treatment processes. Theoretical and experimental work has demonstrated that molecular hydrogen, which is either produced or consumed in the majority of degradation steps, is of primary importance. Thus, it has been suggested by a number of researchers that studies of the steady state and transient behavior of the anaerobic process should include measurement of H_2 concentration. Recently, a microcomputer-

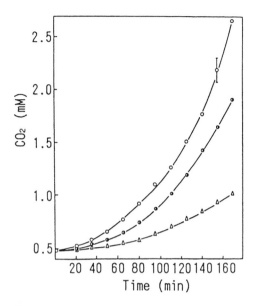

Figure 11. Bacterial response to 5 ppm of (◑) Cu and (△) Cd ions, compared to the control (○) that received no metal.

based system for monitoring and control of the anaerobic wastewater treatment process was described by Slater et al.[3] As shown in Figure 12, the system is composed of three major sections: feed metering, reactors, and gas and recycle stream monitoring. The feed metering section consists of refrigerated tanks connected to metering pumps that can be computer controlled. Control of the organic loading and hydraulic detention time can be done independently. The monitoring system includes a gas chromatograph that measures CH_4, CO_2, H_2, and CO concentrations in the gas phase every 2 min, a second gas chromatograph that measures volatile acids (C_1 to C_4) every 12 min, and sensors for continuous measurement of temperature, pH, ORP, and gas production rate. The monitoring system is controlled by two microcomputers using commercial data acquisition hardware and software. One computer acquires data from the reactor temperature, pH, ORP, gas flow rate, and turbidity sensors and controls the flow rate of the feed pump. The second computer controls the sampling valves for the two gas chromatographs and gathers data from the detectors. As chromatograms are acquired, this computer also determines peak areas and chemical compositions. Ricker[29] shows how this type of information can be used to estimate the response of the system to other types of transient disturbances, such as a step in the feed flow rate. In general, H_2 concentration is very sensitive to small imbalances in the production and consumption of hydrogen, and the response pattern is specific to the type of disturbance affecting the system, e.g., an increase in organic loading, a toxic event, etc. The ability to observe such qualitative patterns during operations may be beneficial in process control.

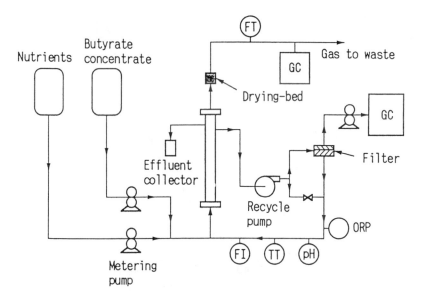

Figure 12. Schematic of the reactor and associated instruments. FT: flow transmitter; GC: gas chromatograph; FI: flow indicator; TT: temperature transmitter; pH: pH transmitter; ORP: ORP transmitter.

5.3.3 ANALYTICAL TECHNIQUES FOR CONTROL OF A CHEMICAL TREATMENT PROCESS

5.3.3.1 Routine Analysis of Mercury by Rapid Batch Method

The discharge of mercury to water courses is strictly controlled, and its effluent standard is therefore set at a very low concentration (5 ppb). In order to control the treatment process and ensure quality of the effluent, an analytical methodology that combines high speed, sensitivity, and minimal interference to determine mercury in complex wastewater samples is indeed necessary. A batch method utilizing cold-vapor atomic absorption has been most widely recognized as the official method;[30] its procedures are shown in Figure 13. However, this method tends to be time consuming and does not remove the interference from organic substances, iodide, etc. Organic compounds and iodide coexisting in samples may then consume the oxidizing agent during pretreatment, while the residual iodide ion would react with mercury(II) leading to the formation of a stable complex, HgI_4^{2-}, which cannot be reduced by acidic tin(II) solution during the reduction of mercuric ions. Furthermore, oxidized products such as IO_3^- are reduced more easily than mercury(II) under an acid solution condition; achieving high reduction efficiency of mercury is therefore very difficult.

For the rapid determination of mercury in wastewater samples in which organic compounds and iodide coexist, a simple batch method with cold-

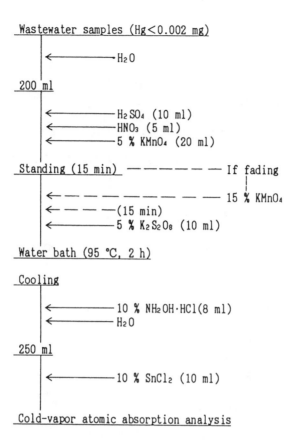

Figure 13. Flow chart of official manual method for COD measurement.

vapor atomic absorption (shown in Figure 14) was studied using alkaline tin(II) as reducing agent.[31,32] Tin(II) in alkaline solution is a stronger reducing agent than that in acidic solution. Consequently, even though the organomercury compounds (such as methyl-, ethylmercury chloride, phenylmercury nitrate, and HgI_4^{2-}) are very stable, they can be easily and completely reduced to metal mercury with alkaline tin(II) in the presence of a certain amount of copper(II) ion as a catalyst. This reduction cannot proceed in the presence of potassium zinc cyanide as a masking agent for silver ion, so a large amount of copper(II) ion was added to resolve this problem. The behavior of copper(II) in the reduction process of mercury has been evaluated as follows:[31]

$$SnO_3^{2-} + 3H^+ + 2e = HSnO_2^- + H_2O \tag{1}$$

$$Cu^{2+} + e \longrightarrow Cu^+ \tag{2}$$

$$2RHgX + e \longrightarrow RHg\cdot + X^- \tag{3}$$

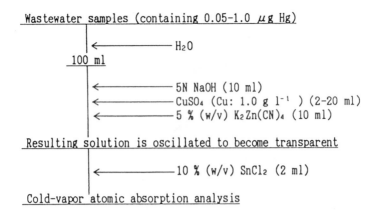

Figure 14. Flow chart of simple and rapid method for COD measurement.

$$2RHg \cdot \longrightarrow RHg^+ + RHg^- \tag{4}$$

$$RHg^+ + H_2O + 2e \rightarrow RH + Hg + OH^- \tag{5}$$

$$RHg^- + H_2O \longrightarrow RH + Hg + OH^- \tag{6}$$

$$RHg \cdot + CN^- \longrightarrow RHgCn + e \tag{7}$$

$$RHg^+ + Cn^- \longrightarrow RHgCN \tag{8}$$

$$RHgCN - Cu^+ \longrightarrow CuCN + RHg^+ \tag{9}$$

Some striking features of this method are its speed and simplicity, because the pretreatment step is omitted, and alkaline tin(II) solution is used as a reducing agent. The determination time required for a sample is much shorter, typically 3 min, than the official manual method which is a long 2 to 3 h, making it suitable for a large number of samples. For the determination of mercury in complicated wastewater samples, an accuracy and precision of better than 3% and a detection limit of 0.1 µg/l can generally be obtained with this method.

5.3.3.2 Computer-Controllable Monitoring System for Chromium Recycling

The greatest amount of chromium waste is produced in chromium electroplating, leather tanning, and material and chemical manufacturing.[33] The wastewater treatment systems in these plants are regarded as a logical point to control the transport of chromium into the water environment. Many pro-

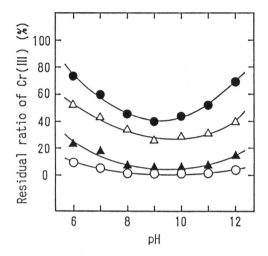

Figure 15. Effect of excess sodium sulfite on precipitation efficiency at different pH. Excess times (W/W₀): (○) 1.0; (▲) 1.2; (△) 1.5; and (●) 2.0.

cedures have been published concerning treatment for wastewater containing chromium(VI) or Cr(VI), but a reduction/coprecipitation method is now widely employed. In this method, sodium sulfite and the like are usually used as an agent for the reduction of Cr(VI) into Cr(III), and sodium hydroxide or lime is used to precipitate Cr(III). A calculation based on thermodynamic data indicates that the concentration of soluble Cr(III) in equilibrium with $Cr(OH)_3$ is about 10^{-9} to 10^{-12} mol/l at pH 7 to 8; therefore, in principle, a high removal efficiency should be readily achieved. However, the concentration of total chromium in treated wastewater often fell short of the effluent standard unless coprecipitation agents were used. This is because there was a conspicuous lack of reliability in the electrode-electrometer combination method, despite which is widely adopted in the process monitoring, so that excess sodium sulfite must be added to assure a safe discharge. The interaction between the insoluble Cr(III) hydroxide and the residual sulfite ion leads to the formation of the basic salt of Cr(III), $[Cr_X(OH)_Y(SO_3)_Z]^{3X-Y-2Z}$, which is dispersed as a dispersion colloid during the precipitation treatment. The effect of excess sodium sulfite on the precipitating efficiencies of Cr(III) was investigated by Zhou et al.[34] As shown in Figure 15, the residual ratio of Cr(III) in clarified fluid increased significantly with increasing excess of sodium sulfite. When the dosage of sodium sulfite (W) reached 200% of the amount (W₀) required for the equivalent reaction with Cr(VI), the residual ratio of Cr(III) increased about 50 to 60%. To precipitate the suspended colloid, coprecipitation reagents such as aluminum and iron(III) salts must usually be added. However, the increase in sludge poses a serious challenge to the possible use of these reagents. In this context, it should be noted that the

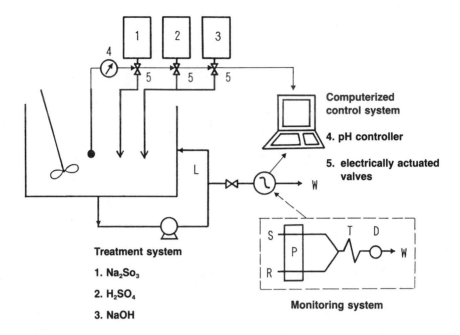

Figure 16. Monitoring system for the process control of wastewater treatment. P: microflow pump; L: fast circulation loop; T: recation tube; D: spectrophotometric detector; R: reagent solution.

control of reducing reagent is important to attain high removal efficiency and to reduce sludge.

An advanced monitoring system for control of the treatment process of wastewater containing chromium(VI) was developed using a flow-based chemical probe.[34] In this system (Figure 16), the acquisition of monitoring information and control of the treatment process is computer-controllable. The application of the system ensures safe discharge of wastewater without the addition of any coprecipitation reagents, and enables recycling of chromium waste because of the minimal contamination of the sludge. This monitoring system could be connected to the treatment process. The wastewater samples and reagent solutions containing potassium iodide and starch were continuously delivered by an improved high precision microflow double-plunger pump, each at a flow rate of 1.8 ml/min, from a fast circulation loop and a reagent reservoir, respectively. The reagent stream was mixed with the sample stream in a PTFE reaction tube (10 cm length, 1.0 mm i.d.), and the mixed stream was then led into the flow-through cell placed in a UV-visible spectrophotometric detector operated at 650 nm. The absorbance change was continuously recorded and was simultaneously adjusted for processs control by a personal computer.

As pointed out, the control of the reducing reagents delivery, namely, the end-point control of redox-treatment is the key in the process control. In fact,

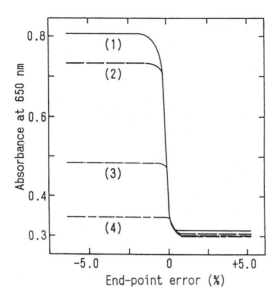

Figure 17. Typical response curve obtained by using the monitoring system. Cr(VI) concentrations: (1) 50; (2) 30; (3) 10; and (4) 1 mg l^{-1}.

the electrode-electrometer combination method has an obvious shortcoming in that it cannot produce a measurable response at the end-point of redox-treatment. An experiment for end-point control was carried out to demonstrate the usefulness of the chemical probe using an iodide-starch solution as reagent stream. Figure 17 illustrates the relationship between the response and the end-point error and shows an encouraging result. In the figure, the end-point error was evaluated using the computer and Equation 11:

$$\text{EPE} = \frac{C_1 \cdot Q \cdot T - C_2 \cdot V}{C_2 \cdot V} \times 100 \ (\%) \tag{11}$$

where
$\quad$ EPE $=$ end-point error (%);
$\quad$ C_1 $=$ equivalent concentration of sodium sulfite;
$\quad$ Q $=$ addition rate of sodium sulfite solution, ml/min;
$\quad$ T $=$ treatment time (min);
$\quad$ C_2 $=$ initial equivalent concentration of Cr(VI);
$\quad$ V $=$ initial wastewater volume, ml

As shown in Figure 17, even the Cr(VI) concentration which goes down to 1.0 mg/l gives an adequate absorbance change at the end point of redox-treatment. These experimental results confirm that the monitoring system is sensitive and accurate in its control of the redox-treatment process.

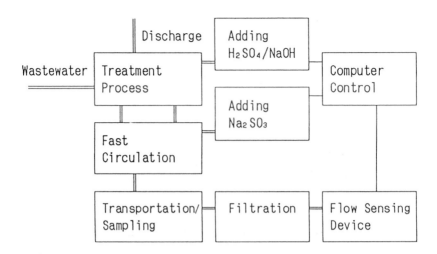

Figure 18. Trial test in industrial scale by application of the monitoring system.

Some important features of the system employing this chemical probe are its miniaturized device, high sensing functions, and dependable accuracy. Generally, a monitoring system cannot be installed close to treatment equipment, and a transport line with a correspondingly longer delay time may be necessary. To diminish this time lag and speed up the response, a so-called fast circulation loop was constructed between the treatment equipment and the sampling head. The wastewater is propelled with increased velocity in the loop, and because of this fast circulation loop, the time lag between sampling and result and the error of end-point control caused by the delay of response is reduced to be negligible. When using a series of three replicate measurements each consisting of 5 individual wastewater measurements, mean error and relative standard deviation (R.S.D.) for the end-point control were ±0.9 and 0.5%, respectively.

A trial test of this monitoring system was conducted in industrial plant scale (Figure 18). During the period of test, the concerted movement of the monitoring system and the manipulator was accurately programmed by a central programming unit running on the personal computer which commanded the manipulator to carry out the treatment process in accordance with the monitoring data. After precipitation treatment, the concentration of residual Cr(III) in wastewater was measured by a colorimetric method that used diphenyl carbazide as a color reagent. The samples were taken from the clarified fluid 24 h after the precipitation treatment. As shown in Figure 19, even though the monitoring system was not periodically recalibrated, it was able to reliably assure that the concentration of total chromium in effluent was lower than 0.5 m/l for a 2 week period.

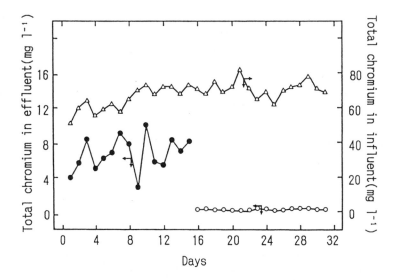

Figure 19. Results of process control under long-term unattended operation by using different monitoring techniques: (○) this system; (●) electrode-electrometer combination.

REFERENCES

1. Korenaga, T., Takahashi, T., Moriwake, T., and Sanuki, S., *Advances in Water Pollution Control,* Pergamon Press, Oxford, 1990, 625.
2. Hong, W.H., Meier, G., and Deininger, R.A., A micro computer interfaced continuous flow toxicity test system, *Water Res., 21,* 1249, 1987.
3. Slater, W.R., Merigh, M., Ricker, N.L., Labib, F., Ferguson, J.F., and Benjamin, M.M., A micro computer-based instrumentation system for anaerobic wastewater treatment process, *Water Res., 24,* 121, 1990.
4. van der Linden, W.E., Flow injection analysis in on-line process control, *Anal. Chim. Acta, 179,* 91, 1986.
5. Gisin, M. and Thommen, C., Industrial process control by flow injection analysis, *Anal. Chim. Acta, 190,* 165, 1986.
6. Ruzicaka, J., Flow injection analysis — a survey of its potential for continuous monitoring of industrial process, *Anal. Chim. Acta, 190,* 155, 1986.
7. Christian, G.D. and Ruzicka, J., Application of FIA in process analysis, paper presented at the 5th Int. Conference on Flow Analysis, Kumamoto, Japan, August 21, 1991.
8. Borman, S.A., Year of the Robot, *Anal. Chem., 57,* 651A, 1985.
9. Schlieper, W.A. and Isenhoure, T.L., Complexometric analysis using an artificial intelligence driven robotic system, *Anal. Chem., 60,* 1142, 1988.
10. Banno, K. and Takahashi, R., Laboratory automation system for the analysis of drugs in biological fluids, *Anal. Sci., 7,* 511, 1991.

11. Korenaga, T. and Yano, Y., Robotics in automation of environmental water analysis, *Anal. Sci.,* 7, 733, 1991.
12. Testing Methods for Industrial Wastewater, Japanese Industrial Standard Committee, Tokyo, JIS K0102, 1986.
13. Karube, I., Matsunaga, T., Mitsuda, S., and Suzuki, S., Microbial Electrode BOD Sensor, *Biotechnol. Bioeng.,* 19, 1535, 1977.
14. *The Determination of BOD$_5$ by a Microbial Electrode,* Japanese Industrial Standard Committee, Tokyo, Japan Industrial Standards, K3602, 1990.
15. Hikuma, M., Suzuki, H., Yashuda, T., Karube, I., and Suzuki, S., Amperometric estimation of BOD by using living immobilized yeasts, *Eur. J. Appl. Microbiol.,* 8, 289, 1970.
16. Kulys, J. and Kasziamskiene, K., Yeast BOD sensor, *Biotechnol. Bioeng.,* 22, 221, 1980.
17. Riedel, K., Renneberg, R., Kuehn, M., and Scheller, F., A fast estimation of biochemical oxygen demand using microbial sensor, *Appl. Microbiol. Biotechnol.,* 28, 316, 1988.
18. Riedel, K., Renneberg, R., and Liebs, P., An electrochemical method for determination of cell respiration, *J. Basic Microbiol.,* 25, 51, 1985.
19. Riedel, K., Lange, K.P., Stein, H.J., Ott, P., and Scheller, F., A microbial sensor for BOD, *Water Res.,* 24, 883, 1990.
20. Nagashio, N. and Harita, K., BOD quick measuring apparatus, *Nisshin Denki Giho,* 35, 78, 1990.
21. *Standard Methods for the Examination of Waste and Wastewater,* 13th ed., American Public Health Association, Washington, D.C., 1971.
22. Adelman, M.H., *Automation in Analytical Chemistry,* Vol. 1, Technicon Symposia, Mediad, NY, 1967.
23. Korenaga, T. and Ikatsu, H., Continuous-flow injection analysis of aqueous environmental samples for chemical oxygen demand, *Analyst London,* 106, 653, 1981.
24. Korenaga, T. and Ikatsu, H., The determination of chemical oxygen demand in wastewater with dichromate by flow injection analysis, *Anal. Chim. Acta,* 141, 301, 1982.
25. Hungate, R.E., ARO11 tube method for cultivation of strict anaerobes, *Methods Microbiol.,* 3B, 117, 1969.
26. Walker, J.D., Effect of chemicals on microorganisms, *J. Water Pollut. Control Fed.,* 60, 1106, 1988.
27. Jardim, W.F., Guimaraes, J.R., and de Faria, L.C., Short-term toxicity test using *Escherichia coli:* monitoring CO_2 production by flow injection analysis, *Water Res.,* 24, 351, 1990.
28. Dorward, E.J. and Barisas, B.G., Acute toxicity screening of water pollutants using a bacterial electrode, *Environ. Sci. Technol.,* 18, 967, 1984.
29. Ricker, N.L., The use of biased least-squares estimators for parameters in discrete-time impulse-response models, *I & EC Res.,* 27, 343, 1988.
30. *Testing Method for Industrial Wastewater,* Japanese Industrial Standard Committee, Tokyo, Japan Industrial Standards K0101, 1986.
31. Izawa, M., Korenaga, T., and Ikatsu, H., A simple determination of mercury by cold-vapor atomic absorption spectrophotometry in alkaline the samples coexisting organic substances, *Eisei Kagaku,* 34, 475, 1988.

32. Korenaga, T., Izawa, M., Noguchi, H., and Takahashi, T., Development of a continuous monitoring system for the determination of total mercury based on gas-liquid separation using coaxial microporous tubing, *Bunseki Kagaku,* 41, 17, 1992.
33. Francoise, C., Richard, C.F., and Bourg, A.C.M., Aqueous geochemistry of chromium: review, *Water Res.,* 25, 807, 1991.
34. Zhou, X., Korenaga, T., Takahashi, T., Moriwake, T., and Shinoda, S., A monitoring/control system for treatment of wastewaters containing chromium(VI), *Water Res.,* 27, 1049, 1993.

**Section
6**

**Practical Technology for Control of
Hazardous Wastes**

CHAPTER 6.1

Hazardous Wastes Control Technology

Hiroshi Takatsuki and Shin-ichi Sakai

TABLE OF CONTENTS

0-8493-682-5/94/$0.00 + $.50
© 1994 by CRC Press, Inc.

6.1.1 DEFINITION OF HAZARDOUS WASTES

A variety of substances with inflammable, explosive, and/or toxic characteristics are produced widely today. The properties of these substances are sometimes useful, but are often dangerous and hazardous. All of these compounds, once produced in a variety of products from raw materials, distributed and consumed, are ultimately disposed of as wastes. Some are so-called hazardous materials, which adversely affect human health and/or the environment if handled unproperly. Wastes discharged from research institutions ("laboratory wastes") are small in volume, but are in many cases extremely hazardous. The toxic and other properties of many such wastes are not yet known. These wastes, therefore, must be controlled, recycled, or disposed of properly as hazardous wastes.

The globally accepted definitions of hazardous wastes are found in the "Basel Convention on the control of transboundary movements of hazardous wastes and their disposal".[1] The Basel Convention defines hazardous waste as waste discharged through the streams specified in Convention Annex I, or waste having as constituents specified in the same Annex, excluding those without hazardous characteristics such as explosiveness and inflammability, provided in the Convention Annex III. In other words, wastes subject to the Basel Convention are those discharged through any one of the 18 specified streams from Y1 to Y18, or containing any one of the 27 hazardous substances from Y19 to Y45, and having any one of the 14 hazardous characteristics from H1 to H13. The European Community (EC) has also proposed to define hazardous wastes[2] in terms of the waste streams, hazardous constituents, and hazardous characteristics, as does the Basel Convention. The EC's definitions differ from those of the Basel Convention in the following points. The waste streams are classified into two groups: 18 routes in group 1A, which correspond to the streams specified in the Basel Convention, and 22 routes in group 1B. Waste discharged through a route in group 1A is considered hazardous if it has a hazardous characteristic, while waste discharged through a route in group 1B is not considered hazardous unless it contains a hazardous constituent from C1 to C51. The hazardous constituents from C1 through C51 include the hazardous constituents from Y19 to Y45 provided in the Basel Convention and other substances such as silver compounds and tin compounds. Since the concept of the EC's proposal is, in essence, within the framework of the definitions of the Basel Convention, it can readily be adapted to conform with the Basel Convention.

Meanwhile, the U.S. government has provided a completely different definition for hazardous wastes.[3] According to the definition by the Resource Conservation and Recovery Act (RCRA), waste is considered hazardous if found in the lists of hazardous wastes by substance and by discharge stream, or if it has hazardous characteristics. Specifically, waste is defined as hazardous if it is found in any one of the following lists or if it shows any one of the following hazardous characteristics. The lists: hazardous wastes from nonspecific sources (F List), hazardous wastes from specific sources (K List), acute hazardous chemical wastes (P List), and toxic chemical wastes (U List). The hazardous characteristics: ignitability, corrosivity, reactivity, and toxicity. Forty substances are identified as having toxicity, and wastes containing more than maximum concentration of these forty substances are designated hazardous wastes D004 through D043. Wastes showing ignitability, corrositivity, and reactivity under specified test methods are designated hazardous wastes D001, D002, and D003, respectively. Thus, the definition of hazardous wastes by the U.S. is much more comprehensive than those by the Basel Convention, EC, and Japan.

The former "Waste Disposal and Public Cleansing Law" (hereafter referred to as the "Waste Management Law") of Japan provided standard values for waste contents of 11 specified hazardous constituents regarding their final disposal methods. Combustion residue, sludge, waste acid, and alkali in which the content of any of the specified hazardous constituents exceeded the standard value were considered hazardous wastes. The hazardous constituents were determined from the standpoint of their toxicity. However, there is an increasing number of cases in which human health or the living environment is affected by hazardous substances leaked from waste treatment processes or disposal sites, or fire or explosion attributable to wastes, or harmful gas generated by these accidents. To tackle this situation, the Waste Management Law was amended in 1992. The amended law includes hazardous characteristics, such as explosiveness, toxicity, and infectiousness, in the concept of wastes under special control.[4] In the process of formulating this amended law, the hazardous characteristics to be considered in designating wastes under special control were defined as described in Table 1. The amended Waste Management Law, enacted in July 1992, designates the following wastes as wastes under special control (Table 2).

Table 1 Hazardous Characteristics for Hazardous Waste Management in Japan

1. **Explosive wastes:** Wastes, which as a result of chemical reaction, produce or may produce gases at such a temperature, pressure, and speed as to cause damage to the surroundings; wastes that explode or may explode under the influence of fire or friction.
2. **Flammable wastes:** Liquid wastes with a low flammable point, or solid wastes that may easily burn in normal handling, ignite through friction, or contribute to igniting other substances. Explosive wastes are not included in flammable wastes.
3. **Reactive wastes:** Wastes having one of the characteristics noted below that damages or may damage human health or the living environment, or wastes, which as a result of chemical reaction, generate toxic gas in dangerous quantities.
   - *Oxidizing characteristic:* The characteristic of causing, or contributing to, the combustion of other materials by yielding oxygen.
   - *Spontaneous combustion characteristic:* The characteristic that is liable to spontaneous heating under normal conditions or to heating up on contact with air and being then liable to ignite.
   - *Characteristic of emitting flammable gases in contact with water:* The characteristic, which by interaction with water, is liable to become spontaneously flammable or give off flammable gases in dangerous quantities.
4. **Corrosive wastes:** Liquid or water-soluble wastes that are or may be corrosive due to their high acidity or basicity.
5. **Toxic wastes:** Wastes falling under one of the following groups below.
   - Wastes containing substances, which if swallowed or inhaled or by skin contact, induce acute or chronic toxicity, thus causing possible death or serious injury to human health.
   - Wastes which damage or may damage the living tissues of a human body through contact.
   - Wastes containing substances that are or may be toxic to the environment.
6. **Infectious wastes:** Wastes which contain or may contain, or bear or may bear on their surface, viable microorganisms that can cause infectious diseases.

6.1.2 PRINCIPLE FOR HAZARDOUS WASTE CONTROL TECHNOLOGY[5]

Wastes, once discharged, undergo treatment (such as incineration) and may be disposed of in landfill sites. Such treatment and disposal can lead to environmental pollution. In selecting a hazardous waste processing technology, therefore, it is essential to ensure that the processed waste will not adversely influence human health or the living environment on either a short- or long-term basis. In this regard, before applying a hazardous waste processing technology on a commercial basis, it must be verified, at least at the pilot plant level, that the technology will make the wastes nonhazardous and/ or stable. In the future, wastes will have more and more varieties of char-

Table 2 Wastes under Special Control by the Waste Management Law of Japan

General Wastes under Special Control

PCB-using parts: Parts with PCBs removed from abandoned air-conditioners, television receivers, and microwave ovens which belong to general wastes.
Fly ash discharged from municipal waste incineration: Fly ash collected in dust separators installed in waste incineration plants with a capacity of 5 t or more per day that discharge fly ash and incinerated ash separately.
Infectious general wastes: General wastes which indicate or may indicate an infectious nature, such as blood-attached items discharged from hospitals or other medical institutions.

Industrial Wastes under Special Control

Highly combustible waste oil: Waste gasoline, kerosene, and light oils discharged as industrial wastes.
Strong acids and strong alkalis: Waste acids with a pH of 2.0 or less, and waste alkalis with a pH of 12.5 or more.
Infectious industrial wastes: Industrial wastes that indicate or may indicate an infectious nature, such as blood and used injection needles discharged from hospitals or other medical institutions
Specified hazardous industrial wastes
1. *Waste PCBs and PCB-polluted wastes:* Waste PCBs and PCB-polluted wastes specified in the Waste Management Law before amendment.
2. *Asbestos wastes:* Friable sprayed-on asbestos and asbestos-containing heat insulating materials removed from buildings, and plastic sheets resulting from the above demolition works. Friable asbestos wastes collected in the dust separators installed in working sites having soot or dust emitting facilities specified in the Air Pollution Control Law.
3. *Hazardous industrial wastes with specified substances:* Hazardous industrial wastes from specified waste sources exceeding the specified standard values of Hg, Cd, Pb, O-P, Cr(VI), As, CN, PCB, TCE, and PCE, which requires landfill in isolated final disposal sites, and combustion residue or dust generated by incineration of industrial wastes exceeding the specified values.

acteristics and discharge streams. It is also essential, therefore, to promote technological development to process such various wastes. Simultaneously, it is also necessary to establish an evaluation system for verifying whether the developed techniques and facilities are optimum or not from the standpoint of availability in the real society. From this standpoint, it is desirable to commence research and development (R & D) for laboratory waste processing and controlling technologies from beaker-scale experiments and to contribute to technological development on the pilot and commercial plant level. For heterogeneous wastes composed of a variety of constituents, a large-scale processing system may not be appropriate. Research institutes must investigate regarding a control method for such wastes.

Figure 1 shows the framework of the hazardous waste management system in relation to waste sources. The figure shows the priority of measures to be taken in the total framework, including hazardous waste generation sources. Specifically, besides the management system for generated hazardous wastes, the first priority measure is to minimize hazardous waste generation itself by employing "clean technology" in various industrial processes — refining and converting of raw materials, and processing and assembling products. We have a wide range of options for achieving this goal, such as changing the production process or changing the product composition. With laboratory wastes, however, it is basically impossible to apply this clean technology concept in every case. It is necessary to keep in mind that research activities in laboratories often require the use of hazardous substances. The second priority measure is to recover/recycle hazardous wastes in the same process that produced them. In-plant resource recovery/recycling would not only improve the total yield of the process and reduce the amount of wastes produced, but would also introduce the possibility of utilizing recovered substances as raw materials for other processes on an exchange or trade basis. Some hazardous wastes are necessarily and unavoidably generated by social activities. Needless to say, an appropriate processing system should be established for these wastes. In many cases, however, the clean technology and in-plant recovery/recycling option is worth considering as the most basic measure.

For hazardous wastes, once generated, the third through fifth priority measures are to be taken: the third priority measure is recovery/recycling of the wastes by the off-site treatment/regeneration process; the fourth priority measure is a process of making the wastes detoxified and stable; and the fifth priority measure is storage/management. The most prominent principles for hazardous waste processing are off-site recovery/recycling as the third priority measure, and the next process of making the waste harmless and stable as the fourth priority measure. For this fourth priority measure, higher priority should be put on eliminating hazardous characteristics from the waste. In this sense, it is desirable to employ a process that would change the characteristics fundamentally, such as thermal treatment, e.g., incineration or melting, or chemical treatment. Thermal treatment, under proper conditions, is effective particularly for many organic compounds (such as organochlorine compounds in organic solvents) and is recognized as the most appropriate process for attaining ultimate detoxification. Of course, proper treatment of exhaust gas and other by-products resulting from incineration is also indispensable. On the other hand, in processing metal components that remain as they are at the total amounts, priority is put on solidification/stabilization, as well as on recovery/recycling. Inorganic wastes may be disposed of when it has been confirmed that the by-product residue generated by the detoxification process is nonhazardous or stable and when solidified/stabilized wastes meet specified standards. Otherwise, the fifth priority measure of storage/management is necessary. Concentration or other volume-reduction measures may be taken

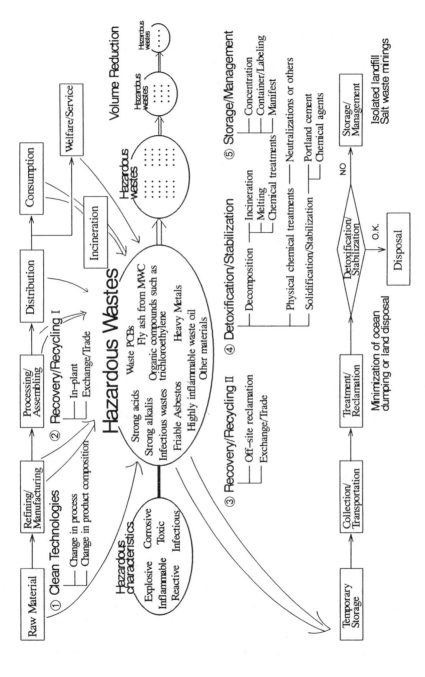

Figure 1. Framework of hazardous waste management in relation to waste source.

prior to storage/management. For storage/management, it is necessary to store wastes in sufficiently strong containers labeled for identification of the contents. These requirements also apply to temporary storage, collection, and transportation of hazardous wastes generated from various activities. The total flow of wastes, from generation to final disposal, including storage/management, should be controlled by the manifest system. An efficient waste management system, including this manifest system, along with the minimization of finally disposed wastes by promoting elimination of hazardous characteristics and stabilization, is an extremely important aspect of hazardous waste management.

Table 3 compares major technologies for processing hazardous wastes. In management of hazardous wastes once generated, priority is given to (1) recovery/recycling, (2) detoxification, (3) stabilization, and (4) storage, in that order. As core technologies for this management, Table 3 provides incineration/thermal decomposition, melting, solidification/stabilization, and disposal in landfill/storage. Each of these elemental technologies will be described in subsequent sections. Since some hazardous wastes contain substances which can be extracted or separated as useful resources, it is desirable to attempt recovery/recycling, insofar as possible. Therefore, processes intended for resource recovery are contained in some elemental technologies.

6.1.3 INCINERATION

Incineration has been widely used as a volume reduction and stabilization method for a long time and, with the accumulated technical experience and data, is one of the most reliable waste processing techniques. Incineration can be applied to municipal solid wastes or infectious wastes. Needless to say, the process assumes that exhaust gas is properly treated and that incinerated ash is safely disposed of in landfills. Desirably, the process should include recovery of heat and resources. The following paragraphs describe the incineration process in terms of (1) principle of incineration, (2) types of incinerators, (3) exhaust gas treatment system, and (4) control of incineration system.

6.1.3.1 Principle of Incineration

The purpose of incineration is to completely decompose wastes by heat and to minimize generation of harmful substances as the secondary product of combustion. To secure efficient combustion, it is necessary to consider the four parameters: (1) temperature, (2) residence time of combustion gas, (3) amount of combustion air (air ratio), and (4) complete mixing (turbulence) of combustion gas. The temperature and combustion gas residence time required for thermal decomposition of waste vary largely, depending on the chemical properties of the waste. Therefore, it is not possible to set combustion conditions suitable to every kind of waste. The properties of the waste must be studied before proper combustion conditions can be determined. The combustion in an incinerator changes sharply when the incinerator is started up or when waste is put into the incinerator. It also changes with the composition of the waste. Since the combustion state in the incinerator is changing at all times, combustion would become incomplete if not controlled carefully. Incomplete combustion may generate organic secondary products. These secondary products are often harmful and contain large amounts of substances which require high temperatures for decomposition. To prevent these secondary products from being discharged into the environment, it is desirable for the incineration plant to have a secondary combustion chamber that achieves complete mixing of combustion gases. Complete combustion requires excess combustion air, the excess differing depending on the physical and chemical properties of the waste. Therefore, the properties of the waste to be incinerated must be taken into consideration when determining the excess air ratio. To maintain the proper excess air ratio, it is also necessary to consider a forced draft fan capable of controlling the combustion air supply, and a double-gate supply unit capable of preventing air leakage into the incinerator.

For waste containing organochlorine compounds, combustion efficiency can be improved by adding a hydrocarbonic compound to the waste to increase the H/Cl ratio, in addition to controlling the above four parameters. In this method, hydrogen atoms and OH radicals generated in the combustion process react with the organochlorine compounds, thus permitting thermal decomposition at a lower combustion temperature than with a smaller H/Cl ratio.

6.1.3.2 Types of Incinerators

There are many types of incinerators available for combustion of wastes. The properties and amount of wastes to be incinerated, and the conditions required for efficient combustion, must be considered in selecting the most suitable type of incinerator. Major types of incinerators are rotary kiln, liquid atomizing incinerator, stoker type incinerator, fluidized-bed incinerator, and fixed-bed incinerator.

6.1.3.2.1 Rotary Kiln

The rotary kiln with a secondary combustion chamber, one of the most typical hazardous waste incinerators, is flexible in waste properties. This type of incinerator is suitable particularly for solid wastes, sludge, and liquid and gaseous waste mixtures.

Table 3 Comparison of Processing Technologies for Hazardous Wastes

	Incineration/Thermal Decomposition	Melting	Solidification/ Stabilization	Landfill/Storage
Elimination effect of hazardous characteristics	Basically high	Extremely high	High for metal wastes	Low for volatile substances; difficult for liquid wastes
Applicable hazardous wastes	Combustible waste oil Waste acid or waste alkali Trichloroethylene and other organochlorine compounds Infectious wastes	Fly ash from municipal waste incineration Asbestos wastes Infectious wastes	Fly ash from municipal waste incineration Asbestos wastes Wastes containing heavy metals Waste acid or waste alkali	Treatment residue Solidified/stabilized wastes
Inapplicable wastes	Incombustible organic substances Wastes with high heavy metal content	Almost none	Organic substances	Reactive substance Toxic substances that are mobile, nondestructive, or accumulative

	Destruction efficiency not yet verified for some components	Practicability has been verified at actual plants	Abundant experience except for process using chelating agent	Abundant experience
Application experience/progress of development				
Recoverable resources	Energy, acid	Energy, metal, construction materials	Construction materials, possibly	None
Cost	Moderate to high	Moderate to high	Moderate	Low to moderate
Affected environmental media	Atmosphere	Atmosphere	Surface water, groundwater	Surface water, groundwater
Problems to be solved	Destruction rate must be determined; Substitute index required; Measures must be taken against incomplete combustion products and by-products	Operation not easy; Safety measures necessary for high temperature operation; Linkage with resource recovery system desirable	Incomplete prevention of leaching; Weathering and deterioration	Long-term stability of structures; Service life of liners

6.1.3.2.2 Liquid Atomizing Incinerator

Liquid waste, such as waste solvent and waste oil, can be incinerated by liquid atomizing incinerator. This type of incinerator has a furnace with a high speed rotary burner (approximately 10,000 rpm).

6.1.3.2.3 Stoker Type Incinerator

In general, the stoker type incinerator contains a flat or stepped grate. Having a closely meshed grate, it is suitable for incineration of wood, fiber, or medical wastes.

6.1.3.2.4 Fluidized-Bed Incinerator

The fluidized-bed incinerator provides effective and efficient combustion due to its continuously moving fluidized bed. This incinerator is particularly effective for sludge and solid wastes of pulverized size.

6.1.3.2.5 Fixed-Bed Incinerator

The fixed-bed incinerator can be used for a variety of wastes, including those which cannot be put on a rotary hearth for physical reasons. This incinerator is suitable particularly for gaseous or liquid wastes and sludge.

6.1.3.3 Exhaust Gas Treatment System

An exhaust gas cooling unit must be installed to cool combustion gas to a specified temperature before it is led to the exhaust gas treatment system. In order to prevent emission of harmful substances, such as dioxin, it is necessary to cool the combustion exhaust gas to 200°C or lower as quickly as possible. Since the dust concentration changes with the type of the incinerator used, different types of dust separators must be installed, depending on the incineration plant conditions. The dust concentration is expected to conform to the standard values specified in the Air Pollution Control Law. Accordingly, it is necessary to install a dust separator which meets those standards. A waste incinerator may emit various harmful gases of high concentration, including hydrogen chloride and NO_x. To eliminate these harmful substances from the exhaust gas, a wet or dry exhaust gas treatment system must be installed.

Some wastes have high calorific value. The waste heat resulting from combustion of such wastes can be utilized as energy. Thus, we should do our utmost to ensure the maximum utilization of energy from waste, while saving processing expenses and suppressing CO_2 emission; installation of a dryer upstream of the incineration process is one such measure. At the same time, however, it must not be forgotten that the primary purpose of incineration is not recovery of energy but decomposition of wastes.

6.1.3.4 Control of Incineration System

Mass flow control, operational control, and analytic control are necessary prior to accepting wastes in an incineration plant. Mass flow control is important for the management of material balance in the incineration plant, for reporting to pertinent authorities, and for determining the waste processing cost. Operational control includes preliminary check of wastes, sampling, and general qualitative or semiquantitative tests. Since comprehensive analysis of wastes is not possible because of time limitations, tests are carried out on particular items. Among the important items for analysis are (1) flammability, (2) chlorine and sulfur contents, and (3) pH value. Analytic control checks sample wastes in detail; the wastes are analyzed comprehensively as to calorific value, moisture content, concentrations of heavy metals, and other harmful substances.

To prevent environmental pollution by combustion, it is necessary to continuously monitor the combustion state. In this regard, it is desirable to continuously measure the CO and O_2 concentrations in exhaust gas, which are accurate indices of the combustion state. The CO concentration is an important index of the combustion state; combustion is more complete the lower the CO concentration. The O_2 concentration is an important index of the excess air ratio during the combustion process.

6.1.4 MELTING

The melting process heats and melts waste to reduce its volume by using combustion heat of fuel or electric heat energy. At a high temperature of 1200 to 1500°C, the organic substances in the waste gasify, burn, and decompose, while the inorganic substances melt to become vitrified slag. Most melting processes developed are targeted mainly for sewage sludge and its incinerated ash.[6] Some have been developed for melting asbestos wastes and fly ash.[7] Figure 2 shows three melting furnaces for sludge and their processing systems.[8] These furnaces are also used for incinerated ash.

Since the melting process involves high temperature and phase change of the wastes to be solidified, safety measures and exhaust gas cleaning measures should be considered. Each part of the melting system is operated at considerably high temperature. In particular, the components around the discharge port of molten slag or metal become very hot, so their designs must incorporate measures for preventing steam explosion. In the waste melting process, it is necessary to treat dust and exhaust gases of HCl, NO_x, and SO_x generated by combustion of wastes. The dust collected in the exhaust gas treatment

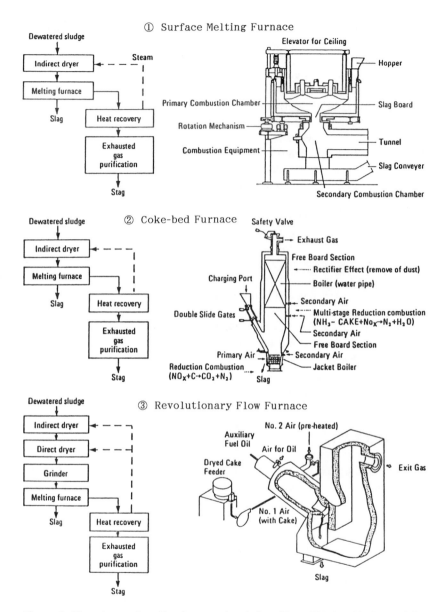

Figure 2. Three types of melting furnaces for sludge. (From Shimizu, K., State of the arts on the sludge incineration plants in Japan, Proc. 3rd Kyoto University — KAIST Joint Seminar/Workshop on Civil Eng., 1990, 238. With permission.)

system of a melting furnace contains heavy metals in enriched form, such as Pb and Cd, which will evaporate at high temperature. Therefore, the dust would best be returned as resources to mines, or treated separately with heavy metal stabilizer. The wastewater generated in the melting system includes that from water cooling the system, that discharged from the exhaust gas cleaning unit, and other miscellaneous wastewater. Such wastewater may contain hazardous compounds or heavy metals. Therefore, it is also necessary to treat such wastewater by an appropriate method. If sludge produced by wastewater treatment contains hazardous substances in amounts exceeding the standard values, it must be dewatered and returned to the melting furnace for retreatment or stabilized by solidification in concrete.

6.1.5 SOLIDIFICATION/STABILIZATION

The primary purpose of solidification/stabilization processes currently used or under development is to prevent leaching of hazardous substances, including heavy metals. These processes include solidification in portland cement, solidification in asphalt, and stabilization by chemicals.

6.1.5.1 Solidification in Portland Cement

Portland cement and water are added to wastes for granulation or molding, in order to prevent hazardous substances from leaching or scattering from collected dust or various kinds of sludge. This process has problems of volume increase of wastes due to the cement added, and leaching of salts ($CaCl_2$, $CaSO_4$, etc.) from dust.

6.1.5.2 Solidification in Asphalt

This process has been developed, for hazardous sludge, radioactive wastes, and ash resulting from waste incineration. Asphalt solidification incurs about 1.5 to 2 times as high a running cost as cement solidification. In addition, asphalt is troublesome to handle. This process is not popular due to these disadvantages.

6.1.5.3 Stabilization by Chemicals

Waste is stabilized by adding chemicals and, if necessary, a pH controlling agent. Specifically, a chemical agent such as a heavy metal fixation or coagulation agent and, if necessary, a pH controlling agent are added to the waste before moisturizing and kneading in order to prevent hazardous substances from leaching. Liquid chelate agents or sodium hydrogen sulfide is used as the heavy metal fixative.

6.1.6 PHYSICAL AND CHEMICAL TREATMENT

In addition to incineration processes for hazardous wastes, there are many other processes available; for example, heavy metal extraction by acid, photochemical (ultraviolet rays) decomposition, chemical decomposition, ozone decomposition, and the biological process. The ferrite process[9,10] for heavy metals containing waste will be discussed in another chapter.

Fly ash, from which heavy metals such as Cd and Pb have been extracted using an acidic scrubber wastewater of hydrochloric acid (pH 0.3 to 1.2), is compressed into pellets and heat treated in an incinerator to stabilize it.[11] Tests now being conducted to verify the practicability of this process have shown that about 1-h agitation of a mixture of fly ash and scrubber wastewater extracts about 90% of Cd, 60 to 70% of Zn, about 20% of Pb, and 20 to 25% of Cu from the fly ash. In addition, heat treatment of fly ash pellets in an incinerator causes dioxin in the fly ash to be thermally decomposed, so that heavy metals are not leached. Thus, this process is believed to be effective in obtaining stable fly ash.

It is reported that decomposition by ultraviolet ray irradiation of organochlorine compounds in water, such as trichloroethylene, tetrachloroethylene, 1,1,1-trichloroethane, carbon tetrachloride, and chloroform, can be decomposed into CO_2, HCl, or other inorganic substances.[12] The report states that the OH-radical generated from water functions to extract Cl from organochlorine compounds, thus initiating an oxidation reaction. Presently, laboratory tests are being conducted on this photochemical decomposition process, using a small pilot apparatus. Development has not yet advanced to the level of practical-scale testing.

Two different chemical decomposition processes have been developed for organochlorine compounds: one using metallic sodium, the other using alkaline polyethylene glycol. In the former process,[13] metallic sodium is allowed to disperse in fine granular form into waste oil containing organochlorine compounds (such as dioxin) so as to eliminate the chlorine atoms of the organochlorine compounds. Since this process uses metallic sodium, its application is limited to waste oil. The chemical decomposition process using alkaline polyethylene glycol[14] relies on the reaction of replacing the chlorine atoms of organochlorine compounds by alkoxide, a product of the reaction between alcohol or glycol and potassium hydroxide or sodium hydroxide. This process is effective only for dechlorination.

REFERENCES

1. 1989 Basel Convention on the control of transboundary movements of hazardous wastes and their disposal, *J. Environ. Law,* 1, 255, 1989.
2. The Council of European Community, Council Directive of December 12, 1991 on Hazardous Waste, 91/689/EEC, 1991.
3. U.S. Environmental Protection Agency, 20 CFR, Part 261, Identification and listing of hazardous waste, 1991.
4. Kaburagi, Y., Waste under special control, *Waste Manage. Res.,* 3, 192, 1992 (in Japanese).

5. Sakai, S., The definitions of hazardous wastes and their management, *Waste Manage. Res.*, 3, 202, 1992 (in Japanese).
6. Takeda, N., Hiraoka, M., Sakai, S., Kitai, K., and Tsunemi, T., Sewage sludge melting process by coke-bed furnace: system development and application, *Water Sci. Technol.*, 21, 925, 1989.
7. Sakai, S., Takatsuki, H., Hiraoka, M., and Tsunemi, T., Sludge melting process with hazardous asbestos wastes, *Water Sci. Technol.*, 23, 2029, 1991.
8. Shimizu, K., State of the arts on the sludge incineration plants in Japan, Proc. 3rd Kyoto University — KAIST Joint Seminar/Workshop on Civil Eng., 1990, 238.
9. Tamaura, Y., Katsura, T., Rojarayanont, S., Yoshida, T., and Abe, H., Ferrite process; heavy metal ions treatment system, *Water Sci. Technol.*, 23, 1893, 1991.
10. Takatsuki, H. and Sakai, S., Treatment of and resource recovery from laboratory hazardous wastes, UNEP-Asia and Pacific Regional Workshop on Hazardous Waste Minimization and Reduction, 389, 1990.
11. Vehlow, J., Braun, H., Hroch, K., Schneider, J., and Vogg, H., Semi-industrial testing of the 3R-process, Proc. Int. Conf. Municipal Waste Combustion, Hollywood, FL, 4B-11, 1989.
12. Pelizzeti, E., Pramauro, E., Minero, C., and Serpone, N., Sunlight photocatalytic degradation of organic pollutants in aquatic systems, *Waste Manage.*, 10, 65, 1990.
13. Bilger, E., Detoxification of organic liquids using sodium/dehalogenation of harmful substances, *Contam. Soil,* 88, 943, 1988.
14. Kornel, A., Rogers, C.J., and Sparks, H.L., KEPG application from the laboratory to Guam, PB90-127200, 1989, 460.

CHAPTER 6.2

Treatment of Wastewater Mixed with Metallic and Inorganic Compounds

Mitsunobu Kitamura

TABLE OF CONTENTS

0-8493-682-5/94/$0.00 + $.50

6.2.1 INTRODUCTION

As the health of a human being is negatively affected by heavy metals, the handling of these metals and their appropriate disposal are an important matter. The subject of disposal has tended to change with the development of advanced technology, though industrial pollution in rivers by heavy metals has shown signs of improvement in recent years. Therefore, we consider it necessary to develop a higher technology for disposal options.

Wastewater mixed with metallic and inorganic compounds is produced from educational research and medical activity and includes heavy metals, mercury, cyanide, and fluoride. Depending on the composition of the wastewater and the manner of its disposal, various methods, such as the ferrite process, precipitate aggregation, electrolytic floating, the iron powder method, and solar evaporation have been developed. For example, using the precipitate aggregation (neutralization) method, most of the heavy metals in the wastewater can be removed as hydroxides by alkali reagents. Actual disposition can best be done by using most appropriate method or combination of methods for the contents of the wastewater.

The treatment process of wastewater mixed with metallic and inorganic compounds is shown in Figure 1. The unitary operations of reaction and disposition are as follows. In the ferrite process of a unitary operation, heavy metals in the wastewater can be removed while ferrite (triiron tetraoxide; magnetite) is produced by oxidation of ferrous salt. This process has the strong point that heavy metal ion is not easily eluted from the produced precipitate. In the neutralization method, most heavy metals in the wastewater can be removed as hydroxides by alkali reagents. This method is the fundamental and most economical technique for the treatment of heavy metals. In the electrolytic floating method, chemically stable compounds that have been formed with heavy metal ions are captured as they float with bubbles in wastewater, and most of those ions can be removed. The iron powder method is a modification of the neutralization method using specifically made iron powder. In solar evaporation, wastewater containing heavy metals is condensed and dried under a heat source like sunshine. This method is successful for wastewater containing high concentrations of special metals and chemical compounds.

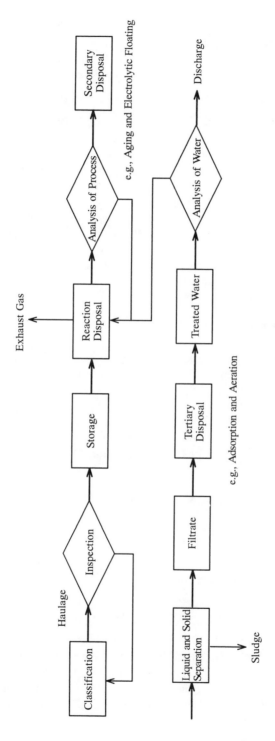

Figure 1. Treatment process of wastewater mixed with metallic and inorganic compounds.

In universities of Japan, these various methods have been used properly. Facilities for disposal of inorganic wastewater from universities were established during the 1970s, and the total amount of disposal of wastewater has averaged 100,000 liters per university since that time.

6.2.2 FERRITE PROCESS

Inorganic wastewater produced from experiments contains many harmful heavy metals in various forms. The ferrite process was the first method developed to remove heavy metals from wastewater. This process involves the addition of ferrous ions to wastewater, oxidation by oxygen in air, the formation of ferrite that is a water-insoluble oxide, and the precipitate separation. Heavy metals are disposed under the same condition and precipitate. As the precipitate is ferromagnetic, it can be separated by a magnet and recovered. Many universities in Japan have used this process for disposal of inorganic wastewater because the recovered ferrite sludge can be reused as magnetic material.

6.2.2.1 Disposal Principle

The ferrite has a spinel structure of tetragonal system and is a crystal complex composed of iron oxide and oxide of divalent metal (M); it is chemically formulated as MFe_2O_4 or $MO \cdot Fe_2O_3$. Moreover, even a uni-, tri-, or tetravalent metal forms ferrite with the crystal structure. The spinel crystal structure of ferrite is composed of two sublattices, as in Figure 2. A sublattice of regular tetrahedron is enclosed by four oxygen atoms, and B sublattice of regular octahedron is enclosed by six oxygen atoms.

Iron(II) hydroxide [$Fe(OH)_2$] is precipitated by adding sodium hydroxide (NaOH) in an aqueous solution of ferrous ions, and triiron tetraoxide (Fe_3O_4) is produced by the air oxidation of $Fe(OH)_2$.

$$Fe^{2+} + 2OH^- \rightarrow Fe(OH)_2 \rightarrow Fe_3O_4$$

In the presence of divalent ions (M^{2+}),

$$(3 - x)Fe^{2+} + xM^{2+} + 6OH^- \rightarrow M_xFe_{3-x}(OH)_6 \rightarrow M_xFe_{3-x}O_4$$

The hydroxide is precipitated as an intermediate formation according to the above formula. After the dissolution of hydroxide, the oxidation, and the crystallization, black ferrite of the complex oxide with a spinel structure is produced. When the spinel structure is formed, metal ions strongly joined in the ferrite lattice are not easily dissolved in water. The conditions for forming

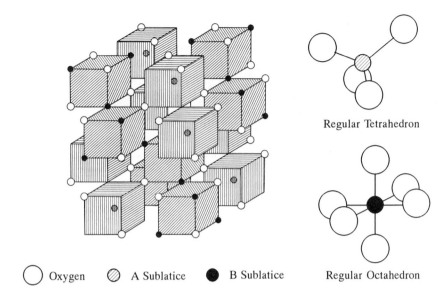

○ Oxygen	⊘ A Sublatice	● B Sublatice

Regular Tetrahedron

Regular Octahedron

Figure 2. The spinal crystal structure.

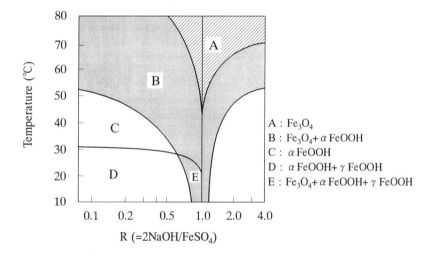

A : Fe_3O_4
B : $Fe_3O_4 + \alpha\,FeOOH$
C : $\alpha\,FeOOH$
D : $\alpha\,FeOOH + \gamma\,FeOOH$
E : $Fe_3O_4 + \alpha\,FeOOH + \gamma\,FeOOH$

Figure 3. The conditions for forming ferrite. (Modified from Kiyama, M., *Bull. Chem. Soc. Jpn.*, 47, 1646, 1974.)

ferrite are oxidation temperature (T), air rate, molar ratio (R) of alkali to ferrous salt, and pH (Kiyama, M., *Bull. Chem. Soc., Japan,* 47, 1646, 1974). As is evident from Figure 3, black Fe_3O_4 with ferromagnetism is formed in region A. The conditions for forming ferrite are R = 1, pH 9 to 10 and T = 65 to 70°C. Ferrite containing heavy metal ions is also formed under the

Table 1 Metal Ions that Construct Ferrite

Valence	Metal Ions
I	Li, Cu, Ag
II	Mg, Ca, Mn, Fe, Co, Ni, Cu, Zn, Cd, Sn
III	Al, Ti, V, Cr, Mn, Fe, Ga, Rh, In, Sb
IV	Ti, V, Mn, Ge, Sn, Mo, W
V	V, As, Sb
VI	Mo, W, Cr

Table 2 Classification of Wastewater for Ferrite Process

Classification	Object	Remarks
1. Mercury	Inorganic mercury and organic mercury	Except metal mercury, amalgam
2. Cyanide	Free cyanide	Except organic cyanide, complex
3. Fluoride	Fluoride	
4. Phosphate	Phosphate	
5. Heavy metal	Heavy metals and chromic acid mixture	Except compounds hazardous to human health
6. Acid	HCl, H_2SO_4, HNO_3, etc.	Does not contain heavy metals
7. Alkali	NaOH, K_2CO_3, etc.	Does not contain heavy metals

same conditions. In other regions, yellow and brown oxyhydroxides with nonferromagnetism are produced as by-products. Table 1 shows various metal ions that construct the ferrite with iron ion. Targets of the ferrite process are many toxic heavy metals (e.g., coppers, lead, iron, nickel, chromium, cadmium, manganese, and mercury) contained in the wastewater. Mercury, however, cannot be disposed of by this ferrite process.

6.2.2.2 Wastewater Disposal System

6.2.2.2.1 Classification of Wastewater

Inorganic wastewaters discharged from universities contain not only heavy metals but also organic compounds. As these wastewaters often contain chemical compounds that are difficult to dispose of by the ferrite process, we must treat them before the disposal by the ferrite process. When various compounds having different properties are contained in wastewater, their disposal is complicated. To dispose of wastewater smoothly, it is therefore necessary to classify it at the spot where it was discharged. It is best to classify wastewaters from research activity and the experiment in universities according to the classification in Table 2.

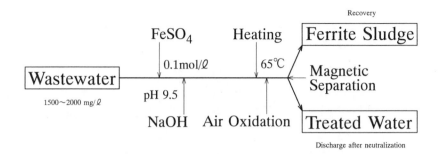

Figure 4. Diagram of disposal by the ferrite process.

6.2.2.2.2 Process of Predisposition

Wastewaters difficult to treat by the ferrite process can be handled in the following way.

Inorganic and organic mercury — Organic mercury and organic compounds contained in wastewater are oxidized by potassium permanganate under 70°C and pH 2 and converted into inorganic mercury and inorganic compounds. After the oxidation, inorganic mercury is adsorbed and removed by chelate resin.

Cyanide (free cyanide) — Free cyanides are oxidized and decomposed by adding sodium hypochlorite to the wastewater at pH above 10.5. The end of the reaction is checked by an oxidation reduction potentiometer (ORP). The reaction is as follows:

$$NaCN + NaOCl \rightarrow NaCNO + NaCl$$
$$2NaCNO + 3NaOCl + H_2O \rightarrow 2NaHCO_3 + 3NaCl + N_2$$

Fluoride and phosphate — Calcium fluorides or calcium phosphate insoluble in water is precipitated by adding calcium chloride and aluminum sulfate to the wastewater. The reaction is as follows:

$$2F^- + CaCl_2 \rightarrow CaF_2 + 2Cl^-$$
$$2PO_4^{4-} + 3CaCl_2 \rightarrow Ca_3(PO_4)_2 + 6Cl^-$$

6.2.2.2.3 The Operation of the Ferrite Process

A block diagram of disposal by the ferrite process is shown in Figure 4. Iron(II) hydroxide and heavy metal ions are coprecipitated at pH 9.5 to 10.5 by adding sodium hydroxide to an aqueous solution of ferrous ions. After the air oxidation of iron(II) hydroxide at 60 to 70°C, ferrite is produced as the

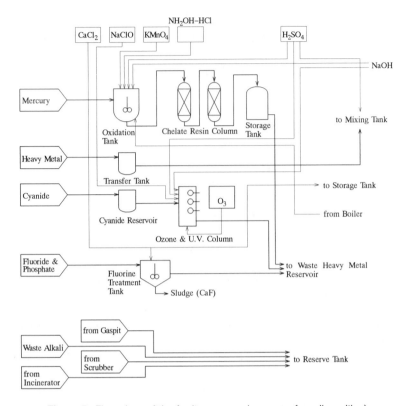

Figure 5. Flow chart of the ferrite process (process of predisposition).

precipitate, and since this is ferromagnetic, it can be separated from the treated water by magnetic separation. The ferrite sludge is recovered and the treated water is released after neutralization. Figure 5 shows a model of disposal by the ferrite process.

This disposal plant is operated automatically using the sequence controller and the sensors. These sensors govern the water level of tanks, the pH of the mixing tank, reaction tank, and neutralization tank, and the temperature and ORP of the reaction tank. A monitoring system using a computer has also been developed.

Preparation — Wastewater can be easily disposed of by the ferrite process after the predisposition of mercury, cyanide, fluoride, and phosphate. Before employing the ferrite process, it is necessary to test carefully whether or not a good quality ferrite can be produced from this wastewater under the usual conditions. If it cannot, the wastewater is disposed of as follows: (1) by adding potassium permanganate or a large amount of ferrous sulfate, (2) by making a small mixing ratio of the substances interfering with the ferrite reaction, (3) by removal of the interfering substances, or (4) by raising the temperature and pH of the ferrite reaction. After evaluating those results, the conditions of the ferrite process can be determined.

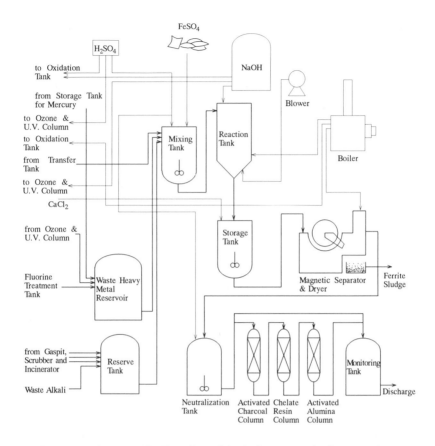

Figure 5 (continued). Flow chart of the ferrite process (main process).

Reaction (ferrite) — Wastewater is prepared to pH 9.5 to 10.5 by adding sodium hydroxide to the reaction vessel, after which the wastewater is oxidized with oxygen in air at 65°C. Through the formation of iron(II) hydroxide, the wastewater forms a precipitate of ferrite particles during the reaction period of 2 h.

Magnetic separation — Ferrite sludge is obtained from the ferrite slurry by separation of liquid and solid phases with a magnetic separator; the water content of the sludge is about 80%. Finally, solid ferrite sludge with a water content of about 50% can be obtained using a filter press and a polymer coagulant.

Final Step — The treated water after the magnetic separation is prepared to pH 7 by sulfuric acid. After chemical analysis, this water is released if safe. If the analytical results do not show it to be so, it is necessary to again subject it to the ferrite process. If only mercury is detected, this water is disposed of by chelate resin or active carbon. The redisposed, treated water is then released if safe.

Table 3 Merits of Ferrite Process

	Features	Grounds
Reaction	Simultaneous disposition without reduction of hexavalent chromium	A kind of coprecipitation method with ferrous sulfate as reducing agent
Separation	Separation of sludge by magnetic force	Ferromagnetic sludge
Sludge	Reclamation disposition and reuse in various fields	Stable solid oxide, specific gravity, magnetism, adsorption, stability

Table 4 Demerits and Countermeasures for Ferrite Process

Demerit	Countermeasure
Wastewater Difficult for Disposal	
Mercury, cyanide, fluoride, and phosphate	Predisposition by chelate resin, NaOCl, or $CaCl_2$
Organic metal, organic complex, and organic compounds with high concentration	Capture of heavy metal by washing combustion gas after burning
Disposal Process	
Ferrous sulfate as disposal reagent	Easy receipt of ferrous sulfate as waste or cheap material from mining industry
Heating for reaction	Addition of a small boiler

6.2.2.3 Disposal Features

The ferrite process is a disposal method with many special merits as shown in Table 3. However, it is not perfect and has several demerits as shown in Table 4; the appropriate countermeasures for these disadvantages are also shown in this table.

Table 5 shows yearly results of water analysis before and after the disposal. The results for mercury, cyanide, fluoride, and phosphate wastewater are also shown. These are for disposal by the ferrite process after predisposal.

Properties of produced ferrite — The total concentration of heavy metals in wastewater from universities is about 5000 mg/l. Ferrite sludge is produced at about 70 kg/1 m^3 of this wastewater. From Mössbauer spectra and X-ray diffraction patterns, it became evident that only ferromagnetic ferrite was produced, and other iron oxides were not produced, though the figures are not shown here. The spinel crystal structure of the ferrite was complete.

Table 5 Result of Disposal

Metals	Wastewater (mg/l)	Treated Water (mg/l)
Hg	0.003 ~ 0.70	<0.001 ~ 0.0045
Cd	1.0 ~ 12	0.004 ~ 0.088
Pb	3.5 ~ 11	<0.01 ~ 0.04
As	0.05 ~ 0.2	0.002 ~ 0.033
Cr	25 ~ 285	0.011 ~ 0.12
Cu	6.7 ~ 53	0.01 ~ 0.49
Zn	11 ~ 72	<0.01 ~ 0.032
Mn	5.6 ~ 82	0.01 ~ 0.07
Fe	53 ~ 83.2	0.20 ~ 1.03
Co	0.4 ~ 1.1	<0.01 ~ 0.04
Ni	3.1 ~ 12	0.02 ~ 0.045

Table 6 Reuse of Ferrite Sludge

Subjects of Reuse		Applications
Magnetic guide	Magnetic guide and magnetic induction	Guide system for blind persons and automated guided vehicle and golf-cart
Electromagnetic wave absorber	Prevention of reflection and leakage for electromagnetic wave	Building, bridge, ship, and electronic oven
Material for vibration damping	Control of mechanical vibration and noises	Vibration-proof base and sound insulating wall
Others	Magnetic fluid, magnetic capture material, and catalyst	Lubrication material, gravity separation, scavenger to recovery of leaked oil

Magnetic curves indicated the hysteresis characteristic of ferromagnetism and highly saturated magnetization.

Recovery and reuse of ferrite sludge — The total amount of ferrite sludge recovered from disposal plants in universities of Japan has been about 30 tons a year since 1975. The water content of this sludge has been about 50%. Ferrite sludge as a waste is also produced by the manufacturing industry and the chemical industry. As shown in Table 6, the sludge has been reused in various fields, especially as material for a magnetic guide system, absorption of broadband electromagnetic wave, and vibration damping.

Accomplishment of disposal — The facilities for disposal by the ferrite process have been established at 3 to 5 universities each year since 1975, and now are operating smoothly on about 35 campuses. The process has been modified by adding an active carbon tour, adsorptive resin tours for mercury and cyanide, or a dehydrator of the sludge. There have been no major breakdowns at all since the establishment of disposal facilities, though the oldest facility has been operating for 15 years. We feel, however, that it will be necessary to reconstruct and renew these disposal plants in a few years because they have been so heavily used.

Present situation and future of the ferrite process — The ferrite process is one of several disposal methods that has attracted special interest recently because the sludge produced coprecipitates with heavy metals, is insoluble in water, and can be reused. As stated, about 35 universities now use this process, and it has been modified according to the accomplishments of disposal for 15 years. It is expected that ferrite sludge is reused as magnetic materials, since the demand for these materials is increasing in chemical and electrical manufacturing. In the future, it will be important to investigate the techniques of reuse so that uses can be expanded.

6.2.3 PRECIPITATE AGGREGATION METHOD

The wastewater discharged from chemical laboratories includes not only heavy metals but also high concentrations of salts or organic compounds. Therefore, it is often necessary to treat the water a second time. The precipitate aggregation method is classical and the most famous of the disposal techniques. The process for removing heavy metals in wastewaters is as follows: (1) heavy metals are precipitated as water-insoluble hydroxides or sulfides, (2) precipitated sludge is separated from the treated water, and (3) heavy metals in the treated water are removed by adsorbing on chelate resins.

6.2.3.1 Disposal Principle

The theoretical foundation for removing heavy metals from wastewater by the precipitate aggregation method has already been established. Most heavy metal hydroxides have minimal solubility in a high pH region; therefore, the hydroxide is precipitated by adding alkali to the wastewater. The equilibrium relationship can be written as

$$M^{n+} + nOH^- \rightleftarrows M(OH)_n \downarrow$$

where M is a metal. The solubility product $K_{sp} = [M^{n+}] \cdot [OH^-]_n$ and the ionic product $K_w = [H^+] \cdot [OH^-]$ are given by

$$[M^{n+}] = K' \cdot [H^+]_n \qquad (1)$$

where $K' = K_{sp}/K_w^n$.

When the metal M forms an amphoteric hydroxide, the equilibrium relationship can be written as

$$M(OH)_n \rightleftarrows H_{n-1} \cdot MO_n^- + H^+$$

if alkali exists in excess. In the equilibrium state for forming hydroxide complex ions, the equilibrium constant K can be written as

$$K = [H^+] \cdot [H_{n-1} \cdot MO_n^-] \tag{2}$$

From Equations 1 and 2, the dissolved heavy metal concentration C_M can be represented as follows:

$$C_M = [M^{n+}] + [H_{n-1} \cdot MO_n^-] = K' \cdot [H^+]^n + K/[H^+] \tag{3}$$

Therefore, in simple systems, the dissolved heavy metal concentration C_M is determined by the pH value. However, the solubility of metal hydroxides changes with the conditions (e.g., temperature, ionic strength, aging time, and coexistent salts). Actual wastewater contains many kinds of coexistent compounds and is very complicated. Therefore, we cannot clarify the removal limit of heavy metals immediately by the value of pH obtained from the articles for solubility and Equation 3.

It is necessary to adjust the value of pH to the most suitable range to remove the heavy metals as hydroxides. We can find the buffer exponent, such as the amount of alkali to be added to the most suitable pH and the degree of difficulty in adjusting pH of the wastewater, by observing the titration curves of the neutralization. Generally, these curves are made up in a preliminary experiment before disposal of the wastewater. Figure 6 shows an example of the titration curve of the neutralization made up after adding iron(III) salt in actual and theoretical wastewaters containing heavy metals. The main factors involved in the precipitate aggregation method are the adjustment of pH of the wastewater and the addition of inorganic coagulating agent. It is necessary for us to be well aware of the effects of these factors to dispose of the wastewater containing heavy metals properly and without incident.

The cautions for heavy metals are as follows. As cadmium (Cd) and manganese (Mn) are heavy metals that can generally be removed only in a high pH region, too much alkali must be added in the disposal of these metals. In addition, chlorine ion (Cl^-) hinders the removal efficiency of coprecipitation by iron(III) hydroxide. pH must be set higher than ordinarily when garbage furnace wastewater containing Cd^{2+} and Cl^- is disposed of by the precipitate aggregation method. Lead (Pb) and zinc (Zn) form a complex

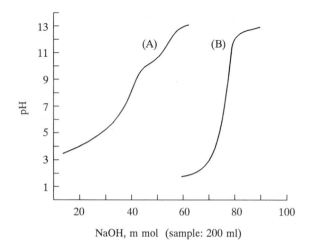

NaOH, m mol (sample: 200 ml)

Figure 6. The titration curve of neutralization. (A) Wastewater mixed with heavy metals and 10 mM of ferric chloride after reduction of Cr^{6+}. (B) Model wastewater mixed with heavy metal and 10 mM of ferric chloride.

hydroxide ion. The coprecipitation effect of these elements is recognized when iron(III) is added and Zn has a more remarkable effect. Copper (Cu) has no remarkable effect when iron(III) is added. Chromium (Cr) has the coprecipitation effect in the acidic region, but dissolves in the alkali region without any iron(III). Cu and Cr are removed successfully if the agent that forms the complex does not exist in the wastewater. Mercury (Hg) is influenced considerably by an anion like a halogen ion that coexists in acidic wastewater. The removal efficiency of Hg on adding iron(III) is remarkable at pH 9 to 10. The removal of arsenic (As)(V) is easier than As(III). Arsenic is removed by coprecipitation with iron(III) hydroxide, after As(V) converted to As(III) by an oxidizing agent such as hypochlorous acid, permanganic acid, chromic acid, or hydrogen peroxide in the predisposition.

6.2.3.2 Wastewater Disposal System

Classification of wastewater — Table 7 shows a simple classification for inorganic wastewater to be disposed of by the coprecipitate neutralization method, the precipitate aggregation method, or the electrolytic floating method. These methods are widely used in the disposal at Japanese universities.

Process of precipitate aggregation — Figure 7 shows an example of the precipitate aggregation process. In this example, categories of a chromic acid mixture, wastewaters containing arsenate and fluoride, and blood test wastewater are added to the classifications shown in Table 7.

Disposal of fluoride — Wastewater containing fluoride discharged from chemical laboratories is mixed with heavy metals and strong acids produced

Table 7 Classification of Wastewater for Precipitate Aggregation Method

Classification	Object
1. Heavy metals	Wastewater containing heavy metals such as Cu, Zn, Fe, Mn, Cr, Cd, and Pb; up to second rinse water for experimental apparatus
2. Mercury	Wastewater containing inorganic mercury; up to fourth rinse water for experimental apparatus
3. Cyanide	Alkaline wastewater containing cyanide and little heavy metal

by the etching of metals and treatment of the metal surfaces. Therefore, it is necessary to dispose not only of fluoride but also of heavy metals as shown in Figure 7. In the disposal by the precipitate aggregation, a method of adding calcium or coagulating by polyvalent metals is used to remove fluorine. The method using calcium employed calcium hydroxide $[Ca(OH)_2]$ or calcium chloride $(CaCl_2)$. Addition of $Ca(OH)_2$ removes fluorine as the precipitate of calcium fluoride (CaF_2) after adding the slurry or the powder of slaked lime to the wastewater and setting pH 8 to 10. The results of disposal using this method alone do not satisfy the standard required for treating water because the precipitate of CaF_2 produced is very small particles and is not coagulative. The $CaCl_2$ method is one of removing fluorine by the addition of $CaCl_2$ after adding the milk of lime in wastewater and setting pH 6 to 8. This means, however, is not appropriate for the disposal of wastewater containing heavy metals that cannot be removed in the neutral region. In the technique of coagulating by polyvalent metals, aluminum hydroxide or iron hydroxide is first produced by mixing the alkali like $Ca(OH)_2$ after adding aluminum sulfate, iron(II) chloride or iron(III) to the wastewater. Then, CaF_2 or F^- is adsorbed on and coprecipitates with this produced hydroxide. This is more successful in the removal of F ion than the method of adding calcium.

6.2.4 ELECTROLYTIC FLOATING METHOD

In the disposal of wastewater requiring the electrolysis, there are three methods: direct, indirect, and electrolysis coagulation. The electrolytic floating method is based on the same principle as the latter. When the composition of the wastewater is simple, it can be disposed of using only the electrolytic floating method. The precipitate aggregation method is usually used as a predisposition process with the electrolytic floating process, however, because wastewater discharged from chemical laboratories is often complicated and contains toxic substances. Figure 8 shows an example of this process in which heavy metals in the wastewater are removed by neutralization as metal

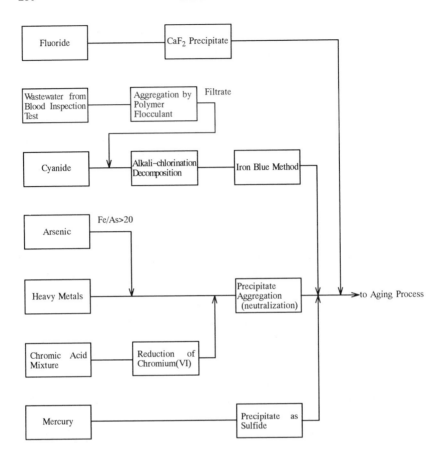

Figure 7. Diagram of disposal by the precipitate aggregation method (precipitate process).

hydroxides with a small solubility product. Practically speaking, it is difficult to dispose of these at the pH value for the precipitation because the solubility product for each metal ion is different. In this process, the wastewater is usually disposed of at pH 8 to 9 where Cd cannot be removed to less than 0.1 mg/l of the standard. Removal of Cd must be done by redisposal at pH above 10.5 by precipitate aggregation, using alkali, acid, and the agent as an active carbon and dehydrating the precipitate. With the electrolytic floating method, Cd can be efficiently removed at a neutral region, and the concentrations of metal ions and organic compounds in the wastewater reduced considerably.

6.2.4.1 Disposal Principle

The heavy metal ions decrease, as shown in Figure 9, when aluminum alloy is dipped in neutral (pH 6 to 9) wastewater containing these ions, and

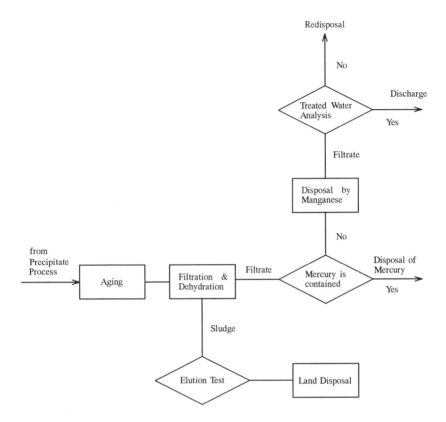

Figure 7 (continued). Diagram of disposal by the precipitate aggregation method (aging process).

direct current is applied to the alloy. Metals plate out on the cathode at high concentration of metal ions in the wastewater. At this concentration, hydrogen ion H^+ is predominantly discharged. Therefore, the concentration of OH^- increases near the cathode and metal ions in the wastewater form hydroxides.

$$2H^+ + 2e^- \rightarrow H_2 \uparrow$$
$$M^{n+} + nOH^- \rightarrow M(OH)_n$$

Aluminum (Al^{3+}) which is dissolved in the wastewater from the anode immediately creates aluminum hydroxide at pH 6 to 9.

$$Al \rightarrow Al^{3+} + 3e^-$$
$$Al^{3+} + H_2O \rightarrow Al(OH)_3 + 3H^+$$

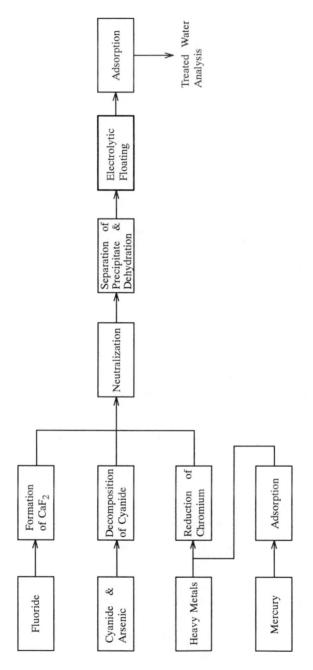

Figure 8. Diagram of disposal by the electrolytic floating method.

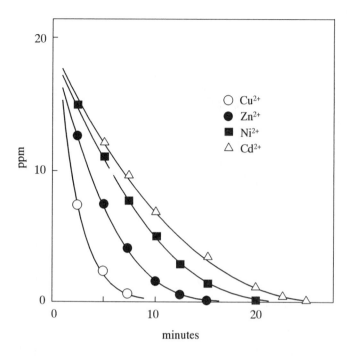

Figure 9. Relation between electrolytic time and concentration of heavy metals.

Colloidal particles of aluminum hydroxide formed near the cathode have a stronger coprecipitation effect than aluminum sulfate or polyaluminum chloride (PAC) which is usually used in the disposal by precipitate aggregation. This is due to aluminum hydroxide which was formed near the cathode and positively charged and metal hydroxide that was formed near the anode and negatively charged. Therefore, an electrostatic attraction improves the efficiency of the removal of heavy metals. Colloidal particles of aluninum hydroxide on which the heavy metals were adsorbed can be scraped out and removed as scum.

The merits of this method are as follows: (1) It is possible to dispose of Cd that cannot be disposed of by the precipitate aggregation method (pH 8 to 9) or the slight traced heavy metals that remain in treated water. (2) Organic compounds like oils can be easily disposed of by the floating separation. (3) The wastewater can be disposed of by dissolving a small amount of aluminum in it, and little sludge is produced because colloidal particles of aluminum hydroxide have high activity. (4) This disposal process does not require a large area to set up the equipment for floating separation. (5) There is no danger of the pipeline being plugged by the recrystallization of the treated water because agents such as slaked lime is not used in the disposal.

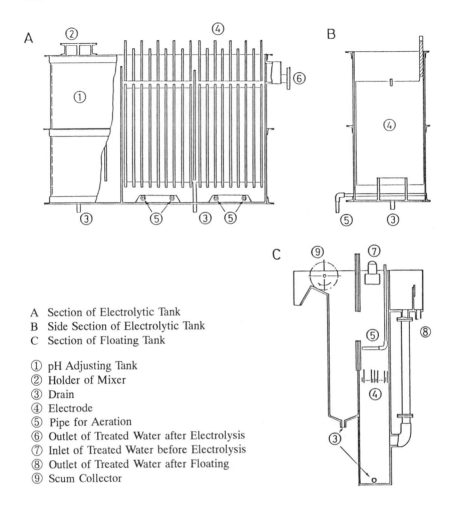

A Section of Electrolytic Tank
B Side Section of Electrolytic Tank
C Section of Floating Tank

① pH Adjusting Tank
② Holder of Mixer
③ Drain
④ Electrode
⑤ Pipe for Aeration
⑥ Outlet of Treated Water after Electrolysis
⑦ Inlet of Treated Water before Electrolysis
⑧ Outlet of Treated Water after Floating
⑨ Scum Collector

Figure 10. General view of the electrolytic tank and floating tank.

6.2.4.2 Wastewater Disposal System

A general view of the electrolytic tank and the floating tank having a capacity of 600 l/h is shown in Figure 10. The pH of the wastewater is adjusted to 7 to 8 in the adjusting tank before transference to the electrolytic tank. In the electrolytic tank, a direct current (5 V, 30 A, and 30 min) is applied between the anode of the aluminum alloy and the cathode of SUS-304 that are arranged alternately and perpendicularly to the direction of the flow. Aluminum, that dissolved in the wastewater when electricity is turned on, is converted into aluminum hydroxide at pH 7 to 8. This hydroxide changes into a water-insoluble colloid by adsorbing heavy metals (e.g., mercury, cadmium, chromium, lead, and copper) and organic compounds. The wastewater is stirred by the aeration from the bottom of the electrolytic tank. The produced colloid is removed into the floating tank together with the wastewater. In the floating tank, the colloid grows into a large flock with the aeration and the addition of a polymer flocculant. Simultaneously, the direct current is applied between the anode and the cathode. The flock of colloid is rapidly floated up on absorption of the gas produced at the two poles. The floated scum is gathered up and removed to the sludge reservoir. The treated water is discharged from the bottom of the floating tank.

6.2.4.3 Disposal Features

Heavy metal ions except mercury of about 20 mg/l can be easily removed from wastewater by the electrolytic floating method. They can be neutralized and dehydrated after the reduction of chromium(VI) in the predisposition process (precipitate aggregation). Therefore, we do not need to consider the concentration of heavy metals when disposal conditions are decided on by the precipitate aggregation process and can predispose the wastewater easily. If cyanide (CN), As, and F have been mixed in the wastewater, it is difficult to predispose it. In this case, we can predispose by adding a proper quantity of iron(II) sulfate or potassium chloride before the neutralization. Usually, it is not necessary to add PAC, aluminum sulfate, or iron(II) sulfate as the flocculant in the predisposal process, because 20 mg/l of heavy metal ions can be easily disposed of in the electrolytic floating process. Therefore, we can reduce the amount of sludge in the predisposal process.

We must be aware not only of the final quality of the treated water, but also of the sludge produced in wastewater disposal. This sludge contains a very small amount of the toxicant even if the sludge is judged to be safe based on the results of a leaching test. We must therefore reduce the amount of sludge to reduce disposal cost and insurance cost at the disposal site. We can reduce this amount produced in the predisposal process by selecting the most suitable conditions and the ability of the electrolytic floating process. Various wastewaters can be safely disposed of by the electrolytic floating and the precipitate aggregation processes as predisposal options.

6.2.5 IRON POWDER METHOD

Inorganic wastewater produced from experiments includes a slight amount of many hazardous substances and requires the following consideration: (1) It contains a variety of hazardous substances of various concentrations. (2) It contains chelate and ammonium complexes that obstruct the formation of the precipitate of metal ions. (3) It contains manganese(VII) or chromium(VI) that can be disposed of after reduction. (4) It contains organic compounds that obstruct the disposal and As, Hg, or free cyanide that must be classified before disposal. The iron powder method was developed as the method of disposal for inorganic wastewater; the wastewater need not be classified, and hazardous substances can be disposed of collectively. This method is one type of precipitate aggregation method which uses a specially made iron powder. The chemical oxygen demand (COD) in liquid wastes from photograph developing or liquid wastes from disinfectants at a hospital (e.g., formalin) can be reduced by such iron powders. These powders have differing contents of the powder used for the disposal of heavy metals.

6.2.5.1 Disposal Principle

In the iron powder method, heavy metals are removed by a series of reactions. When the pH of the wastewater is as acidic 3 or less, the following reducing reaction occurs:

$$2H_2CrO_4 + 2Fe + 6H_2SO_4 \rightarrow Cr_2(SO_4)_3 + Fe_2(SO_4)_3 + 8H_2O$$

A part of the iron powder is dissolved in acid. Cr^{6+} is changed to Cr^{3+} easily by the reducing power in this dissolution and the power of the produced Fe^{2+}. Therefore, Cr can be removed as a hydroxide using the correct pH. Mn^{7+} can be removed in the same way, because it is not necessary to reduce it by predisposal.

When the iron powder is added to acidic wastewater containing copper ions, Cu^{2+} with its low ionization tendency is separated as a metallic copper by the following electrochemical reaction:

$$Cu^{2+} + Fe \rightarrow Cu + Fe^{2+}$$

This reaction proceeds rapidly because the iron powder with a large surface area constantly produces a new surface with the stirring. Cu, Hg, and Ag (of which the standard electrode potential is high) can be removed by this reaction.

When heavy metals are disposed of by the iron powder method, the metals are adsorbed on the surface of the iron powder having an activity. The power of the adsorption on this powder is very strong in this reaction. Heavy metals are adsorbed on the surface and the inside of the porous iron powder; this fact became evident from an electron micrograph of the precipitate with this powder after the reaction. Heavy metals are also adsorbed on iron hydroxide in the precipitate. Several percentages of iron powders are dissolved and most of them remain as metallic iron, though there are some differences in pH or the concentration of the wastewater.

In disposal of the wastewater, part of the iron powder is dissolved as Fe^{2+}. Fe^{2+} is changed into Fe^{3+} by air oxidation after pH is raised, and the mononucleus complex of iron is produced by the hydrolysis. Following the reaction, the solution is changed into a sol and the sol precipitates as a polymer complex containing H_2O_2, O, and OH. At the same time, iron oxyhydroxides such as α-FeOOH and γ-FeOOH are produced as the precipitate. Heavy metals are removed completely by this active iron precipitate produced because these reactions proceed uniformly.

At low concentration of heavy metal ions, most heavy metals can be removed because the neutralizing reaction takes place with the addition of iron powders. However, at high concentration, it is better to neutralize the wastewater by alkali and produce hydroxides because the reaction from the added powder is insufficient. Fe is removed as hydroxide after the iron powder that was dissolved as Fe^{2+} is quickly changed into Fe^{3+} with the raising of pH. At the same time, As, Sb, and Pb can be removed by the coprecipitation effect of $Fe(OH)_3$.

H_2O_2 completely decomposes the compounds (COD) which are then easily oxidized because of the oxidation power of H_2O_2 and the catalytic action of an iron powder coated by copper. This iron powder can be removed completely upon neutralization after the oxidizing reaction.

6.2.5.2 Wastewater Disposal System

Figure 11 shows the basic disposal system for heavy metals by the iron powder method. The iron powder (5 g/l) is added to the wastewater after adjustment of the pH to 2 to 2.5. The pH rises to about 6 with the proceeding of the reaction when stirred. Then, the wastewater, to which polymer flocculant (1 mg/l) has been added and pH set to 9 to 9.5 by adding alkali, is stirred. The precipitate obtained is separated from the treated water by a filter press. The wastewater containing cyanide and heavy metals is disposed of by the disposal process for heavy metal after decomposition of cyanide by the alkali-chlorination method (Figure 11). Figure 12 shows the disposal process for COD. Wastewater to which iron powder, H_2O_2, and a deforming agent have been added is stirred after the pH and temperature were set at 2 and 30°C, respectively. Alkali is then added, pH of the wastewater set at pH 7 to 10, and it is stirred. After the reaction, the precipitate produced from the wastewater is filtered by adding the flocculant and further stirring.

6.2.5.3 Disposal Features

The drawbacks of the iron powder method for the disposal of wastewater are as follows: (1) The reaction proceeds differently in different kinds of iron powder. (2) It is difficult to dispose of heavy metals when their concentration is high (above 1000 mg/l). (3) Consumption of the iron powder increases when an oxidizing agent is mixed in the wastewater. The merits of this method, on the other hand, are as follows: (1) The classification of wastewater is of only two kinds—heavy metals and free cyanides. (2) It is necessary to pre-dispose for cyanides but not for heavy metals. (3) Silicates in the wastewater

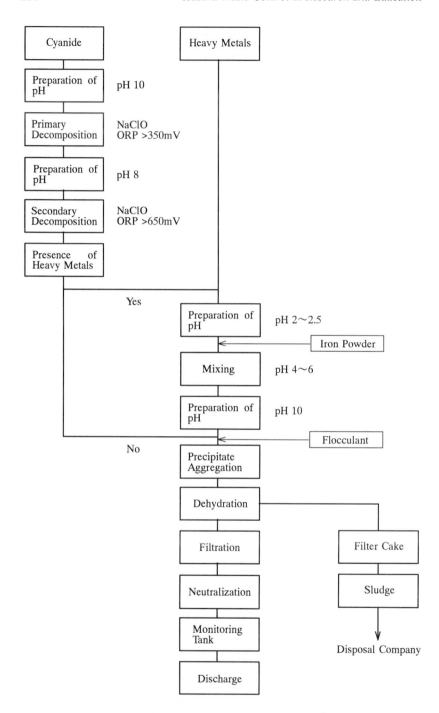

Figure 11. Basic disposal system by the iron powder method.

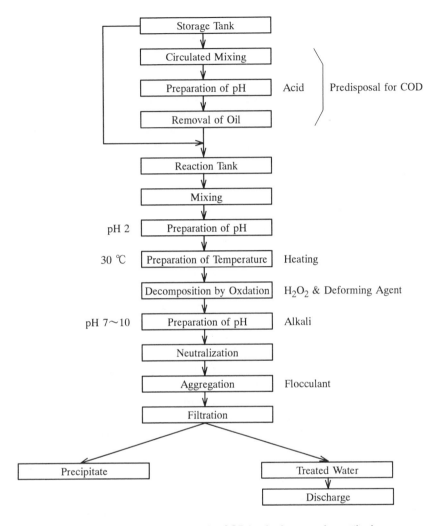

Figure 12. Disposal system for COD by the iron powder method.

can also be removed. (4) Filtration is very easy because the rate of aggregation and sedimentation of the products is large and the iron powder acts as a filter aid. This is one of the most effective disposal methods because the classification is simple, and most of the inorganic wastewater can be disposed of all at the same time.

6.2.6 SOLAR EVAPORATION METHOD

In the ferrite process, the precipitate aggregation method, the electrolytic floating method, and the iron powder method, wastewater is lumped together and subjected to the sedimentation separation of heavy metals. Treated water containing salts is discharged to the sewerage. In contrast, by the solar evaporation method, heavy metals can be recovered individually from each wastewater and reused; the wastewater in each tank (10 to 50 liters) is not diluted, not mixed with any other, and is treated chemically and physically (e.g., neutralization, oxidation-reduction, precipitation, electrolysis, ion-exchange, distillation, and filtration). Then, heavy metal compounds (sludge) or salts are recovered after the treated wastewater is put into the solar evaporating chamber so that water evaporates.

6.2.6.1 Wastewater Disposal System

Solar evaporation chamber and exhaust gas treatment equipment — The solar evaporation chamber is like a greenhouse and is made of glass with an aluminum frame to use solar energy effectively. For example, the chamber has a width of 250 cm and a height of 90 cm with two saucers coated with polytetrafluoroethylene inside it. Each saucer has a width of 122 cm, a depth of 115 cm, and a height of 10 cm. We can dispose of about 240 l of wastewater at a time because the 20 l polypropylene vessels can be placed in two columns and three rows on each saucer. The vessels are 34 cm wide, 50 cm deep, and 20 cm high. The wastewater is dried naturally by air from an air intake duct to which a filter paper is attached, and the chamber is equipped with a draft fan; 20 l of wastewater can be dried in a polypropylene vessel in about 1 month by the evaporation of water. The exhaust gas is discharged into the atmosphere by the draft fan after being washed with water by exhaust gas treatment equipment. This equipment has a width of 60 cm, a depth of 80 cm, and a height of 280 cm. The water in it is reused to send a shower by the circulating pump as circulating water. pH of the wash water can be monitored as the equipment has a pH electrode. At the same time, the conductivity of the wash water is monitored by circulating this water at 10 l/min in the cell to measure conductivity. The results of the monitoring are continuously recorded on each piece of exhaust gas treatment equipment.

Disposal for recovery — In the solar evaporation method, wastewater is grouped into about eight classes:

1. Wastewater containing volatile matter is neutralized and disposed of at the sedimentation tank after the volatile matter is removed by distillation. The precipitate produced is separated from the filtrate at the filter tank and put in a solar evaporation chamber to obtain the sludge. The filtrate evaporates as water vapor in this chamber and becomes salts. The residue is recovered as heavy metal compounds and salts and purified for reuse.

2. Wastewater with acid and alkali is disposed of by solar evaporation after neutralization, precipitation, and filtration.

3. Heavy metal wastewater is disposed of by solar evaporation after neutralization. When it contains oxidizing and reducing compounds, oxidation-reduction or ion-exchange is performed first.

4. In mercury wastewater, mercury ion is recovered as sulfide by the precipitation method and purified for reuse. The filtrate is disposed of in the solar evaporation chamber for mercury. The produced gas is discharged safely after disposal by an adsorbent for mercury.

5. Cyanide wastewater is disposed of in the solar evaporation chamber after the cyanide is decomposed into nitrogen and carbon dioxide by the oxidative decomposition process.

6. In wastewater containing noble metals, these metals are recovered by electrolysis and purified for reuse.

7. In highly concentrated wash water with a chromic acid mixture, chromium ion is recovered as chromium oxide after reduction and neutralization. The filtrate is disposed of in the solar evaporation chamber for salts. In low concentrated wash water, it is reused as pure water after removing chromium ion by the ion-exchange process.

8. Wastewater with the chromic acid mixture is reused by condensation in distillation under reduced pressure. Sulfuric acid is recovered by filtration and ion-exchange and reused. In this system, various monitors grasp rapidly and exactly the mass transfer in the recovery process.

6.2.6.2 Disposal Features

The features of the solar evaporation method are as follows:

1. In this method heavy metals can be recovered individually from each wastewater and reused.

2. The method allows us to see where the responsibility for each investigator lies.

3. Only water vapor and carbon dioxide are discharged when the wastewater is disposed of by this method; no effluent is produced.

4. Few chemicals and little energy are required for disposal by this method, and the cost is less expensive than methods such as the ferrite process.

CHAPTER 6.3

Incineration of Organic Solvent Wastes

Tetsuji Chohji

TABLE OF CONTENTS

6.3.1 INTRODUCTION

A great deal of organic wastewater is released from research, educational, and medical facilities of universities and other institutes. The wastewater released from laboratories shows diversity in its quality and quantity. Organic wastewater contains organic solvent, solvent containing halogens, noncombustible solvent, and dilute organic liquid waste.

These are very hazardous materials and, thus, unwelcome in the environment. Although some of them are known as carcinogenic and acute or chronic toxic substances, the influence of many of them on the human body is not yet understood. It is most desirable, then, that organic wastewater be collected. Some kinds of organic substances may be appropriate for discharge into the sewerage, but these should be restricted to dilute solution with alcohols, proteins, amino acids, sugars, etc. which decompose biologically.

Although it would be best if collected wastewater could be reused, the variety of materials it contains generally makes this difficult. Organic wastewater must then be rendered harmless by the incineration method.

Small amounts of organic solvent can be burned on a metal tray or a bed of sand in the open air or in a draft chamber. However, care must be taken in dealing with a large amount of organic solvent, noncombustible solvent, or dilute organic liquid waste. Incineration is a very valuable method of dealing with laboratory organic wastewater.

6.3.2 BASIC THEORY OF INCINERATION

Two separate processes may occur in the same combustion chamber. The first is pyrolysis, which is the breakdown of materials by heat into vapors, ash, and char. In this process, the organic wastewater becomes small drops and gas, as expected in highly efficient combustion, which is the combination of a material with oxygen. The combustion itself is the second process.

After being preheated by radiant and convection heat in the combustion chamber, the drops of wastewater become vapor. The vapor of the organic substances diffuses from the surface of the drops to the outside and mixes with the ambient air while the gas burns. Figure 1 show schematic model of the process.

Organic wastewater with a higher boiling point burns incompletely because pyrolysis occurs on the surface of drops. When the thermal decomposition temperature of wastewater is higher than the boiling point, soot is produced in the chamber. When the drops are smaller, the rate-determining step is generally a process of gas combustion with diffusion of the surrounding air, while when the drops are larger, the step is a process of combustion of the drops. Actual combustion occurs between these two processes. Drops of organic wastewater are among 10 and 300 μm in size, with the average generally from 50 to 100 μm. Good combustion is best achieved with drops of average small size.

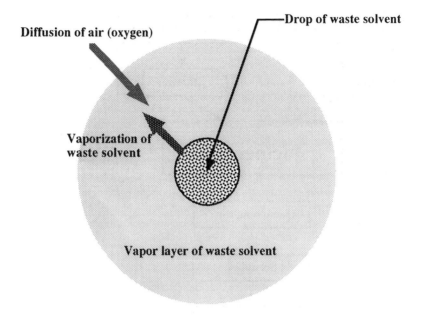

Figure 1. Schematic model of burning drop of solvent waste.

To achieve complete combustion, a good balance of the components of the wastewater and good atomization of the drops is required. Temperature in the combustion chamber must be controlled and a well-balanced amount of air supplied. There are many incinerators with commercial designs that satisfy these conditions.

6.3.3 LABORATORY INCINERATORS

The following three types of incinerators are most commonly used by universities and other institutes: atomizing combustion type, emulsion combustion type, and combustion-in-water type.

6.3.3.1 Atomizing Combustion Type

The most remarkable feature of the atomizing combustion type incinerator is its high speed rotary burner. When organic solvent flows through a rotating hollow axle with a fan, a thin film of the solvent is formed in the axle. At the end of the axle, the film is dispersed by centrifugal force and is atomized by air sprayed from a ring-shaped nozzle. Rotation speed of the axle is usually about 5000 rpm for organic solvent waste combustion. Good atomization of the organic solvent by the burner requires proper adjustment of the flow rate and viscosity of the solvent in the rotating hollow axle of the burner.

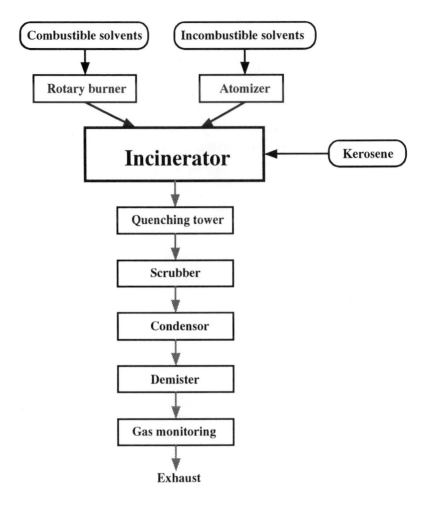

Figure 2. Process of atomizing combustion type incinerator.

The process of the atomizing combustion type incinerator is shown in Figure 2. While the waste organic solvent is fed through a high speed rotary burner into the combustion chamber, diluted organic liquid waste is sprayed from an atomizer at the top of the chamber. The temperature of the incinerator is maintained at about 800°C. Kerosene is used in place of waste organic solvent at the start, as well as if the incinerator temperature falls to under 750°C because of the lack of organic solvent and its calorific power.

The atomizing combustion type system is characterized by the following:

1. Wastewater is generally compressed to 0.3 kg/cm².
2. The opening angle of spray of organic wastewater is controlled by changing the angle of the introductory blade and/or the flow rate of atomizing air.

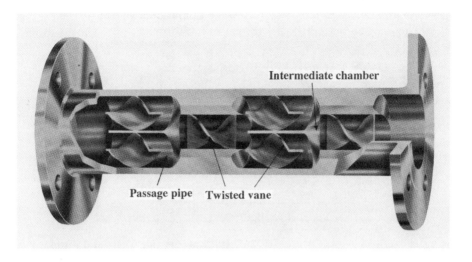

Figure 3. Illustration of in-pipe mixer.

3. The dynamic range over which the wastewater supply can be controlled is relatively large.
4. Highly viscous wastewater is hard to completely atomize.

6.3.3.2 Emulsion Combustion Method

The main feature of the emulsion combustion method is that an emulsion of organic solvent and water is supplied to a combustion chamber to carry on safe and complete pyrolysis. A comparatively stable emulsion of organic solvent and water is obtained with the aid of a static mixer and a nonionic emulsifying agent.

To produce an emulsion, organic solvent and water are mixed using an in-pipe mixer; Figure 3 shows the schematic structure. The in-pipe mixer is made of elements with 180° twisted blades and intermediate chambers. There are two parallel, cylindrical, built-in conduits in the mixer. When two different kinds of liquid are fed together under pressure into the mixer element, the combination is divided into eight phases at the end of the first mixing element. Consequently, the number of divided layers, N, can be approximately expressed by $N = 4^n$, where n is the number of mixer elements; for example, 3 mixing elements produce 128 streams of liquid. This mixer can be used for any viscosity emulsion without mechanical problems and it resists abrasion.

An oil-in-water type of emulsion, as shown in Figure 4, is more common for organic solvent combustion. It can be stocked safely without danger of firing or explosion by a rapid chemical reaction or a runaway polymerization reaction because of the effect of large sensible heat of water. Since there is no accumulation of static electricity in the emulsion, there is also no danger that it will catch fire.

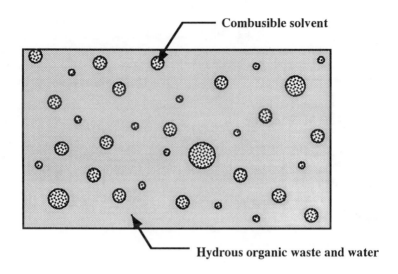

Figure 4. Illustration of oil-in-water type of emulsion.

The process of the emulsion-combustion type incinerator is shown in Figure 5. Organic solvent and water are mixed with a small amount of nonionic emulsifying agent in a mixing tank. The mixture is circulated through a static mixer for a certain time to produce an emulsion. If there is a large amount of diluted organic wastewater, it may be used instead of water to produce an emulsion. The heat quantity of the emulsion must be maintained at 3500 to 7000 kcal/kg of emulsion.

At the start of combustion, the temperature in the chamber is raised by the combustion of fuel oil. The emulsion is sprayed by a nozzle called an atomizer into the pyrolysis chamber when the combustion chamber temperature is over 800°C.

6.3.3.3 Combustion-In-Water Type

A remarkable feature of a combustion-in-water type incinerator is its capability to simultaneously treat both organic and inorganic wastewater. Figure 6 shows a schematic diagram of this incinerator. Organic solvent is sprayed from a nozzle at the top of the incinerator. The flame runs downward into a wastewater tank that is filled with inorganic wastewater. The fact that combustion occurs in the water gives this unit its name. Although usually a rather direct contact evaporator, this incinerator is also used for the combustion of organic solvents.

This method was hitherto utilized primarily for the burning of concentrated solvent, but recently has also been used for the incineration of waste oil. Exhausted gas passes by the liquid, causing the gas to be washed by the thermal exchange and agitation effect of the gas-liquid contact. Harmful

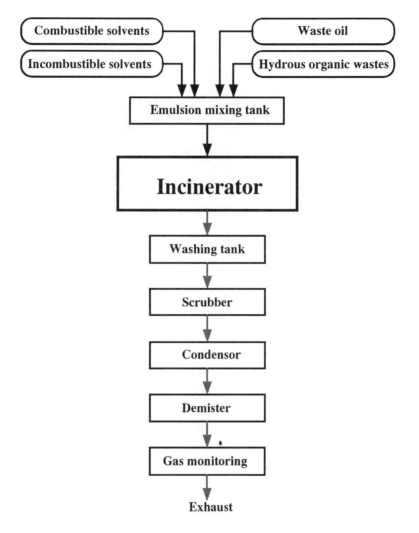

Figure 5. Process of emulsion-combustion type incinerator.

chemical compounds released by combustion and heavy metals present in the waste solvent are removed by the iron hydroxide coprecipitation system.

In a general incinerator, the bottom fire-resistant material is damaged because of the heap of molten salt generated when inorganic salts are put in the combustion furnace. However, there is no such fear with a vertical combustion-in-water type incinerator.

The process of this combustion type is shown in Figure 7. Liquid propane gas (LPG) is used to ignite the unit. Kerosene sprayed from a stabilizing burner is burned to raise the furnace temperature to 800°C. After this preheating, combustible waste solvents are sprayed from the main burner.

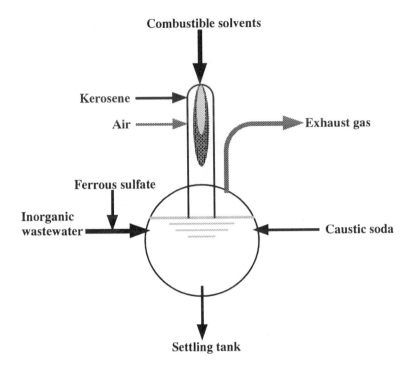

Figure 6. Schematic diagram of combustion-in-water type incinerator.

Furnace temperature is controlled between 800 and 1100°C. When it reaches about 1000°C as a result of the combustion of the combustible waste solvent, noncombustible organic wastewater is sprayed from a burner in the middle of the incinerator. Should the furnace temperature drop because of a lack of heat generated by the combustion, kerosene from the stabilizing burner is again burned.

This combustion system is characterized by the following:

1. Organic and inorganic wastewater are simultaneously treated.
2. The treatment system is more compact because an inorganic reactor is set up inside the combustion furnace.
3. Organic wastewater containing heavy metals can also be burned.
4. No damage is caused to the furnace by a heap of molten salt because of the use of fire-resistant material on the bottom of the unit.

Inorganic wastewater is drained into a tank under the incinerator, a certain amount of ferric chloride is added, and it is adjusted to a neutral pH range.

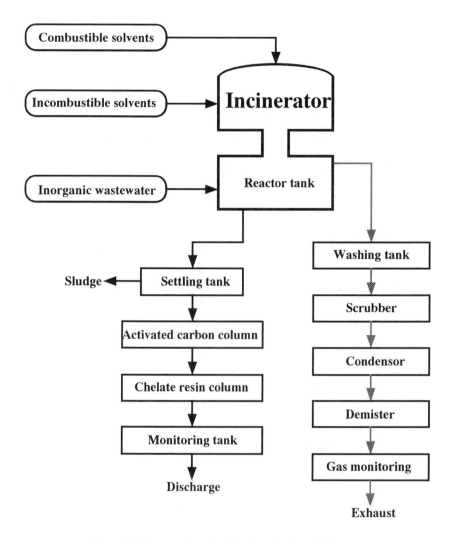

Figure 7. Process of combustion-in-water type incinerator.

6.3.4 ORGANIC SOLVENT WASTES

6.3.4.1 Classification of Wastewater

Organic wastewater has various properties of combustibility (viscosity, halogen content, nitrogen or sulfur, and some water content, and it must be burned wisely and efficiently. It is better not to mix different kinds of wastewater in laboratories, but rather to classify each kind and store it carefully for later disposal. Types of organic wastewater include

1. Nonaqueous waste solvent — organic solvents that are highly combustible.
   - Normal organic solvents: hydrocarbons, alcohols, esters, ketones, aldehydes, organic acids, etc.
   - Organic solvents containing nitrogen: pyridine, aniline, amine, alkides, picoline, formamide, nitriles, etc.
   - Organic solvents containing sulfur: mercaptane, alkylsulfide, thiourea, ABS detergent, etc.
2. Waste oils — waste that contains petroleum products, or animal and vegetable fats and oils; their characteristics are higher combustibility and higher viscosity.
   - Petroleum products: kerosene, fuel oil, machine oil, lubricating oil, etc.
   - Animal and vegetable fats and oils: fats and oils, fatty acids, etc.
3. Noncombustible solvents — primarily waste halogenated solvents. Organic wastewater that contains much of one or more of the following chemicals is considered a noncombustible solvent: carbon tetrachloride, chloroform, chloroethyrenes, chlorobenzenes, etc.
4. Hydrous organic wastes — wastewater that includes aqueous solutions containing water-soluble organic solvents or compounds, and oils containing a great deal of water. Developing solution also falls into this category: an aqueous solution which contains hydroquinone, metol (p-[methylamino]phenol sulfate), phenidone, acetic acid, etc. Fixing solution, however, is not treated as hydrous organic waste because it includes a great deal of silver; this should be recycled.

6.3.4.2 Preparation of Wastewater

6.3.4.2.1 Spontaneous Combustibility

In an ordinary incinerator, temperature in the combustion chamber is kept high by burning spontaneous-combustible wastes, which are nonaqueous waste and/or waste oils. Before beginning burning of the combustible waste, it is advisable to test-burn a little of it in an evaporating dish to learn its combustibility. If it does not burn spontaneously, kerosene may be mixed in, but if it burns explosively or generates much soot, it means that the waste contains explosive or harmful chemicals requiring that care be taken in its combustion.

6.3.4.2.2 Viscosity

Viscosity of wastewater significantly affects its atomization. Waste oil is often over 20 cp in viscosity, which is the maximum value for complete atomization. It is therefore advisable to mix these oils with another solvent such as a nonaqueous waste solvent.

6.3.4.2.3 Contents of Halogen, Nitrogen, and Sulfur

Halogen, nitrogen, and sulfur in wastewater create gas such as HCl, Cl_2, NO_x, and SO_x in a combustion chamber. This acidic gas damages the incinerator material. It is necessary to properly handle these elements to reduce the generation of acidic gas.

6.3.4.2.4 Wastewater Contaminated by Solid Matter

Various solid matter is often present in organic wastewater. Although most of it is precipitates generated by unexpected reactions during storage, they are sometimes small laboratory instruments. The solid matters must be removed by prior filtration or decantation so that it will not clog pipelines and atomizers.

6.3.4.3 Caution

Highly flammable liquids should not be directly incinerated except in special facilities, as they may explode or "flashback" as they are being loaded. An exception is that small amounts may be mixed in a 2% solution in oil, then burned by an atomizer or absorbed on vermiculite. Organic wastewater low in fuel value should be burned with a larger amount of combustible material whenever possible.

Considerable thought should be given to the delivery and storage of waste near the incinerator so that there is no fire hazard or health risk to the operator.

Substances which cannot be combusted must be handled with special care:

1. Explosive materials — chemicals listed in Table 1 are viewed as explosive.
2. Highly reactive compounds — chemicals shown in Table 2 react explosively when they are mixed.
3. PCB (polychlorobiphenyl) — PCB, polychloronaphthalene, and hexachlorobenzene must be carefully stored at laboratories. They are particularly difficult to burn, requiring a temperature of 1200°C for at least 2 s, which is currently achievable by few incinerators.
4. Radioactive substances — these substances are never treated in a general incinerator. The waste of radioactive substances is rigorously controlled by law.
5. Highly flammable substances — as started above, these liquids may explode or flashback as they are being loaded, and special facilities are required for their incineration. Since ethers (diethylether, diisopropylether, tetrahydrofuran, dioxan, dimethoxyethane, etc.) easily make peroxides, they must never be mixed with other solvent waste. Wastes that contain mainly ethers should be diluted with water for storage to protect from electrification or explosion.
6. Carcinogens — some chemicals pose a special risk in the waste disposal operation. These are regularly used substances that often harbor possible cancer hazards, as listed by the International Agency for Research of Cancer (IARC) in Lyon, France.

Table 1 Explosive Chemicals

Acetylene	Picramide
Acetyl peroxide	Picric acid
Ammonium nitrate	Picryl chloride
Ammonium picrate	Picryl sulfonic acid
Benzoyl peroxide	Propargyl bromide
Cumene peroxide	Succinic peroxide
Dinitrophenylhydrazine	Trinitroanisole
Dipicrylamine	Trinitrobenzene sulfonic acid
Dipicryl sulfide	Trinitrobenzoic acid
Ethylene oxide	Trinitrocresol
Lauric peroxide	Trinitronaphthalene
Methyl ethyl ketone peroxide	Trinitrophenol
Nitrogen trifluoride	Trinitroresorcinol
Nitroglycerin	Trinitrotoluene
Nitroguanidine	Urea nitrate
Nitromethane	

Table 2 Some Examples of Chemicals with Possibility of Explosive Reaction by Mixing A and B

A	B
Harogen	Ethers, ethan, acetylene, ethylene imine, harogeno carbons hydrazine, hydroxy amine, etc.
Ammonia	Ethylene oxide, piclic acid, etc.
Potassium, sodium	Carbon tetrachloride
Silver	Oxalic acid, tartaric acid, etc.
Chlorine, bromine	Turpentine oil, benzene, etc.
Perchloric acid	Alcohol, acetic anhydride, etc.
Potassium permanganate	Ethanol, methanol, acetic anhydride, glacial acetic acid, glycerine, ethylene glycol, acetic ethyl, etc.
Hydrogen peroxide	Alcohol, acetone, aniline, nitromethane etc.
Chromic acid	Acetic acid, naphthalene, glycerine, turpentine oil, alcohols, etc.

6.3.5 EXHAUST GAS TREATMENT

Since international discussion of the global environmental problems in May 1988, at the summit in Toronto, the debates about acid rain, depletion of the ozone layer, and increase of global atmospheric temperature due to greenhouse gases have accelerated. Acid rain damages man-made construction as well as the ecosystem on land and in rivers and lakes. Massive use of ozone-depleting chemicals such as CFCs (chlorofluorocarbons) and halons

Table 3 Effective Treatment Method of Pollutants

Pollutants	NO$_x$	SO$_x$	HCl	PAH	Soot	Hg
Pretreatment	G	G	G	B	B	G
Treatment in incinerator	V	G	G	V	V	B
Treatment of exhaust gas	V	V	V	G	G	V

Note: V, very good; G, good; B, bad.

causes skin cancer and disturbance to the ecosystem due to excessive exposure to ultraviolet radiation. Carbon dioxide generated from incinerators has caused atmospheric temperature to increase on a global scale. Within 60 years, the global temperature is predicted to increase by 1.5 to 4.5°C, resulting in a rise in sea level of 30 to 150 cm, flooding of low-laying areas, and destroying the ecosystem. The environmental effects of pollutant emissions from laboratory incinerators cannot be viewed as negligible, no matter how small they may be.

Environmental pollutants released from incinerators are

- Gas: NO$_x$: NO, NO$_2$, N$_2$O; SO$_x$: SO$_2$, SO$_3$; HC: CH$_4$, C$_n$H$_m$Oi, PAHs (polycyclic aromatic hydrocarbons); CO$_2$, etc.
- Liquid: mist, aerosol, submicron particles, etc.
- Solid: soot, metal oxide particles, etc.

Nitrogen oxides are formed from the nitrogen in organic waste (fuel-NO$_x$) and from nitrogen in the combustion air (thermal NO$_x$). The formation of thermal NO$_x$ is dependent on combustion temperature and residence time. If not burning wastewater containing SO$_x$, HCl, heavy metals, etc. there is no necessity of treating gas exhausted from the incinerator. Since harmful materials cannot be eliminated from wastewater, however, these harmful materials are, in fact, often burned with the wastewater. CO, PAH, and soot are generated in a combustion furnace from the reactions of wastewater. Thus, their quantities vary with the operating conditions used. Because they are extremely harmful even in minimal quantities, PAHs must be particularly controlled and decreased as much as possible.

For these reasons a laboratory incinerator is never too small to overlook removal of pollutants from the exhaust gas. There are three general processes for this: pretreatment, treatment-in-furnace, and treatment of the exhaust gas. The quantity of materials which can be decreased by each of the three techniques is listed in Table 3.

A typical system for removing air pollutants is shown in Figure 8. Exhaust gas is passed through a quenching tower and reaches a scrubber. This process reduces the temperature of the gas decrease and removes from the exhaust gas certain acidic gas constituents. The operation of gas scrubbing is often

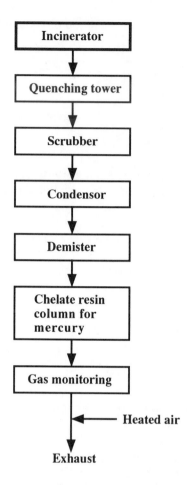

Figure 8. Typical system of removing air pollutants from exhaust gas.

called gas washing or gas absorption. Absorption is a mass transfer operation involving contact between a gas mixture and a liquid. Intimate contact between the two phases is usually obtained by bubbling the gas, spraying the liquid, or using a packed column.

Figure 9 shows three typical devices for quenching and scrubbing exhausted gas. In each of these, the gas and liquid phases may be in continuous contact. Let us describe a packed column as an example of scrubbers. The device consists of a cylindrical column or tower equipped with a gas inlet and distributing space at the bottom; a liquid inlet and distributor at the top; liquid inlet and distributor at the top; liquid and gas outlets at the bottom and top, respectively; and a supported mass of inert solid shapes. The solid shapes are called tower packing.

The most common type of tower packing is the Raschig ring, which is a thin-walled cylinder having a diameter equal to its height. A mass of identical

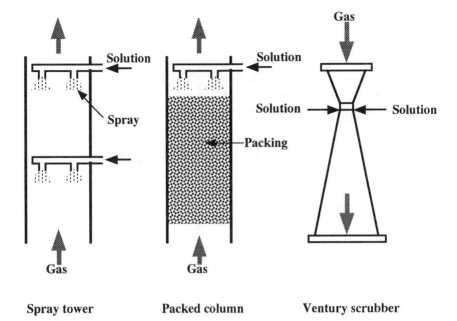

Spray tower   **Packed column**   **Ventury scrubber**

Figure 9. Schematic diagrams of typical devices for quenching and scrubbing exhaust gas.

rings is dumped at random into the tower, and they fill the space above the packing support, which usually is a perforated plate. The liquid used will be referred to as the solution. The solution should be selected for the constituent present in the smallest amount. It is well-known that sulfur dioxide reacts with water to form sulfurous acid, which in turn ionizes. Equilibria involved in the absorption of sulfur dioxide and reaction with water and subsequent ionization are:

$$SO_2 + H_2O \rightleftarrows SO_2 \cdot H_2O$$

$$SO_2 \cdot H_2O \rightleftarrows H^+ + HSO_3^-$$

$$HSO_3^- \rightleftarrows H^+ + SO_3^{2-}$$

According to the above equilibria, the solution should contain alkaline to completely remove sulfur dioxide from the gas phase. Of course, after scrubbing, the solution must be handled by a plant capable of disposing of inorganic liquid wastes.

The second step in removing air pollutants utilizes a demister; this device is used to mechanically separate particles of solid or drops of liquid. Mechanical separation is applicable to heterogeneous mixtures, but not to homogeneous solutions. This method best handles particles larger than 0.1 μm.

Use of a sieve, septum, or membrane, such as a screen or filter, which retains one component and allows others to pass is also common.

Screening is a method of separating particles according to size alone: the undersized pass through the screen openings while the oversized do not. Screening becomes progressively more difficult as the size of the particles is reduced below approximately 100-mesh.

Many methods of mechanical separation are based on the movement of solid particles through a gas. One of these utilizing the movement of particles through gas is a settling chamber. A stream of dust-laden air enters the chamber and flows horizontally at a low velocity through the unit. The dust particles settle under the influence of gravity to the bottom of the chamber, from which they can be removed, and the clean air exists from the chamber.

Both gravitational and centrifugal methods are used to remove particles or drops from gas streams. A gravity settling chamber is useful to remove particles of dust larger than about 325-mesh, although the capacity for smaller particles is minimal and the chamber occupies considerable space. The cyclone separator that uses centrifugal force to amplify the settling rate is used to remove smaller particles and is more compact.

Dust or mist from gas streams can be removed by a variety of methods which depend, entirely or partially, on impingement of the particles against solid surfaces placed in the flowing stream. Because of their inertia, the particles are expected to cross the streamlines of the fluid and to strike and adhere to the solid from which they can subsequently be removed.

Mercury is a challenging material because it has highly toxic components and is thus hazardous in the environment. The presence of mercury components in organic wastewater has recently become a serious problem. These components come from unknown sources and sometimes contaminate the alkaline solution used for washing exhaust gas as well as the exhaust gas itself. Mercury cannot be satisfactorily removed by a scrubber with alkaline water, and an advanced chelate resin treatment process is often used.

Finally, monitoring is required to assure the complete treatment of exhaust gas. Monitoring is usually done to check the content of CO, CO_2, HCl, NO_x, O_2, SO_2, dust, and hydrocarbon. A mercury gas monitor can also be set up on the gas flow line as required.

REFERENCES

1. McCabe, Warren L. and Smith, Julian C., *Unit Operations of Chemical Engineering*, McGraw-Hill, Tokyo, 1956.
2. Pitt, Martin J. and Pitt, Eva, *Handbook of Laboratory Waste Disposal*, Ellis Horwood, West Sussex, England, 1985.
3. Takahashi, Teruo and Korenaga, Takashi, Treatment of organic wastewater, in *Waste Treatment in Universities*, Daigakuto Haikibutsu Shorishisetu Kyogikai, Kyoto, Japan, 1988.
4. Ohtake, Kazutomo, Combustion and environment, *Energy Res.*, 11, 214, 1990.

CHAPTER 6.4

Problems in Waste Treatment Operation

Hideki Tatsumoto

TABLE OF CONTENTS

0-8493-682-5/94/$0.00 + $.50
© 1994 by CRC Press, Inc.

6.4.1 INTRODUCTION

Since 1967, the free discharge and abandonment of waste in Japan has been regulated by the "Environmental Pollution Prevention Act", the "Law Concerning the Disposal and Cleaning of Waste", and the "Sewage Law". The waste created by universities is also subject to these laws. Wastes from educational, research, and medical activities in a university must be appropriately processed by that institution in accordance with these three laws.

The principle generally accepted in Japan is that the polluter is responsible for treating the waste materials he or she creates. This applies to all professors, researchers, and students in the institution. However, the treatment of waste takes a great deal of time and requires special techniques; it is, thus, neither efficient nor economical in the process of education and research. Many universities in Japan have tended to set up central waste treatment facilities.

In general, the waste discharged from universities, research institutions, and hospitals has the following peculiarities:

1. The wastewater contains multiple substances, although the quantity is less than that from an industrial firm.
2. The quantity of the waste discharged from such institutions is not necessarily constant, but is actually rather variable.
3. The sources of emission are spread over a wide area.
4. The waste of medical units may contain bacteria, viruses, and other germs which have been directly discharged into it.
5. The waste of other units may contain new types of chemical compounds developed to meet a new technology, but there may not yet be a method developed to render them harmless.

Therefore, the methods of treating waste materials must be variable, depending on the components and the quantity generated by the departments of an institution.

The kinds and volumes of wastes differ greatly according to the size and composition of the institution or research laboratory concerned, but can be generally divided into inorganic and organic wastes.[1-3] For wastewater that includes inorganic materials, some of the following treatment methods are applicable: the coagulation (or precipitation) activated carbon (or chelate resin) adsorption method, iron powder method, electrolytic flotation method, and ferritization method. For wastewater in which there are organic substances, the spray combustion method or emulsion combustion method is applicable. Combination of these two latter techniques can be used to simultaneously treat both inorganic and organic wastewaters followed by the coagulation/precipitation method.

6.4.2 CLASSIFICATION OF WASTES

According to their derivation, experimental wastes differ in the requirement for their disposal and must be separated and labeled in storage so that the disposed method used will be the one which will be rendered them the most harmless. An example of the separation is indicated in Table 1. For storage, the physical and chemical characteristics of a waste must be identified; waste may be mixed if it is known that no dangerous reaction will result. Mixing a variety of wastes in the same waste fluid container poses the possibility of explosion, combustion, generation of dangerous gases, or putrefaction.

Waste which contains poisonous or hazardous material must be labeled and separated and should not be mixed with general waste material. Filters which have accumulated sedimentation (such as heavy metals, bottles, and pieces of glass to which poisonous materials have adhered) should not be mixed with other general waste containers. Poisonous and other hazardous elements must be separated by rinsing, etc. General waste should also be sorted into combustible solids, metals, and glass and each type stored in a separate container.

A waste and the water used to rinse its container should be clearly labeled when sent to storage. The suggested standard is to include the water from at least two rinses. For mercury, the water from four rinses should be considered hazardous and labeled for storage. When the concentration of a particular waste or the poison potential is high, or the stated standard is felt to be too low for the material in a container, the standards for mercury waste effluent should be applied.

6.4.3 GENERAL PRECAUTIONS RELATIVE TO COLLECTION AND STORAGE

Collection, storage, and handling of wastes produced by experiments shall be conducted with the following precautions in mind:

1. It is preferable to separately treat and dispose of strong wastes produced by experiments as soon as possible. After identification, waste for which the same method of disposal can be used should be kept together.
2. Complex ions, chelating reagents, etc. which contain components difficult to dispose of should be kept with wastes which contain the same type of materials.

Table 1 Classification of Hazardous Laboratory Waste

Code	Classification	Waste Classification for Storage	Container Color	Cautionary Steps to be Taken Prior to Storage and Disposal Treatment Areas
A	Mercury wastes	Inorganic and organic mercury	Blue	1 All wastes containing any amount of mercury should be classified as mercury waste. 2. Tools or instruments, filters, and filter paper, etc. that have contact with mercury should not be rinsed with water. Up to the fourth rinsing, the rinse fluid waste shall be retained. 3. Metallic mercury, mercuric compound, sediment waste, and amalgams should be separately identified for storage. 4. When cyanide is included, the alkali value should be over pH 10 and clearly labeled. 5. Clearly identify any other heavy metals that are contained.
B	Cyanide wastes	Free cyanides and cyanide complexes	Purple	1. Waste should be alkaline over pH 10. 2. Any metals included should be identified.
C	Wastes of fluoride, phosphoric acid	Inorganic fluoride compounds, phosphoric acid and inorganic compounds	Light blue	1. Division between fluorides and phosphoric acid should be clearly identified. 2. Any heavy metals included should be identified.
D	Wastes including heavy metals	Poisonous metals, Cr, Cd, Pb, Zn, Fe, Mn, As, etc.	Yellow	1. Elements from atomic number 21 (scandium) through 83 (bismuth) and compounds thereof to be placed and clearly identified before stored. 2. Tallium and osmium and their compounds should be stored separately.
E	Mixed chromic acid wastes	Wastes of chromic acid and sulfuric acid	Orange	1. These wastes should be separated from those in category D when stored and placed in separate identified storage from D classification. 2. Note that polyethylene containers are not appropriate for long-term storage.

F	Combustible wastes	Combustible organic solvents, petroleum products, vegetable and animal fats and oil, etc.	Red	1. Beware of fire. 2. Besides the normal organic agents (hexane, alcohol, acetone, ethylene, benzene, etc.), organic acids such as pyridine, DMF, DMSO, CS_2, etc. should be clearly identified. 3. Strong ether compounds (Et_2O, THF, etc.) to be disposed of immediately at point of origin. 4. Kerosene, heavy oil, machine oil, grease, cutting tool lubricants, etc. shall be made free of solids. 5. High viscosity items to be diluted with solvent.
G	Nonflammable wastes	Water inclusive of dilutant wastes	Green	1. Wastes of water solutions of organic compounds (amines, organic acid, etc.), degradable resistant cyanic mixtures, organic cyanide compounds, chelate compounds, antibiotics, disinfectants, photographic materials (exclusive of fixers), etc. should have contents identified.
		Silicone oils	Yellowish green	2. Silicone oils should be placed in separate storage and identified.
H	Halogen wastes	Organic halogen solutions ($CHCl_3$, CCl_4, PhCl, etc.)	Black	1. Halogens should be stored separately from F category wastes.
I	Organic fluoride wastes	Organic fluorides (CF_3COOH, PhF, etc.)	Gray	1. Organic fluorides should be stored separately from F and H category wastes.
J	Poisonous or hazardous solid wastes	Filter papers used with poisonous hazardous solids; papers and residuals of lids filtering wastes, oil sludge, etc.	Polybuckets or with blue	1. Papers should be divided by content and clearly identified. Particular caution should be taken to identify any mercuric matter. 2. Experimental animal carcasses should be placed in double layered polyethylene bags and frozen; inquire at disposal centers for detailed instructions.
K	Others			1. Inquire at disposal centers for detailed instructions.

3. The following wastes must never be mixed together:
   - Organic matter and peroxides
   - Cyanides, sulfides, hypochlorites, and acid
   - Volatile acids such as hydrochloric acid, fluoric acid, and nonvolatile acid
   - Acids such as sulfuric acid, sulfonic acid, oxalic acid, polyphosphoric acid, and others
   - Ammonium salts, volatile amines, and alkalis.
4. Unbreakable, noncorrosive containers must be used for storage in the laboratory and clearly labeled with the contents and components before storage in a safe place. Special precautions are imperative for wastes that are highly poisonous or hazardous.
5. Wastes of melcaptans which generate highly obnoxious fumes, wastes of amines, cyanides, and phosphines which generate poisonous gases, and those of highly combustible carbon bisulfide, ether, etc. must be stored appropriately in leakproof containers for early disposal.
6. Wastes which contain peroxides and highly explosive nitroglycerine should be handled with great care and disposed of promptly.
7. Waste which contains radioactive material should be collected separately and in accordance with the laws governing such material, then disposed of under strict supervision to assure against leakage.
8. Combustible wastes (paper, wood, textile, animal residue, and garbage) should be incinerated as soon as possible or stored where it can be tightly contained.
9. Nonflammable wastes (metal, glass, ashes, trash, soil and sand, pottery, cement, empty cans, bottles, etc.) should be segregated and securely stored.
10. Used dry cells and phosphorescent tubes should be stored in respectively designated places.

6.4.4 INORGANIC WASTES

6.4.4.1 Mercuric Wastes

If the mercury in a waste is organic mercury or metal, sulfides, or colloidal material, a complex and complete treatment is difficult. These wastes are treated by chelate resins adsorption which are exclusively for mercury and have either a dithiocarbamic acid base ($-NCS_2H$) or a thiol base ($-SH$). At the time of treatment it is important to pay due attention to the mercury concentration, adjustment of the pH of the waste, and the chemical state of the mercuric ions when subjected to adsorption. Without pretreatment, direct adsorption by chelate resins cannot be achieved for an effluent standard of less than (0.005 mg/l). One pretreatment is to induce decomposition to a sulfuric acid in an oxidation tank, adding potassium permanganate, and the heat in the tank will then cause oxides to decompose. The organic mercury and mercuric compound or other existing organic material in the waste is decomposed, and conditions are ready for adsorption by the chelate resin.

After oxidation decomposition and adjustment of the pH of the waste to around 6, it is passed through a chelate resin column at SV2-10 (volume of the waste should be 2 to 10 times that of the resin) and transferred to a storage tank. The treated waste is then analyzed for mercury concentration by an atomic adsorption spectrophotometer, and if it is found to be below permissible effluent standards, it is removed to a heavy metal storage tank and given a ferritization treatment. If the effluent standard is not met, the fluid is once again recycled through the chelate resin once again. Chelate resin which has adsorbed the mercury is transported to the Itomuka mining plant, the only place in Japan equipped with a facility to refine mercury. Caution must be exercised here as during treatment mercury analysis is undertaken, and when treatment is insufficient, it is then necessary to determine what the most appropriate step may be. Mercury contained in an alcoholic waste compound is first submerged in a chelate resin to remove the mercury, and the remaining waste is incinerated; the resulting mercuric vapor is reclaimed in an alkaline liquid and returned once again to the oxidation decomposition tank.

Mercuric compounds in wastes of known mercuric content are not a problem, but when the mercury is contained in a chemical, care must be taken not to erroneously place the material in storage under an incorrect classification, such as:

- Mercury containing chemicals or pharmaceutical compounds are mercurochrome, thimerosal, roycomycin tablets, etc.
- Mercury containing reagents are Nessler's reagent, Meyer's reagent, Million's reagent, Linter's reagent, Nass reagent, etc.

6.4.4.2 Cyanide Wastes

Treatment of free cyanide wastes (under 1000 mg/l concentration as CN^-) is done by alkaline chlorination. The progress of the reaction is confirmed by employing the ion electrode method for cyanide ion at the time of conclusion of the reaction, employing an oxidation reduction potentiometer.

The oxidative decomposition method employing calcium hypochlorite cannot treat cyano complex salts such as potassium hexacyanoferrate(III), (potassium ferricyanide), or sodium pentacyanonitrosilferrate(III). Such wastes must be separated, and the various noncombustible components they produce should not be mixed with other free cyanide wastes. When an experiment makes it inevitable that the wastes are mixed upon emission, they shall be sorted into free cyanides and clearly labeled with the compounds and their respective concentrations. In the same manner, when free cyanide wastes are mixed with organic solvents they should be divided according to the free cyanide and identified for compound and concentration, respectively.

Since they are not uniform with respect to compounds and concentration, with the exception of the free cyanide wastes, it is preferable to dispose of them through incineration. For appropriate treatment, the pH of each waste should be adjusted to over 10 and the contents clearly defined.

6.4.4.3 Fluorides and Phosphoric Wastes

When the fluoride ions formulate mixes such as boren fluorides and fluoro silicates, Equation 1 does not progress sufficiently, and even with the activated alumina adsorption method, elimination through adsorption of complex fluoride ions is difficult.[4] As shown in Equations 2 and 3, aluminum ions are added to the waste, and after causing the fluoride ions to formulate an aluminum complex, calcium chloride is added.

$$2F^- + CaCl_2 \rightarrow CaF_2 + 2Cl^- \tag{1}$$

$$3BF_4^- + 2Al(OH)_3 \rightarrow 2AlF_6^{3-} + 3H_3BO_3 \tag{2}$$

$$AlF_6^{3-} + 3Ca^{2+} \rightarrow 3CaF_2 + Al^{3+} \tag{3}$$

The discharge of wastes mixed with fluoride ions in formaldehyde and alcohol solvents is conspicuous, but after the fluoride ions are forced to sink in the sediment as much as possible in the reaction shown in Equation 1 for separation, the remaining clear liquid is subjected to incineration disposal. During the sedimentation process, a large volume of formaldehyde and alcohol is adsorbed so that the sediment can also be incinerated in the solids furnace.

In organic fluoride compounds, such as trifluoroacetic acid, the Equation 1 reaction does not progress as a primary treatment after incineration, and then as secondary treatment, the fluoride ions in the scrubber water are treated with Equation 1 or by activated alumina adsorption.

In the same way, phosphoric wastes are treated in accordance with the reaction of calcium chloride

$$2PO_4^{3-} + 3CaCO_3 \rightarrow Ca_3(PO_4)_2 + 6Cl^- \tag{4}$$

and eliminated as sediment of insoluble calcium phosphate.

For wastes containing polymerized phosphatic salts such as tripolyphosphates, the Equation 4 reaction does not progress so that after using sulfuric acid for acidification of them, boiling for 2 to 3 h, and then breaking down the thiophosphatic acid by adding water, the reaction in Equation 4 occurs. Wastes containing red phosphorus, hydrogen phosphate, phosphorus sulfide, etc. are made alkaline and dissolved in a solution of sodium hypochlorite for oxidation to obtain the reaction in Equation 4.

Fluoride and phosphoric wastes do not respond to the same reaction formulas which makes treatment difficult. The sorting out and identifying collection methods are the same, but it is preferable to place the wastes in separate containers for proper identification and separation.

In the same manner, for other wastes in fluorides or in phosphoric acid related compounds, the contents should be clearly identified.

In ferritic conversion, not all heavy metals are unconditionally converted to ferritic compounds. When heavy metals are present in a stable form similar to metal chelates, they are left outside of the ferritic conversion reactions. Copper amines, cacodylic acid (dimethylarsinic acid), cyano complexes [such as potassium hexacyanoferrate(II) (potassium ferrocyanide)], and EDTA complexes of heavy metals are representative. Furthermore, negative ion materials such as fluorides and phosphoric acids not only cannot, in principle, become ferritic compounds, but they obstruct the ferritic reaction of heavy metal ions; also, the coexistence of organic materials affects ferritic reactions.

6.4.5 ORGANIC WASTES

6.4.5.1 Problems in Organic Wastes

As stated, organic wastes resulting from experiments are incinerated in all cases. The most corrosive halogen group of wastes, pyridine wastes that have a very nauseating odor, and many others are disposed in this manner, so that the equipment and facilities employed very soon sustain corrosion damage and are worn from friction and the generation of salts, etc. In the furnaces where fluids are burned, within 4 to 5 years of operation at 10 weeks of burning per year, the furnace walls suffer severe damage. There is also extensive corrosion in the washing tower, suction fans, and ducts. Pinholes appear in the piping, and the couplings of rotary burners and mechanical seals of pumps become so damaged that they are inoperative. Major problems encountered in a waste treatment plant are listed in Table 2.

Methods used to cope with such damage are fourfold:

(1) The concentration of the waste fluids handled can be monitored by predicting or estimating the degree of heat generated by combustion and keeping the temperature at the most appropriate degree. A high concentration of halogen wastes can be diluted to keep the salt content under 3%, and fluids that are combustible at particularly high temperatures can be mixed with other wastes that burn at a much lower temperature. This action exercises a degree of control which should lengthen the life of the furnace used.

(2) The organic wastes, i.e., benzene, ethanol, carbon tetrachloride, MIBK (methyl isobutyl ketone), and others, are used to extract inorganic compounds; therefore, inorganic compounds are often found in organic wastes. Caution must thus be employed to avoid such admixtures as much as possible, particularly the admixture of mercury.

(3) Depending upon the content of organic wastes, with the passage of time, heat is generated, and in many cases, polymerization or solidification occurs. When mixing such wastes, the possible changes in these admixtures should therefore be observed for a specified time to note any physical changes that occur.

(4) Combustion disposal usually results in the emission of gas, HCl, SO_x, and NO_x, which are washed with NaOH solution. Waste from the gas and

Table 2 Examples of Problems Encountered in a Laboratory Waste Treatment Plant[5,6]

Type of Problem	Conditions	Countermeasures
Classification of items for which content description is insufficient and wastes that include sediments	• Inorganic fluorides and phosphatic types of wastes are involved but allowed to burn freely • Separated but then mixed together • Wastes labeled as benzene, etc. but sediments remained also	• Spray combustion disposal • Scrubber waste disposal by fluorides and phosphatic acids • Before disposal, sediment separated out by cloth and net filtering
Organic halogen-type wastes	• Chlorine-related solvents such as methylene chloride and chloroform wastes	• Combustion results in a strong hydrogen chloride gas being generated; dilute with kerosene or other combustible waste fluids prior to combustion
Wastes containing highly concentrated fluorides (contained in organic wastes)	• Fluoride ions contained 330–2000 mg/l in organic wastes (not noted in delivery charts) • Operating frequency of waste gas in washing tower rose • Salt concentration during scrubber rose	• Dilute before combustion, (notwithstanding treated volume increase) • Measure concentration of fluoride ions prior to disposal • Salt concentration predicted by watching changes in specific gravity of the scrubber waste and changing scrubber water frequently

Troublesome reaction occurring in the the receiving tanks	• When MIBK 80%, less than 5% benzene was mixed in the receiving tank with other wastes, within 30 min temperature in the tank rose to 80°C, polymerization reaction occurred, and sediments generated; combustion terminated, resulting in plugging of pipes, pumps, flowmeter, and combustion nozzles	• Sediment removed and dissolved with DMF; pipes and pumps dismantled and washed with DMF; solids and rinse fluids incinerated in solids furnace; the wastes subjected to prior beaker tests, but failed to predict the reaction that occurred 30 min later; emissions generated must be supplied with appropriate information
Appearance of cyanide from waste solutions containing no cyanides	• When mixtures of several wastewaters from a hospital were treated by an automatic alkali-chlorination process, cyanide concentrations of 0.064 mg/l were produced, in spite of the fact that none of the wastewaters contained any cyanide	• Amounts of oxidizing agents, pH, and range of ORP[a] should be carefully adjusted in treating wastewater containing chemical compounds from which cyanides may be formed

[a] ORP, oxidation-reduction potential

admixture (scrubber water) is treated as inorganic waste. This scrubber water is characterized by a high content of fluoride and inorganic salts. The component in this water should be determined, and the most appropriate disposal method then selected. In controlling the high salt content of scrubber water, sufficient alkali must be employed in the washing of the emission gases, but use of too much alkali will cause the cooling tower spray to plug up the spray screens and the excessive salt will damage the pumps. Thus, caution is necessary.

6.4.5.2 Photographic Developers and Fixatives

Waste from photographic developers and fixatives (G category) contains a large amount of sodium salts. These adhere to the furnace parts and cause them to deteriorate. These wastes are the subject of current studies on a new method known as the wet disposal method.

6.4.5.3 Combustible Wastes

In the disposal of wastes in the F category (organic combustible wastes) and the H category (organic halogen-related wastes), the detailed precautions mentioned earlier should be taken; but in actual fact, disposal is not as easy as it appears. The most frequent reason for this is that detailed information on the content of the waste on the delivery chart provided by the laboratory generator does not always coincide with the actual contents. The most prevalent case is failure to mention the presence of carbon tetrachloride, methylene chloride, chloroform, etc. in F category wastes. These chloride-related solvents, in actual fact, are often in the amount of 20 to 40% or more of the total content. This failure is believed to be due to the fact that within laboratories there is far too little awareness of the necessity of listing all of the contents.

There is one more precaution necessary regarding F category wastes: the admixture of highly volatile substances like diethyl ether and carbon bisulfide. The disposal of these elements by combustion in spray furnaces is exceedingly dangerous.

6.4.5.4 Nonflammable Waste Liquids

Nonflammable waste liquids (G category) contain the most chemical materials that must be separated and identified. Therefore, depending upon the content of the waste, various steps must be followed in the methods of combustion employed. When halogen-related organic solvents are included, combustion is similar to the H category (organic halogen-related wastes), while if fluoride ions are among the contents, combustion must be done in small batches for effective control.

6.4.6 SOLID WASTES

The J category solid wastes has only one division of waste to be identified, but the form is exceedingly varied. Incineration is the method of disposing of generated ash. This is ash that can be mixed with cement to make concrete, as well as other kinds which require repeated rinsing prior to their use in concrete. If the waste generator sorts the waste into respective polyethylene bagged categories, the subsequent sorting into groups for disposal is simplified. Insufficient sorting will result in rinsing that it is not necessary for some types, but which will be done anyway. During the last few years, there has been considerable improvement in the attention paid by waste generators in Japan to properly identify the contents of the J category wastes. The acceptability of the presorting of generated waste determines, to a great degree, the amount of extra work required and is directly reflected in the resulting rates charged for disposal.

6.4.7 SPECIAL WASTES

The following enumerated cases require extra precautions and should be disposed in special waste treatment facilities located in only a few places.

1. Insufficiently sorted wastes and those of unidentified content requiring extra handling:
   - Radioactive material
   - Organism-adhering material
   - Material containing explosives
   - Metallic mercury
   - Be, Se, Te, Os, and their compounds
   - Highly hazardous material (nickel carbonyl, alkylaluminum, etc.)
   - PCB
   - Strong carcinogenic materials
   - Strong types of ether
   - Photographic fixatives (developers in G category)
   - Other nondisposable material (nonflammable solids), poisonous and hazardous materials, and material included in Table 1
2. Easily incinerated waste (paper, sawdust, shavings, textile waste, vegetation waste, kitchen and restaurant garbage) should be quickly incinerated or retained in a sheltered place where it will not be scattered. Filter paper, chemical wrapping, tissue paper, and spent activated carbon, etc. to which poisonous materials have adhered should be separately collected and incinerated in disposal facilities (refer to Table 1).
3. Nonflammable waste (metal, glass, ashes, porcelain, cement rubble, empty cans, empty bottles, etc.), after proper sorting into respective areas, should be stored where they will not be scattered.
4. Used dry cells and fluorescent tubing shall be discarded in places specified for retention.

5. Incineration
   - Filter paper and chemical wrapping to which poisonous materials have adhered, tissue paper, used activated carbon, plastic containers, etc. should not be discarded in garbage bins but collected separately and disposed by incineration in special disposal facilities.
   - Incineration of waste in places other than disposal facilities requires prior approval of the Environment Preservation Council or some other governing body.
   - Explosive wastes require a safe outside location for incineration in small quantities. When igniting the material, oil-soaked waste on the end of a long pole should be used from upwind. The fire should be properly monitored until it is completely out.
   - Metals salts that react harshly with water should be stored in metal containers filled with oil and finally burned all at once. These wastes should be placed in a steel container with oil-impregnated wastes and charcoal, which is then ignited. The generated smoke is dangerous, so care should be taken to not allow it to touch exposed skin; it should also not be inhaled. Cooled ashes should be placed in water in small batches to be neutralized and discarded.
   - Combustible waste should be incinerated as soon as possible in furnaces.

REFERENCES

1. Sanders, D.H., Hazardous wastes in academic labs, *Chem. Eng. News,* 3, 21, 1986.
2. O'Sullivan, D.A., Environmental concerns gain prominence in Europe, *Chem. Eng. News,* 27, 7, 1989.
3. Barnhart, B.J., The disposal of hazardous wastes, *Environ. Sci. Technol.,* 12(10), 1132, 1978.
4. Environment, Safety, Earth, Environmental Science Committee, University of Tokyo, 1991.
5. Suzuki, S. and Tatsumoto, H., Recent situation of waste treatment problems in Japanese national universities, in Proc. Symp. Air Waste Management Association, 83rd Annual Meeting and Exhibition, Pittsburgh, PA, June 24–29, 1990.
6. Tatsumoto, H. and Hattori, T., Appearance of cyanide from waste solutions containing no cyanides, *Environ. Tech. Lett.,* 9, 1431, 1988.

Section
7

New Trends in Hazardous Waste Technology

Ecological Planning in Japan:
A Tokyo Bay Area Example

Harvey A. Shapiro

TABLE OF CONTENTS

0-8493-682-5/94/$0.00 + $.50
© 1994 by CRC Press, Inc.

7.1.1 ECOLOGICAL PLANNING IN JAPAN

7.1.1.1 Introduction into Japan

Ecological planning was developed in the West in the early 1960s by McHarg and others,[1] and it was introduced into Japan a decade later by the author and his colleagues.[2,3] It entered Japanese academia in April of 1971 at Osaka Geijutsu University's then newly established Department of Environmental Planning, the first such degree-granting program at a university in

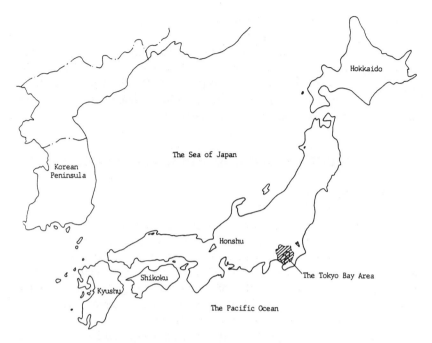

Figure 1. The location of the Tokyo Bay Area in Japan.

Japan. Ecological planning went commercial with the establishment of a planning consulting firm, the Regional Planning Team Inc. (RPT), in Tokyo in January, 1974.[4] However, research on its application to Japan dates from 1969 with the author's thesis at the University of Pennsylvania.[5] Research on its application to the People's Republic of China has recently been done by the author.[6] It is now being practiced in at least nine Asian nations, as well as in Japan.[7]

7.1.1.1.1 Definition

Ecological planning has been defined by McHarg as an instrument for revealing regions as interacting and dynamic natural systems having intrinsic opportunities for, and constraints on, all future uses. The goals of such planning are derived from the region as proffered by locations where all or most propitious factors exist in the presence of none, ideally, or few detrimental ones for all prospective uses. What constitutes ''detrimental'' or ''propitious'' is derived from the prospective uses and the value system of the initiating person or group.[8]

7.1.1.1.1.1 The Tokyo Bay Area in Brief

Ecological planning is an approach based on this view that has been developed and applied worldwide, in this case, to the Tokyo Bay area located in eastern Japan (Figure 1). It is one of Japan's four most important and productive urbanized/industrialized estuarine ecosystems and the site of the nation's capital, Tokyo, located on its northwestern shore. About 25% of Japan's 123.3 million people[9] live in the so-called "Capital Region" centered on the Bay, with 15.5 million living in the 16 cities and towns that line the bayshore and make up the study region (Figure 2).[10] Despite being the smallest of the previously mentioned four major urbanized estuaries, it is by far the most intensively used, and abused, ecosystem of them all.

7.1.2 THE TOKYO BAY AREA ECOLOGICAL PLANNING STUDY

7.1.2.1 Methodology

The approach to the research presented here is shown schematically in Figure 3. Briefly, it involves defining the study area as ecologically as possible, in this case to include the entire bay and all of its surrounding cities and towns which, with the exception of Tokyo, encompass entire river systems. Next, an ecological inventory of both existing natural and sociocultural information for the entire study area is done. These geographic factors are then mapped at a convenient common scale (in this case, 1:500,000). These natural and human geographical factors are then interpreted individually and in combination. Using an overlay method, they are then combined to develop a number of safety/hazard, health/sickness, and amenity criteria considered essential for bay area environmental management and planning, as well as for delineating the "coastal zone" of the region. The concept of "coastal zone" is used here as a tool to link land and water ecosystems throughout the region.[11] Unfortunately, there is no legal requirement to define the coastal zone in Japan,[12] and so land and water environments tend to be planned for and managed separately to the detriment of both. Once the coastal zone is defined, it becomes part of the basis for contemplating environmental management and ecological planning strategies for various future regional scenarios.

7.1.2.1.1 Characteristics of the Tokyo Bay Area

7.1.2.1.1.1 Geology
The terrestrial geology of the northern and central part of the study region is dominated by thick volcanic ash and loam deposits from volcanic activity mainly to the west, especially from Mt. Fuji. These deposits are cut by rivers along which formed broad, water-abundant alluvial sand, gravel, and clay deposits. There, great foundation instability with high vulnerability to seismicity are characteristic with a similar high vulnerability to severe flooding

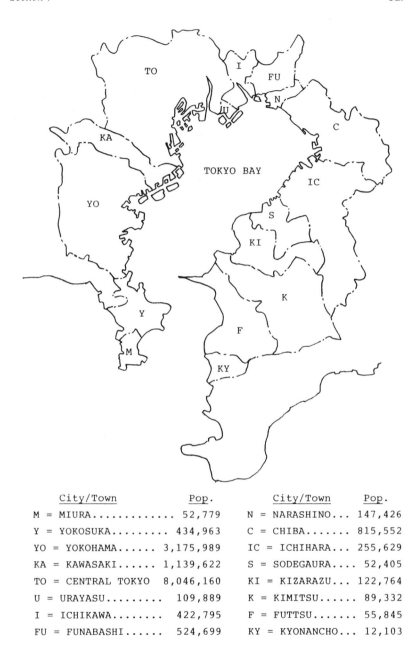

City/Town	Pop.		City/Town	Pop.
M = MIURA	52,779		N = NARASHINO	147,426
Y = YOKOSUKA	434,963		C = CHIBA	815,552
YO = YOKOHAMA	3,175,989		IC = ICHIHARA	255,629
KA = KAWASAKI	1,139,622		S = SODEGAURA	52,405
TO = CENTRAL TOKYO	8,046,160		KI = KIZARAZU	122,764
U = URAYASU	109,889		K = KIMITSU	89,332
I = ICHIKAWA	422,795		F = FUTTSU	55,845
FU = FUNABASHI	524,699		KY = KYONANCHO	12,103

Figure 2. Tokyo Bay area coastal cities (populations as of 1991). (From *Local Government Yearbook, 1991,* Dai-ich-hosei, Tokyo, Japan, 1991.)

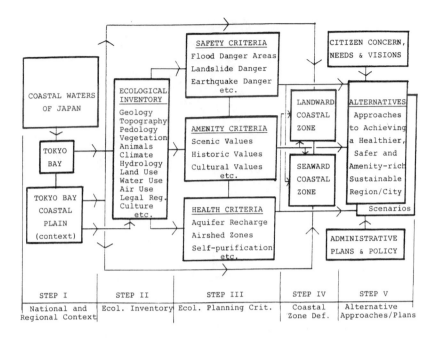

Figure 3. Flow chart of the research methodology.

along rivers, large and small. These alluvial areas coincide closely with the areas of most recent natural land-making having been under water until as recently as the last glacial age in what was Old Tokyo Bay.[13] They also happen to be the most intensively urbanized parts of the region, an extremely dangerous place for urbanization from the viewpoint of natural hazards. The southern part of the bay area is dominated by porous diluvial terraces and older, erodable, tertiary sandstone, mudstone, and gravelstone mountainous areas. These are remnants of the glacial age uplands that surrounded Old Tokyo Bay. They are cut by lateral faults, some of which are active.[14]

The geology of the bay bottom is a critical factor in any bay area plan, especially considering continuing landfill trends along the bay's shoreline. Despite its importance, information on bay bottom geology is incomplete, especially in the center. Nearshore deposits out to about 10 m in depth are generally sand of various textures mixed with muds near the mouths of the rivers that drain the Kanto plain. These nearshore sand and mud strata are extensions of terrestrial alluvial deposits and tidelands, 5 to 15 m thick in the inner bay area to 15 to 20 m thick in the area near the mouth of the bay. These nearshore surface strata are underlain by thick silt strata up to 30 m deep in the inner bay. Inner bay geology at depths greater than 10 m is mainly mud.[15] Deposits at the mouth of the bay are a complex mixture of sands, muds, and rock deposited in or around the valley bottom of the ancient Tokyo River that used to flow there. These deposits are constantly moved about by the currents flowing into and out of the bay. On the whole then, the geology of the bay bottom is unstable and highly dynamic with implications for fishing, marine transport, and other uses, including landfilling.

7.1.2.1.1.2 Topography

The Tokyo Bay area has been formed mainly by volcanic activity and sediment deposition due to the erosion-sedimentation process. Starting from the bay entrance, the topography along most of the Uraga Channel leading into the bay from the Pacific Ocean is mostly steep rocky land with 100 to 200 m high coastal terraces on both sides. The coastline there has been strongly influenced by vertical faulting. In the bay area itself, there are 20 to 30 m high diluvial terraces at the coast on the southwest, but there are only low terraces behind the coastal plain almost everywhere else. The inflowing rivers have formed the alluvial lowlands, at the mouths of which may be seen deltaic landforms and broad tidelands, where they have survived. The result is a highly variable coast in what would otherwise be a rather simple, sandy beach shoreline.[16] However, landfilling and coastal structures have obliterated most of the natural landform, leaving only that near the mouth of the bay and a couple of small inner bay tidelands off the north coast. The most conspicuous natural feature is the protruding sandy Futtsu peninsula.[17] While protecting the inner bay from rough ocean waves, it constricts the bay entrance, resulting in navigation difficulties.

Bay bottom topography reveals broad shallow tidelands at the north end connected to terrestrial alluvial plain deltas, but as already mentioned, these highly productive shallows have been mostly lost to coastal landfill. The inner bay bottom slopes gently to the southwest to an almost flat area in the center. From there it narrows and deepens quickly through the ancient Tokyo River channel. The maximum inner bay depth is 56.7 m at this location.[18] However, the average depth of the bay as a whole is only 15 m. The coastline of the inner bay is 170 km in length.[19] It is still seen as one of Japan's typical, highly undulating coastlines, despite deformation from landfills.[20]

7.1.2.1.1.3 Hydrology

The surface area of Tokyo Bay north of the Futtsu peninsula is about 1200 km^2. This is about 75% of the original area, as the rest has been lost to landfill.[21] The volume of the water in the bay is 18.3 km^3, and the tidal difference is about 2 m.[22] Due to the narrowness of the bay's mouth, the water exchange rate is low, making the estuary highly vulnerable to eutrophication.[23] Bay and ocean water mix just north of the Futtsu peninsula. The exact location of this mixing depends greatly upon wind speed and direction. Winter winds are strong and tend to keep ocean water offshore with top and bottom layers mixing vertically. In the summer, winds blow the heavier bottom-flowing ocean waters toward the inner bay coast under the lighter surface waters, with weak mixing near the shore.[24] This has important implications for commercial fishing as well as for water pollution.

A total of 13 major rivers and an equal number of minor rivers carry fresh water into the bay. They carry a total of over 10.1 billion m³ of fresh water into the bay every year at an average rate of 3000 m³/s. The Edo and Ara Rivers, which drain most of the Kanto plain, carry over half of the in-flow. Since they carry wastewater from the most highly urbanized part of the bay area, their role as the major source of bay water pollution is obvious. The fact that these two rivers flow into a part of the bay where mixing is poor only aggravates the situation. Bay water salinity is lower near the coast and in the inner bay than elsewhere, and nutrient buildup is higher there, a major cause of eutrophication and red tide outbreaks which have occurred throughout the bay area since 1954.[25] In addition, in the summer, an oxygen-poor water layer develops on the bottom (below 10 m in depth) resulting in cold weather "blue tides".[26] The combination of surface-layer red tides and bottom-layer oxygen depletion-caused blue tides has resulted in numerous severe fish kills as well as extensive damage to shellfish.

7.1.2.1.1.4 Climate

There is considerable difference in the temperatures in the bay area at any one time. It is hotter on the west coast than on the east in the summer and colder there in the winter than the east coast. Due to the presence of the bay, coastal areas are generally warmer in the winter than inland areas, but in the summer, temperatures are almost as hot as inland. In addition, winter temperatures are higher near the mouth of the bay than in the inner bay area. In the summer, bay mouth area temperatures are lower than in the bay area. The difference between east and west coast temperatures is the result of a seasonal wind. Summer winds blow from the southeast and south carrying hot air from the inner bay over the western coast, and they blow warm air from over the Pacific Ocean and Tokyo Bay inland. The edge of the sea breeze front has been estimated to extend 10 to 14 km inland. At the leading edge of this front of "airshed", a stagnant air pocket often develops resulting in both unbearably hot temperatures and high air pollution concentrations. In the winter, strong, cold, dry winds blow in from the north and northeast. These heat up as they cross the bay, carrying dampness across the west coast, and making it relatively warmer and more humid in the winter.

Precipitation in the bay area is relatively low, about 1200 mm per year, as compared with the heavier rainfalls, over 2000 mm per year, on the Boso peninsula outer Pacific coast. However, because of the relative scarcity in rainfall throughout the region, Tokyo Bay has been referred to as the Seto Inland Sea of the Kanto District.[27]

7.1.2.1.1.5 Pedology

Gley soils up to 50 cm thick can still be found on the alluvial lowlands and coastal plains that have not yet been completely urbanized or paved over. They have been used for rice farming, reflecting their high natural productivity and intrinsic suitability for that use. On the alluvial fan terraces, Ando soils prevail to a depth of 25 cm or more, while light-colored Ando soils are also found, 25 cm or less deep. These soils have low water tables and are well drained. Both Andos are good for growing forests and are moderately suited for crops. Finally, dry, brown forest soils are widely distributed on the high terraces on the south side of the bay area where tertiary granites are highly weathered. The resulting soils are acidic and not very good for forestry. The other major "soil" area is the coastal landfills. The materials used to make these fills, to be discussed later, vary from garbage to toxic industrial waste and sand-pumped, dredged, bay bottom sediments, as well as construction waste from throughout the region. Many of these fills are covered with a thin layer of soil dressing carried from various inland sites. These displaced soils have little natural productivity, so vegetation is hard to grow and maintain.[28]

As for the sea bottom, the productivity of the bottom muds varies greatly with the depth and quality of the seawater. As a highly enclosed estuarine embayment, Tokyo Bay is a *nutrient sink* and thus a place of naturally high biological productivity, especially in its shallowest portions less than 20 m deep. Muds in areas less than 10 m in depth are also critical in terms of the bay's self-purification capacity, especially tidelands.[29] The productivity in these shallows must be given high priority for protection in any plan for the bay area if the *sustainability* of the bay as a viable ecosystem is to be maintained.

7.1.2.1.1.6 Flora and Fauna

The southeast coast of Chiba Prefecture is generally covered with cryptomeria forest, and oaks and chestnuts are found in the mountainous highlands of the peninsula. The rest of the nonurbanized portion of the region shows the predominance of agricultural vegetation in the low terracelands and rice paddies on the alluvial lowlands. These reflect the high water tables in the lowlands.

Despite their presence, upland forests are not very effective in preventing erosion or flooding in times of heavy rainfall. The lowland paddyfields can serve to hold excess flood waters temporarily. In urban areas, they can also help to serve as fire stops if they are large enough (at least 100 m on a side).[30] Unlike cryptomeria forests, oak-chestnut forest does offer good flood water-holding capacity and erosion-prevention ability, as well as a desirable habitat for various forms of wildlife.[31]

Plants and animals in the bay itself are distributed in four places: the coast, the bay surface, in the water column, and on the bottom.[32] Tidal and shallow water organisms exist where the water is relatively clean and rich in oxygen. Bay bottom organisms work to clean and purify the bay water. Nearshore areas are vital spawning and breeding grounds for fish, especially in the winter and spring. Other species that spawn in the inner bay or in the outer ocean come in to grow in the highly dynamic and productive bay mouth area. The bay has long been favorable for seaweed aquaculture, especially near river mouths at a depth of 5 to 10 m. Most fin fish species in the inner bay are found in the shallows at less than 10 m in depth. Ocean fish can be found at 30 to 70 m in depth near the bay's mouth where rocky/gravelly conditions occur.[33]

The coast was abundant in sea birds, and it still is where natural tidelands remain. In the summer and fall, migrating snipes and plovers can be seen, while in the winter, sea gulls and ducks are seen.[34] Coastal landfill has destroyed most of these birds' feeding grounds. The importance of protecting what little natural habitat remains seems important both for the fauna and for the bay as a living ecosystem. These "remains" can be guidelines for the restoration of the bay's natural habitat in the future.

7.1.2.1.1.7 Present Land Use

The coastal plains of the study region are almost completely covered with intensive urbanization, separated from the bay by large-scale industrial and port uses located on vast amounts of coastal landfill. Inland area on the east side of the bay are still used for agriculture with forests in the hinterlands. On the Tokyo and west side of the bay, little open space remains. Health, safety, and welfare consideration are reasons enough to protect natural open space and expand it wherever possible.

Much of the bay itself is in designate port areas in which almost all of the region's industrial uses are located.[35] The industrial plants of the Tokyo Bay area produce one third of Japan's industrial output. Furthermore, 40% of the nation's oil is imported through Tokyo Bay ports in over 2000 giant tankers annually. The result is the world's most congested[36] and certainly one of the most dangerous embayments. This is because many ships that ply the bay's waters carry volatile liquid natural gas (LNG) and liquid propane gas (LPG). A maritime collision of one or more of those vessels in the crowded bay would not only result in a severe oil spill, but also terrible explosions and the leakage of large amounts of toxic materials into the bay from the numerous storage tanks that line its shore. It has been estimated that should an LNG tanker explode in the bay, depending upon the wind speed and direction, intense heat and flames could spread up to 5000 m from the source of the fire engulfing everything in its way in a catastrophe beyond imagination.[37] The radius of danger for an LPG explosion could easily reach 2000 m (Figure 4). Fighting such fires would be all but impossible.[38]

In addition to commercial ports, there are several fishing ports, mostly near the bay's mouth in proximity to coastal seaweed aquaculture grounds. Bay area fishing grounds vary from season to season and are often close to heavily travelled commercial shipping lanes. Furthermore, several sports fishing and yacht harbors are located in the bay area, too. All of these uses going on at the same time explain even more the extreme value of the bay for so many users, as well as the extreme potential danger that exists there as users compete for its limited and every-shrinking surface area and coastline.

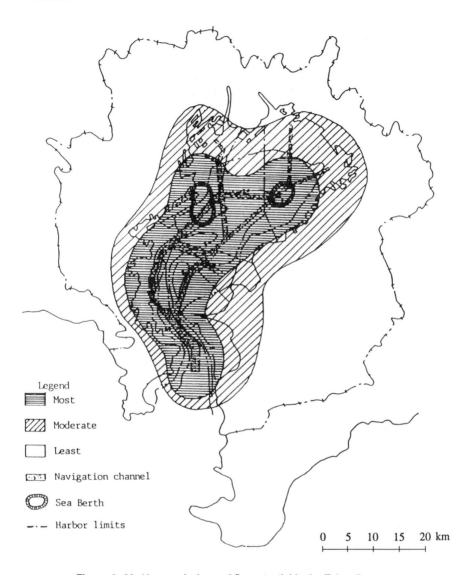

Figure 4. Maritime explosion and fire potential in the Tokyo Bay area.

7.1.2.1.2 *Some Hazardous Wastes in Tokyo Bay*

7.1.2.1.2.1 *Dioxin*

Japan presently uses some 23,000 types of toxic chemicals and adds about 300 more annually, according to Japan's Environment Agency calculations. Of these, dioxin is the most toxic and has been detected in both Tokyo and Osaka Bays.[39] A 1985 to 1986 Environmental Agency survey revealed a concentration of 0.0002 ppb of dioxin in Tokyo Bay, a level the Welfare

Heavy Metal	Amount of Build-up (tons/yr)	Outflow Amount (tons/yr)	Conc. in Bottom Muds (ppm)
Zinc	560	110	900
Copper	100	48	140
Lead	71	7.3	120
Mercury	0.67	0.96	1.75

Source[41]

Figure 5. Major heavy metals in Tokyo Bay in one year (1980 to 1981).

Ministry said is not dangerous to human health. However, U.S. standards for dioxin are said to be a hundred times stricter than Japan's for both man and fish. Suspected sources of dioxin in the bay are agricultural fertilizers and pesticides, tainted muds and sludge, and incinerators and sewage treatment plants, many of which are located on coastal landfill.[40] Those fills are themselves often filled with hazardous materials that a large earthquake or seawall structural failure could result in further leakage into the bay. The hazard to human and marine life's health comes not from the tiny concentrations detected, but from the fact that if such toxic chemicals enter the food chain, they can build up to toxic levels in the fish that people eat.

7.1.2.1.2.2 Heavy Metals

In addition to dioxin, numerous heavy metals flow into Tokyo Bay every year from landward sources. These include zinc, copper, lead, and mercury (Figure 5). Most of these hazardous materials settle on the bottom and build up in this enclosed estuarine water body. As with dioxin, if they get into the food chain, they pose a significant source of toxic pollution. They come mainly from landfill-based industrial sources that must be strictly regulated to reduce/eliminate the risk of fatal heavy metal poisoning,[41] such as that which occurred in Minamata Bay in Kyushu in western Japan in the early 1960s.[42]

7.1.2.1.3 Sea Level Rise and Tokyo Bay

7.1.2.1.3.1 Global Warming and Sea Level Rise

Numerous global environmental problems are now occurring. Among these, "global warming has emerged as perhaps the most serious environmental threat of the 21st century, and only by taking action *now* can we insure that future generations will not be put at risk".[43] Sea level is expected to rise considerably in the years and decades ahead due to global warming caused by increasing amounts of greenhouse gases, such as CO_2, CH_4, N_2O, and chlorofluorocarbons (CFCs).[44] The sea level rise (SLR) over the last century ranges from 8 to 30 cm with 12 cm generally accepted as the "true" global SLR.[45] Researchers have estimated a probable doubling of CO_2 by the year 2050.[46] This could produce a global temperature increase of 1.5 to 5.0°C, which includes the so-called "2°C concensus", the average probable temperature increase.[47] Such a global warming could produce a SLR of anywhere from 50 to over 350 cm by the year 2100.[48] Factors that will ultimately determine global SLR include population growth, climate change, ocean heat absorption, glacial melting, and the amount of greenhouse gases in the atmosphere.

7.1.2.1.3.2 The Influence of Tectonic Activity

One more factor that is important in places like the Pacific Rim is site-specific tectonic movements, as well as man-caused subsidence, from the over-pumping of groundwater. These are particularly relevant factors in Japan. A study of data over a 50-year period from 89 tidal gauge records for Japan and Korea reveal that the maximum rates of tectonic submergence occur along the eastern coasts of Hokkaido (Japan's northern-most island) and Honshu (Japan's largest island), along the southeastern coasts of Honshu, Kyushu (Japan's large western island), Shikoku (Japan's large southern island), and the Inland Sea enclosed by these three islands. The northwest coast of Japan shows no relative change in SLR or even a slight emergence. The maximum rates of submergence reached 24 mm per year at Toga on Japan's central Pacific coast (1965 to 1977). A tectonic submergence of 4 mm per year in the inner bay area (40 cm per 100 years) to 8 mm per year in the outer bay area (80 cm per 100 years) was observed in the Tokyo Bay region. This is greater than the average submergence of 3.1 mm per year for Japan as a whole. This Pacific-side subsidence in the bay area coincides with the areas of subduction of the Pacific and Philippine plates beneath the Eurasian plate.[49] When these tectonic submergence values are added to the global average SLR of 50 to 350 cm by 2100, a range of areas of potential inundation can be plotted as shown in Figure 6.[50,55] Almost all coastal landfills, including those filled with hazardous materials, would be submerged and/or damaged by high and increasingly violent seas. The result would be then, if not before, the leakage of toxic materials from the fills into the rising sea, pollution of the living resources and health hazards via the food chain, and the ultimate destruction of Tokyo Bay as a living ecosystem. A giant earthquake could have similar impacts sooner and more suddenly.

7.1.2.1.4 Landfill for Trash Disposal

Trash disposal by dumping it in small inland fills dates back to the Edo Period (1600 to 1867) in Japan. It was not until 1957 that giant coastal landfills were built for that purpose. Trash and toxic waste from large cities and

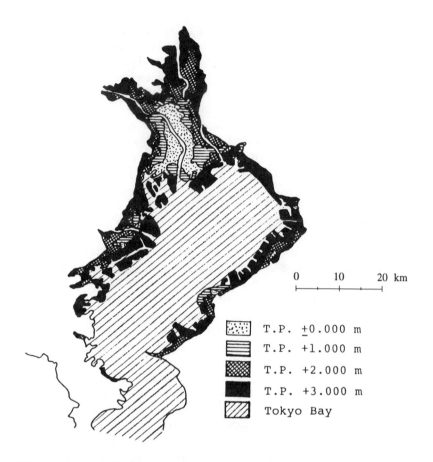

0   10   20 km

░░░░	T.P. ±0.000 m
☰	T.P. +1.000 m
▓▓	T.P. +2.000 m
██	T.P. +3.000 m
▨	Tokyo Bay

Figure 6. Estimated potential inundation for a sea level rise from 1 to 3 m. (From Japan Environment Agency General Office, *Dictionary for the Global Environment,* Chao-hoki, Tokyo, 1990. With permission.)

industries, in particular, increased dramatically along with economic growth in the 1960s and 1970s, and it became difficult to find places to dump them inland, so they began using coastal fill sites. To date, a total of some 600 ha of fills with a volume of 109 million m³ have been built in the Tokyo Bay area for this purpose at a cost of about ¥192 billion (about 1.5 billion in U.S. dollars). These final trash disposal fill sites are called "Phoenix Plans" named after the mythological bird that lived in the Arabian Desert for hundreds of years, finally consuming itself in fire only to rise renewed from the ashes to start another long life (a symbol of immortality).[51] It's a beautiful name for a not-so beautiful use.

In 1980, the Welfare and Transportation Ministries proposed building some 1200 ha of such Phoenix Plan sites between 1986 and 1995 in the Tokyo Bay area alone, only about half of which have been built so far. The latest bay area Phoenix Plan calls for 500 to 600 ha more of such sites to be built between 1996 and 2005. These fills *temporarily* "solve" the problems of final trash disposal and land shortage, as well as being good for the construction industry, but they cause numerous problems. Among others, they often leak, polluting the surrounding seawater; they sink when built on or even when not built on; they are extremely vulnerable to seismic activity; they destroy valuable coastal ecosystems; and they help reduce the natural self-purification capacity of the bay. Those filled with garbage release methane gas as the garbage decays, providing a new energy source but also contributing to global warming, as well as posing another explosion danger in this densely used bay. The incineration of plastics and other such nonbiodegradable wastes emits toxic gases, polluting the air and returning to the earth and water as acid rain threatening to pollute the soil as well as both surface and groundwater. Then there is the smell and unsightliness that repels people, but attracts rats and vermin. Also, there is the air pollution, noise, and vibration related to garbage trucks, as well as the potential for maritime accidents between garbage vessels and hazardous/volatile material-bearing ships in the bay.[52] Only about 30% of the waste dumped in Tokyo Bay fills is from general sources, while 70% is from construction and industrial sources, i.e., toxic, including PCBs, cyanide, asbestos, radioactive materials, and new kinds of wastes from high tech-related sources.[53] Planning for the future of the bay must take these problems into consideration.

7.1.2.2 Toward Criteria for Ecological Planning in Tokyo Bay

Since health, safety, and welfare/amenity should be among the common goal of all land-use planners, these are considered to be appropriate objectives to ecological planning in the Tokyo Bay study area as well. One or two representing each category will be discussed briefly and their maps shown.

7.1.2.2.1 Safety/Hazard-Related Criteria

7.1.2.2.1.1 Earthquake Hazard Potential
Over the ages, the Tokyo Bay-centered region has experienced numerous earthquakes, the largest of tremors were well over 7 on the Richter scale. Being located primarily on deep, unconsolidated, unstable, alluvial deposits, the most highly urbanized areas, including Tokyo and Yokohama, etc., remain extremely vulnerable to earthquake damage from seismic-caused liquefaction, subsidence, and fire and building collapse. The landfills, which now encircle most of the bay, can be seen as a man-made extension of these hazardous conditions and perhaps even more hazardous in a big earthquake.

The only active surface faults in the region are located at the very southern edge of the study area on the Boso peninsula. This area is known to be highly vulnerable to landslides.[54] The diluvial and other terraces are generally considered to be stable and more resistant to seismicity than the alluvial plains.

As for the bay itself, the bottom sediments in the inner bay area are extensions of the unstable alluvial plains and are considered to be equally unstable and vulnerable to seismicity. Ironically, this happens to be where all of the coastal landfills with their giant industrial complexes, oil storage tanks, and port facilities, etc. are concentrated. In addition, areas less than 20 m in depth would be subject to an increase in wave height should a tidal wave hit the bay in a great quake, with areas less than 10 m in depth being particularly vulnerable to inundation. These shallow areas are thus judged to be more hazardous to maritime vessels and coastal structures located there than in deeper areas. The results of these considerations are shown in Figure 7.

7.1.2.2.1.2 Flood Hazard Potential

The land-water interface poses an actual or potential risk of flooding. In Japan, these areas almost always coincide with the alluvial lowlands and coastal plains. In this study, all of the alluvial lowlands are judged to have high flood risk, with areas along the rivers having the highest risk. In addition, water-formed topography, such as deltas and regularly water-covered tidelands, are likewise judged to have very high intrinsic risk of flooding. In contrast, terracelands and steeply sloping areas are seen as having the least relative risk to severe flooding except along streams and dry phantam drainage gullies.

Despite flood protection works, the proximity of coastal fills to the sea is seen as exposing them to constant flood risk, as well as eventual failure of the seawalls and leakage into the sea of fill material, including toxic wastes. Furthermore, considering the prospect for long-term sea level rise due to global warming (and natural subduction) of as much as 3 m by 2100,[55] the risk of such areas to coastal flood damage is likely to increase over time. Thus, at the bay area scale, all coastal landfills were judged to be extremely vulnerable to flooding. Figure 8 shows the result of this evaluation and includes the entire study region, from the watershed to and including the entire bay. Such an evaluation is considered essential for regional planning.

7.1.2.2.2 *Amenity-Related Criteria*

7.1.2.2.2.1 Scenic Values

Topographic, hydrologic, and land-use factors were used to do a preliminary evaluation of the scenic values in the study area (Figure 9). Terrestrial areas along clean rivers and cleaner portions of the bay were evaluated to be of high scenic value, while those along rivers in urban areas, as well as those facing onto the dirty waters of Tokyo Bay, particularly in the ports of Tokyo, Kawasaki, Yokohama, and Chiba, were given a somewhat lower rank. The highly urbanized/industrialized coastal lowlands and landfills were given the lowest relative rating, despite a few exceptions which could not be interpreted

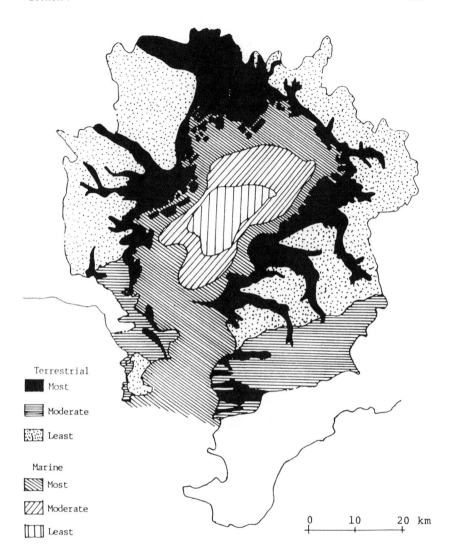

Terrestrial
■ Most
▤ Moderate
▦ Least

Marine
▨ Most
▥ Moderate
▥ Least

0   10   20 km

Figure 7. Vulnerability to seismicity and Tsunami (tidal wave) hazard.

at the scale of the study. In contrast, terracelands were evaluated to be of moderate scenic value, while steeply sloping and high relief areas were given the highest value. Improvement of low and moderately ranked scenic areas should be one goal of planning for the area.

The surface area of the bay near its mouth was given the highest scenic value due to the vast expanse of clean ocean visible there. The view of ships passing there enhance this value, but poor water quality in other such maritime areas resulted in a somewhat lower scenic value. Improvement of water quality in the region could substantially improve its scenic value. Finally, despite

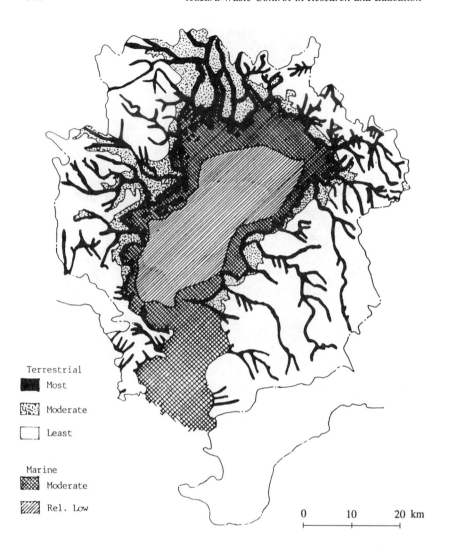

Figure 8. Vulnerability to flooding.

their proximity to ports and river mouths, the remaining inner bay tidelands off of the Funabashi-Urayasu coast were highly evaluated due to their remaining naturalness and the seasonal visitation of sealife and birds using the East Asian Flyway.[56] The importance of preserving these vital, yet endangered, habitats cannot be overemphasized, not only for aesthetic reasons but for biological ones as well.

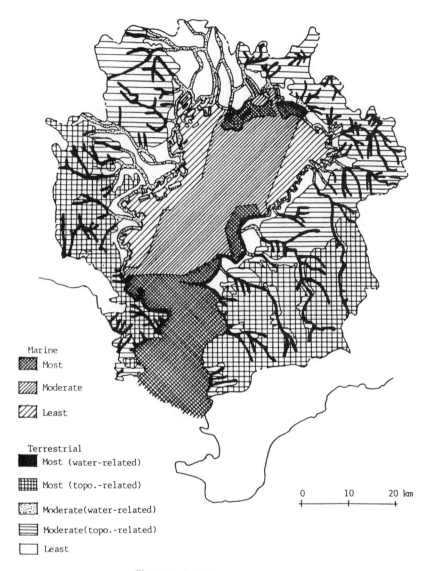

Figure 9. Relative scenic values.

7.1.2.2.3 *Health and Welfare Criteria*

7.1.2.2.3.1 *Biological Productivity*

The ability of the natural environment to produce and sustain plant and animal life is seen as an indicator of a healthy nutrient cycle. Soils information

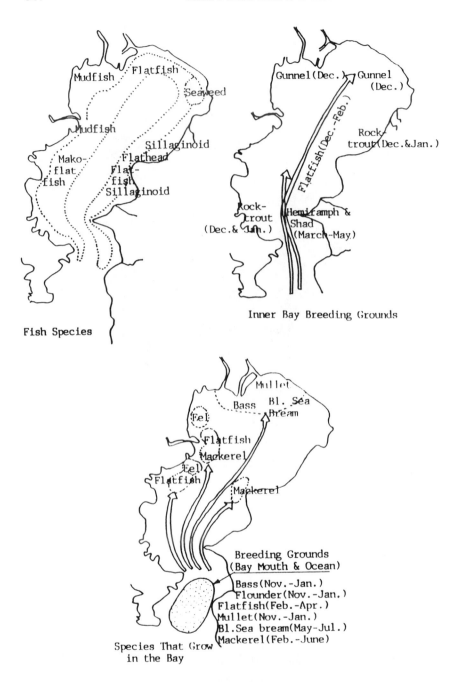

Figure 10. Fish species in Tokyo Bay.

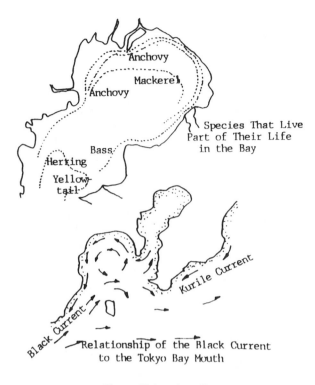

Figure 10 (continued).

was used to make the terrestrial evaluation of biological productivity. Areas with high natural productivity for agriculture and forests were seen as having high biological productivity. Ando soils and lowland soils predominate in this category, but those in urban areas have mostly been lost to paving, etc. Forest and gleyed soils dominate the next lower level of productivity, with immature soils and landfill soils making up the least productive category in terms of agriculture or planting.

The bay is considered to have tertiary treatment capacity and also high nutrient storage capacity due to its enclosed character.[57] Both of these characteristics presume a fully functioning shallows area; however, the fact is that much of this area has already been lost to landfill and dredging, reducing the bay's self-purification capacity greatly. Nevertheless, these shallows, less than 10 m in depth, continue to be highly productive or potentially so due to the permeability of sunlight and supply of nutrients from both land and sea sources. The bay mouth area is likewise considered to be highly productive in terms of living marine resources from within and from outside the bay, fed by the Black Current, as shown in Figure 10.

Water areas 10 to 20 m deep are judged to be relatively productive, but somewhat less so than shallower areas, and deeper areas are seen as being

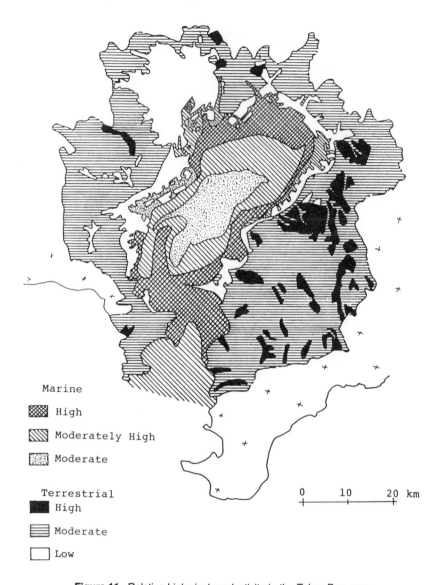

Figure 11. Relative biological productivity in the Tokyo Bay area.

the least productive of them all. However, being part of this estuarine embayment, none of the bay (except filled areas) is seen as unproductive biologically, and so the minimum evaluation is only moderate, as shown in Figure 11. However, due to its enclosed nature, the bay's ecosystem could be extensively damaged from a major oil spill or leak of toxins into the bay's waters. Every effort must be made to avoid this possibility.

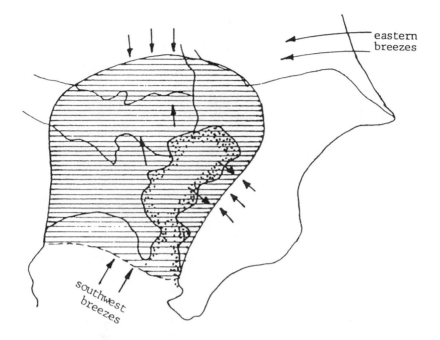

Figure 12. Tokyo Bay region airshed (vulnerability to air pollution). (From Mitsudera, M., *Integrated Ecological Studies in Baycoast Cities, Part I*, Numata, M., Ed., Seibunsha, Chiba, Japan, 1979, 23. With permission.)

7.1.2.2.3.2 Airshed (Air Pollution Vulnerability)

The concentration of people, industry, and landfill in this relatively small embayment necessarily results in a large amount of air pollution and heat.[58] Due to the wind patterns in the region, an airshed is formed, as shown in Figure 12. When winds blow from the south along the bayshore, NO_2 can still be measured at a height of 1000 m, with high concentrations inland. The center of this concentration moves out over the bay when winds blow from the north, especially in the winter.[59] The wind velocity weakens from the coast to the surrounding hills, but due to the expanse of the Kanto plain, a great heat island of polluted air develops. The strategic distribution of vegetated open space with compatible uses throughout the region could increase greatly the natural self-purification of the air and the break-up of the heat island to some extent.[60]

7.1.2.3 The Future of the Bay Area

7.1.2.3.1 Existing Plans for the Bay Area

The Tokyo Bay area remains the site of many large development projects, most involving or related to landfill. Among others, a 1200-ha garbage dump fill (Phoenix Plan) off the coast of Ichikawa City, a 686-ha fill for a sewage treatment plant and housing nearby, a 337-ha fill to expand Haneda Airport in Tokyo Port, and a 186-ha of fill for housing in Yokohama Port are planned. In addition, there are several transportation projects, including the continuation of the coastal road and two road-bridges across the bay, etc.[61] Instead of looking forward, many of these projects appear to look back to the dreams of gigantic landfill projects that made up what was known as the ''Neo-Tokyo Plan'' proposed by the business community in 1958 to fill two thirds of the bay for industry and urban uses.[62] That plan was never fully implemented, but remnants of it are still being done or proposed over three decades later. That plan is said to have ignored the problems and potential of land subsidence and the vulnerability of sea bottom sediments to seismicity, etc.[63] The last version of this plan is the so-called ''Group 2025 Plan for the Capital Region'' with a completion date of the year 2025, as the name of the group implies. Among others, it calls for the construction of a new 30,000-ha landfill island in the middle of the inner bay area to house and provide employment for 5,000,000 people. The Ministry of International Trade and Industry (MITI) has proposed its so called ''Cosmopolis Concept'' for the bay which calls for no less than four 10,000-ha new islands, three by landfill and one to float.[64] The list goes on, but it is clear that the *sustainability* of the bay as a viable living ecosystem is not the goal of any of these plans. Sustainable alternatives are called for now before it is too late.

7.1.2.3.2 A Recreation-Oriented Concept

This proposal takes into account Japan's national trend toward greater leisure time and foresees an evergrowing demand for recreational opportunities, especially marine-oriented ones. This will mean that coastal access be protected where it still exists and restored where it has been lost or restricted so that the region's people can take full advantage of coastal recreation and new leisure opportunities. Such opportunities will, of course, require clean water, natural coastal conditions, facilities designed to blend with the environment, marinas, and sports fishing opportunities far beyond anything now available or imagined for the bay. Such requirements will place considerable limitations on other uses that must then coexist compatibly with recreation, including industrial, urban, and commercial fishing uses, etc. The environmental needs of commercial fishing overlap those of recreation, and only careful planning can reduce/avoid conflicts. As for industry, nothing short of nonpolluting types are seen as being compatible with both recreation and commercial fishing. Strict design standards for industry will be needed to protect scenic values. Such a concept should result in *sustainability* now sorely lacking in the area.

7.1.2.3.3 A Commercial Fishing-Oriented Concept

This scenario is similar in many ways to the previous one, but differs in that the needs for healthy and prosperous commercial fishing in the bay are given top priority. The prospects of climate changes and various related sociopolitical developments in food producing countries could well result in a considerable decline in importable foodstuffs.[65] This would mean that Japan must increase its food self-sufficiency rate considerably. To do this, its agricultural and domestic fisheries resources will have to be available and in healthy/clean condition for ready use. The present trend in Japan is, on the contrary, toward destruction of both its valuable agricultural land by urbanization and soil pollution, nearshore fishing grounds with landfill and pollution, and forestland from lack of proper management. The study area is no exception to this trend. If commercial fishing is to be given top priority, protection and restoration of the bay's biological productivity must be fundamental to any regional plan. The shallows must be restored and protected, and water quality must be improved. The safety of the bay for commercial fishing operations must be insured, and the need for all other uses to respect the needs of commercial fishing will be essential. As before, industry must be nonpolluting (including non-fill types) and recreational opportunities must be planned so as not to have negative impacts on commercial fishing operations. Here again, the need for careful ecological planning, which includes a thorough compatibility evaluation, will be indispensible. Needless to say, living marine resources are highly vulnerable to toxins, so strict regulation, even prohibition, on their use in the bay area will be essential to protect these valuable resources from harm.

7.1.2.4 Coastal Zone Delineation

Before establishing regional environmental guidelines for all prospective land and water uses in the bay area, it is essential to develop a means by which actions on land and water can be controlled so as to positively influence each other and coexist sustainably. This means seeing land and water environments as a unity and not as separate, as is the present tendency in Japan.[66] To do this, the *coastal zone concept* was used. In brief,

"The coastal zone is a horizontal and vertical interface between air, land and water in which human activities may be significantly influenced by and/or significantly influence the natural and social processes operating there . . . "[67]

Since the Tokyo Bay study area is not in itself an ecologically enclosed unit, it is difficult to define the entire study area as a coastal zone. Thus, in this study, the water-related criteria developed earlier were overlayed to delineate

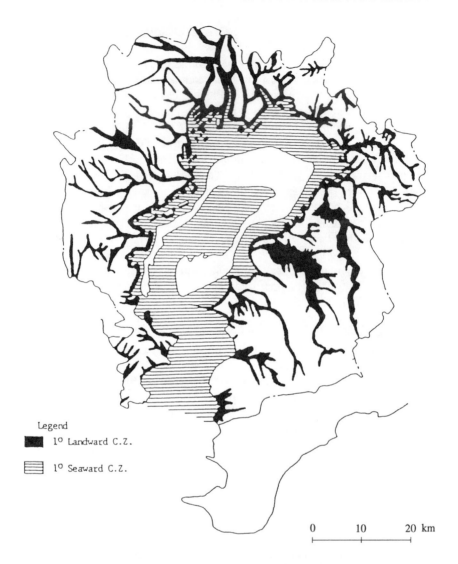

Figure 13. Primary coastal zone of the Tokyo Bay area.

a preliminary coastal zone. The factors used in this process included: flood hazard potential, scenic values, potential biological productivity, and aquifer recharge areas. Two other factors, maritime explosion and fire hazard potential, and air pollution vulnerability were also given consideration. The result of this exercise is shown in Figure 13. All prospective actions within this zone will be regulated so as to protect, restore and/or enhance the quality of the coastal zone environment of which the bay and its surrounding land area are integral parts.

		TERRESTRIAL								COASTAL								MARINE							Remarks
Land Uses / Evaluation Criteria		Housing	Industry	Agriculture	Forestry	Fac. Rect.	Nat. Rect.	Energy Fac.	Other	Housing	Industry	Port Fac.	Fishing	Fac. Rect.	Nat. Rect.	Energy	Other	Landfill	Housing	Harbor Fac.	Comm. Fishing	Recr. Boating	Windsurfing	Other	
SAFETY	Earthquake/Tsunami Hazards	X	X	+	+	O	+	X		X	X	X	O	O	+	X		X	X	O	O	O	O		
	Flood/High-tide Hazards	X	X	O	+	O	+	X		X	X	O	O	X	+	X		X	X	O	O	O	O		
	Landslide Hazards (due to ppt)	X	X	O	+	X	+	X		X	X	X	O	X	+	X		NA							
	Explosion/Fire Hazards	X	X	+	+	X	+	X		X	X	X	O	X	+	X		X	X	X	X	X	X		
	Other																								
WELFARE	Aesthetic/Scenic Values	O	X	O	+	O	+	X		O	X	X	O	O	+	X		X	O	X	O	+	+		
	Scientific Values	X	X	X	O	O	O	X		X	X	X	O	O	O	X		X	X	X	O	O	+		
	Cultural/Historic Values	O	X	O	O	O	+	X		X	X	X	O	O	+	X		X	X	X	O	+	+		
	Educational Values	O	X	O	O	O	+	X		X	X	X	O	O	+	X		X	X	O	O	O	O		
	Other																								
HEALTH	Aquifer Recharge Areas	X	X	O	O	O	+	X		X	X	X	+	O	+	X		X	X	O	O	+	+		
	Biologically Productive Areas	X	X	X	O	X	O	X		X	X	X	O	+	O	+	X	X	X	X	+	+	+		
	Air Pollution Pot.(Airshed)	X	X	O	+	O	+	X		X	X	O	+	O	+	X		X	X	O	+	+	+		
	Other																								

Key
Severe Conditions= X
Moderate Conditions= O
Limited Conditions= +
Not Applicable= NA

Figure 14. Land and water use guidelines for prospective uses in the Tokyo Bay area.

7.1.2.5 Toward a Basis for Land- and Water-Use Guidelines

The final step in this study was to develop some general guidelines for all prospective uses in the study area that might be included in any bay area regional plan. The bay area has already been defined in terms of safety, amenity, and health/welfare criteria, which are among those considered essential for ecological planning and environmental management throughout the study area, terrestrial, marine, and coastal areas all included. A broad variety of land and water uses are possibly being done, including housing, industry, agriculture/fishing/forestry, recreation, energy generation, and others. Each must be carefully planned, from its location through its site preparation, construction, and daily use phases so as not to create irreversible negative impacts on the environment or surrounding uses. The data shown in Figure 14 indicate a set of general conditions, in terms of severity and cost, which are considered to be necessary to achieve the overall objectives of *sustainable development,* that is "development that meets the needs of the present without compromising the ability of future generations to meet their own needs." This means, in other words, improving the quality of human life while living within the carrying capacity of supporting ecosystems.[68]

Depending upon the exact location and specific plans for each use, detailed performance standards can be developed for each development phase to fit any future scenario. This holds true for hazardous wastes as well, the strict legally binding regulation and handling of which must be controlled so as to protect life, property, and the environment in both short and long terms. Anything less would be contrary to the objective of *sustainability*. Recently announced plans to develop domestic legislation to compliment the Basel Convention on the international transport of toxic waste is a step in the right direction[69] and should have positive impacts on the Tokyo Bay area.

REFERENCES

1. Landscape Architecture Research Office, Graduate School of Design, Harvard University, *Three Approaches to Environmental Analysis*, The Conservation Foundation, Washington, D.C., 1967, chap. 3.
2. Shapiro, H.A., Isobe, Y., Ito, T., and RPT Inc., Ecological planning: its method and application, Part I, *Kenchiku Bunka*, 30, 43, 1975.
3. Shapiro, H.A., Isobe, Y., and RPT Inc., Ecological planning: its method and application, Part II, *Kenchiku Bunka*, 32, 23, 1977.
4. Shapiro, H.A., What happened in the introduction of the McHargian Method to Japan, *Landscape Architecture*, 69, 575, 1979.
5. Shapiro, H.A., Yodo Region Ecological Planning Study, University of Pennsylvania, Graduate School of Fine Arts, unpublished, 1970.
6. Shapiro, H.A., Research on environmental planning for coastal urban ecosystems, Part III: a Chinese case study, in *Integrated Studies in Urban Ecosystems as the Basis for Urban Planning III)*, Obara, H., Ed., Seibunsha, Chiba, Japan, 1988, 115.
7. Shapiro, H.A., Environmental planning in Asia, in *Environmental Resources Discussion Meetings Report, No. 3*, Okayama University Environmental Management Center, Ed., Kowa, Okayama, 1991, 22.
8. McHarg, I.L., Human ecological planning at Pennsylvania, *Landscape Planning*, 8, 109, 1981.
9. Tsuneta Yano Memorial Society, Ed., *Nippon: A Charted Survey of Japan, 1990–1991*, Kokusaisha, Tokyo, 1990, 26.
10. Local Government Research Association Data Center, Ed., *Local Government Yearbook, 1991*, Dai-ichi-hosei, Tokyo, 1991, 245.
11. Shapiro, H.A., The coastal zone atlas as a potential tool for citizen participation in Japanese coastal area planning, in *Proc. Int. Symp. Ocean Space Utilization, Vol. 2*, Kato, W., Gerwick, B.C., Homma, M., Lenschow, R., and Magoon, O.T., Eds., Springer-Verlag, Tokyo, Berlin, 1985, 58.
12. Shapiro, H.A., Coastal area management in Japan: an overview, *Coastal Zone Manage. J.*, 12, 19, 1984.
13. Kikuchi, T., *History of Tokyo Bay*, Dai-nihon-tosho, Tokyo, 1974, 44.
14. Emura, M., Ed., *The Active Faults of Japan*, Tokyo University Press, 1980, 139.
15. Kikuchi, T., *History of Tokyo Bay*, Dai-nihon-tosho, Tokyo, 1974, 25.
16. Kikuchi, T., *History of Tokyo Bay*, Dai-nihon-tosho, Tokyo, 1974, 18.
17. Hamada, R. and Utsugawa, T., Relief, soil and soil bacteria in the Futtsu Cuspate area along the Tokyo Bay coast, in *Chiba Baycoast Cities Project, Part III*, Numata, M., Ed., Seibunsha, Chiba, Japan, 1981, 25.
18. Kikuchi, T., *History of Tokyo Bay*, Dai-nihon-tosho, Tokyo, 1974, 22.

19. Tokyo Bay Citizens Council, *Tokyo Bay Observation Guidebook No. 1,* Tokyo Bay Citizens Council Office, Funabashi, Japan, 1985, 2.

20. Kaizuka, S., Naruse, Y., and Ota, Y., *Japan's Plains and Coast,* Iwanami-shoten, 1985, 97.

21. Japan Scientists Association, Ed., *Tokyo Bay,* Otsuka-shoten, 1979, 111.

22. Tokyo Bay Research Group, A Diagnosis of Tokyo Bay, in *Tokai Area Fishery Research Report C, No. 21,* Tokyo Bay Research Center, Tokyo, 1978, 6.

23. Tokyo Bay Citizens Council, *Tokyo Bay Observation Guidebook No. 1,* Tokyo Bay Citizens Council Office, Funabashi, Japan, 1985, 1.

24. Kikuchi, T., *History of Tokyo Bay,* Dai-nihon-tosho, Tokyo, 1974, 33.

25. Tokyo Bay Research Group, A Diagnosis of Tokyo Bay, in *Tokai Area Fishery Research Report C, No. 21,* Tokyo Bay Research Center, Tokyo, 1978, 15.

26. Tokyo Bay Research Group, A Diagnosis of Tokyo Bay, in *Tokai Area Fishery Research Report C, No. 21,* Tokyo Bay Research Center, Tokyo, 1978, 15.

27. Kikuchi, T., *History of Tokyo Bay,* Dai-nihon-tosho, Tokyo, 1974, 27.

28. Sakagami, K. and Hamada, R., Soils of Landfill Along Tokyo Bay, in *Integrated Ecological Studies in Baycoast Cities, Part I,* Numata, M. Ed., Sei-bunsha, Chiba, Japan, 1979, 18.

29. Tokyo Bay Fishermen, unpublished interview, 1985.

30. Ministry of Construction, *Development of a Method for Urban Fire Prevention Policy,* Ministry of Construction, Tokyo, 1982, 227.

31. Regional Planning Team Inc., *Chiba Prefecture Environmental Management Plan Basic Survey,* Chiba Prefectural Government, Chiba, Japan, 1982, 85.

32. Tokyo Bay Research Group, A Diagnosis of Tokyo Bay, in *Tokai Area Fishery Research Report C, No. 21,* Tokyo Bay Research Center, Tokyo, 1978, 37.

33. Japan Scientists Association, Ed., *Tokyo Bay,* Otsuka-shoten, 1979, 9.

34. Tokyo Bay Citizens Council, *Tokyo Bay Observation Guidebook No. 1,* Tokyo Bay Citizens Council Office, Funabashi, Japan, 1985, 11.

35. Shapiro, H.A., The landfilled coast of Japan's inland sea, in *Ocean Yearbook 7,* Borgese, E., Ginsburg, N., and Morgan, J.R., Eds., University of Chicago Press, Chicago, IL, 1989, 294.

36. Tajiri, M., *The Uncontrolled Development of the Sea,* Iwanami-shinsho, Tokyo, 1983, 3.

37. Tajiri, M., *The Uncontrolled Development of the Sea,* Iwanami-shinsho, Tokyo, 1983, 25.

38. Tajiri, M., *The Uncontrolled Development of the Sea,* Iwanami-shinsho, Tokyo, 1983, 3.

39. Murata, T., The Unknown Hazard Pollutants in Tokyo Bay, in *The Protection and Restoration of Tokyo Bay,* Tajiri, M., Ed., Nihon-hyoronsha, Tokyo, 1989, 183.

40. Murata, T., The Unknown Hazard Pollutants in Tokyo Bay, in *The Protection and Restoration of Tokyo Bay,* Tajiri, M., Ed., Nihon-hyoronsha, Tokyo, 1989, 184.

41. Akiyama, K., This is Happening to the Tokyo Bay Ecosystem, in *The Protection and Restoration of Tokyo Bay,* Tajiri, M., Ed., Nihon-hyoronsha, Tokyo, 1989, 210.

42. Ruddle, N. and Reich, M., *Island of Dreams,* Autumn Press, New York, 1975, chap. 4.

43. Booth, W., Action urged against global warming: scientists appeal for curbs on gases, in *Washington Post,* February 2, 1990.

44. Hoffman, J.S., Titus, J.G., and Keyes, D., *Projecting Future Sea Level Rise,* U.S. Environmental Protection Agency, Washington, D.C., 1983, chap. 1.

45. Jones, C.B., *Sea Level Rise: Assessing the Scientific Debate,* Pacific Basin Development Council and University of Hawaii Social Science Research Institute, 1989, 4.

46. Wuebbles, D.J., A Proposed Reference Set of Scenarios for Radioactively Active Atmospheric Constituents, U.S. Department of Energy, Washington, D.C., 1984, chap. 1.

47. Jones, C.B., *Sea Level Rise: Assessing the Scientific Debate,* Pacific Basin Development Council and University of Hawaii Social Science Research Institute, 1989, 10.

48. Jones, C.B., *Sea Level Rise: Assessing the Scientific Debate,* Pacific Basin Development Council and University of Hawaii Social Science Research Institute, 1989, 10.

49. Aubrey, D. and Enery, K., Relative sea level rise of Japan from tide-gaige records, in *Geol. Soc. Am. Bull.,* 97, 194, 1986.

50. Japan Environment Agency General Office, Ed., *Dictionary for the Global Environment,* Chuo-hoki, Tokyo, 1990, 36.

51. Guralink, D.B., Ed., *Websters New World Dictionary,* Wm. Collins and World, Cleveland, 1976, 1070.

52. Fujiwara, Y., Landfill and trash: what should be done?, in *The Protection and Restoration of Tokyo Bay,* Tajiri, M., Ed., Nihon-hyoronsha, Tokyo, 1989, 249.

53. Murozaki, M., Thinking about the trash problem, *Setonaikai,* 24, 12, 1991.

54. Kikuchi, T., *History of Tokyo Bay,* Dai-nihon-tosho, Tokyo, 1974, 27.

55. Japan Environment Agency, Tokyo Bay: Toward Its Protection and Creation, Ministry of Finance, Tokyo, 1990, 23.

56. Friends of the Earth, Japan, International Wetland Symposium, 1977-Tokyo, *Japan Environmental Monitor,* 5-2, 4, 1992.

57. Clark, J., Coastal Ecosystems, The Conservation Foundation, Washington, D.C., 1974, chap. 1.

58. Japan Environment Agency, *Tokyo Bay: Toward Its Protection and Creation,* Ministry of Finance, Tokyo, 1990, 42.

59. Mitsudera, M., The urban atmosphere along the Tokyo Bay Coast, in *Integrated Ecological Studies in Baycoast Cities, Part I,* Numata, M., Ed., Seibunsha, Chiba, Japan, 1979, 23.

60. McHarg, I.L., *Design With Nature,* Natural History Press, Garden City, NJ, 1971, 64.

61. Tokyo Bay Citizens Council, *Tokyo Bay Observation Guidebook No. 1,* Tokyo Bay Citizens Council Office, Funabashi, Japan, 1985, 13.

62. Kikuchi, T., *History of Tokyo Bay,* Dai-nihon-tosho, Tokyo, 1974, 195.

63. Kikuchi, T., *History of Tokyo Bay,* Dai-nihon-tosho, Tokyo, 1974, 196.

64. Fujiwara, Y., Landfill development in Tokyo Bay: how it occurred, in *The Protection and Restoration of Tokyo Bay,* Tajiri, M., Ed., Nihon–hyoronsha, 1989, 78.

65. Kupfer, D. and Karimanzira, R., Agriculture, forestry and other human activities, in *Climate Change*, Intergovernmental Panel on Climate Change, Island Press, Washington, D.C. and Covelo, California, 1991, 73.

66. Shapiro, H.A., Comprehensive environmental management and planning for the bayshore cities in Osaka Basin, in *Chiba Baycoast Cities Project, Part III*, Numata, M., Ed., Seibunsha, Chiba, Japan, 1981, 128.

67. Shapiro, H.A., Comprehensive environmental management and planning for the bayshore cities in Osaka Basin, in *Chiba Baycoast Cities Project, Part III*, Numata, M., Ed., Seibunsha, Chiba, Japan, 1981, 128.

68. I.U.C.N., U.N.E.P., and W.W.F., *Caring for the Earth*, I.U.C.N., Switzerland, 1991, 10.

69. Anon., Procedure proposed for restricting the import and export of hazardous waste, *Asahi Shimbun*, 3 June, 1992, 2.

CHAPTER 7.2

Technology Transfer of Wastewater Treatment Facility to a Developing Country

Toshikazu Akita

TABLE OF CONTENTS

7.2.1 INTRODUCTION

Japan's GNP (gross national product) reached 410 trillion yen in 1990, and the country has long been one of the most economically advanced countries in Asia. However, it is still mentioned that Japan should make further efforts to become a country in which politics, economy, and culture are well balanced. This is because environmental problems, for example, are no longer domestic concerns within a single country; their global effects are expanding as the economic size of a country expands, whether the effects of that expansion are good or not. It is true that coexisting on the same planet are advanced countries, which consume a huge amount of energy and discharge waste, and lesser developed countries, which have difficulty even in meeting the daily demand for food. Therefore, it is now important for all human beings on the earth to make efforts toward coexistence and coprosperity and to preserve a sound environment, not only now, but also in the future.

The fact that Japan's ODA (Official Development Assistance) became the largest in the world in 1990 indicates the level of Japan's aid to developing countries. However, like Japan, which has made serious mistakes as well as great achievements as a result of having put so much weight on economic development, developing countries are now about to follow the same course of exploding populations in the cities and pollution of water, air, and soil in parallel with their own hasty development. Economic development and environmental problems are as closely related as are two wheels of a car, but the history of the world clearly shows that countermeasures for environmental problems have always been taken, not before, but after the problems have arisen. In this, Japan was no exception. Moreover, the gap between the regulations and the reality can often be surprisingly observed. It is becoming increasingly important for Japan, as a country having advanced pollution control technology, to provide economic and technical aid, such as educating personnel, constructing relevant facilities, and transferring technology in order to prevent further expansion of environmental destruction.

NEC Environment Engineering Ltd. has been developing and marketing integrated wastewater treatment systems based on the heavy metal ion removal technology (Ferrite Treatment Technology) developed in 1972. As a result, the ferrite treatment method has been the method most utilized at wastewater treatment facilities in universities and research institutes in Japan since 1975. In addition to the technique's technical superiority, a well-organized support system for operation and maintenance of the facilities is considered a key factor contributing to the acquisition of such a high market share, despite its slight complexity. Though wastewater from universities and research institutes characteristically varies in nature and type, hard-to-treat wastewater can be handled by the ferrite treatment method, and technical service staff are always available for instructions. We consider knowledge acquired on-site as the most valuable. In addition, research was also conducted to develop technologies for making effective use of sludge (a by-product of ferrite) formed as a result of the ferrite treatment; the results of this research were presented by T. Tsuji in 1984 and drew great attention from people in the field.

We are making every effort to preserve the environment by introducing this technology not only in Japan but also abroad. As an example, discussed herein is a case of technology transfer, including the design, construction and operation, and management of ferrite treatment facilities, carried out under an agreement between a Korean engineering company and NEC Environment Engineering Ltd.

7.2.2 A CASE IN SOUTH KOREA

7.2.2.1 Present Situation in South Korea

South Korea achieved economic development by rapid industrialization and urbanization in the 1960s, but was faced with serious pollution problems in the 1970s. After the environmental preservation law went into effect in 1977, policies designed to balance the economy with the environment were implemented in the 1980s. South Korea is now regarded as a leader among the Newly Industrialized Countries. It also initiated improvement and/or installation of pollution control facilities at universities and research institutes during the 1980s, and it was at that time that many people from the government, universities, and industries visited Japan to learn about pollution control technologies.

At the request of the South Koreans, leading figures from Japanese universities, which had succeeded in the systematic operation of pollution control facilities, visited that country to lecture or provide instructions there.

Currently, the South Korean Environment Agency, which is the regulatory body for environmental affairs, has about 1000 staff members at its head office, branch offices, and research institutes. According to a 1988 statistical survey conducted by South Korea's Ministry of Education, the total number of universities and college is about 430, of which about 20% are national universities, while the total number of students stands at 795,000, of which approximately 25% are students of national universities.

7.2.2.2 Progress of Technical Assistance to the University of Seoul

Various technologies having different features are employed at wastewater facilities to treat various types of wastewater from universities and research institutes.

The ferrite treatment process, as discussed in Chapter 6.2, is a relatively new technology derived from electronic material production techniques. Since the first facility was delivered in 1975, a number of others have been supplied to research institutes at many universities. From around that time, many people from the South Korean government and private sector started visiting Japan to study Japanese regulations relevant to environmental protection and to inspect pollution control facilities. Enterprises, eager to acquire pollution control skills, sent their engineers to Japan for long periods of time to study them at universities and companies. It was about that time that ferrite treatment, which was becoming a major heavy metal wastewater treatment method in Japan, drew the attention of a Korean water treatment engineering company, which in 1980 was granted a license by our company to introduce the technique in South Korea. NEC had successfully commercialized the facilities utilizing the method. As a result of joint efforts by Japanese and Korean engineering companies, a wastewater treatment facility was delivered to the University of Seoul in 1982 and has been operating smoothly since then. For wastewater treatment at universities in particular, it is important to establish a good organization and to carry out the operations of the treatment facility and train personnel concerned in accordance with each university's basic concept of environmental control. We explained to the South Koreans that it is necessary for an engineering company not merely to design and construct a facility but also to be able to make comprehensive proposals on the planning, organization, operation, and personnel training. It has proved to be what our South Korean counterparts needed.

7.2.2.3 Actual Technical Cooperation

The environmental control program at the University of Seoul, which is the largest academic institution in South Korea, was carefully planned from 1980 or even before so that it could be the model for other education and research institutes in that country.

Following repeated technical reviews by both the Japanese and Koreans, the adoption of the ferrite treatment method, which was the method most employed at the time (1980), was chosen. The facility was then designed and constructed jointly with our company which was familiar with developing treatment systems utilizing this technique.

After receiving information on the various faculties at the University of Seoul — the number of students, relevant regulations, etc. from the Koreans — the actual design and construction were begun. The basic design, an engineering flow sheet, layout drawings, reference civil drawings, and other documents were prepared with the initiative taken by the Japanese side while at the same time providing instructions to Korean engineers. Working drawings were prepared mainly by the Koreans and were checked by the Japanese experts. After a review of the design from various aspects, a basic treatment flow sheet very similar to those adopted at Japanese universities was selected. The overall treatment can be roughly divided into two processes: (1) the organic wastewater treatment process and (2) the inorganic wastewater treatment process. Waste scrubbing water from the organic wastewater incineration process is introduced to and treated in the inorganic wastewater treatment process and then discharged. However, even though the separate collection, followed by the separate treatment of organic and inorganic wastewater, is considered preferable from the viewpoint of treatment efficiency, the faithful practice of separate wastewater collection is not always followed, based on our experiences at several Japanese universities. For these reasons, some universities are now adopting a method by which both organic wastewater and inorganic wastewater are treated together by atomized incineration, and the waste scrubbing water is treated by the ferrite method before being discharged.

Actual construction work at the University of Seoul was performed by the Korean engineering company, and just before the completion of the work, a design engineer, a construction engineer, and an operation and management engineer were dispatched from Japan to conduct a final check of the construction, to run a general test of the facility, and also to instruct the Korean counterparts at the same time. The university had not yet established its organization for the actual operation and management of the facility, so discussions with the university were made through a professor who was expected to become chairman of the environmental safety committee. It is regrettable that we could not directly instruct the personnel who would be in charge of the operation since we have always found that direct instruction is the most effective way. However, it is due to the efforts of both our Korean counterparts and the university that the facility at the University of Seoul was described in an academic journal (AAWT, July 13, 1989) as a fine example of a large wastewater treatment facility functioning successfully in a Korean university.

Among the features of wastewater from universities and research institutes are small amounts, various kinds, high concentration, and large variations in annual effluent. In addition, the number of new types of pollutants in the wastewater are increasing year by year. We consider it of great importance to provide not only the facility itself, but also the operation and management methods, to deal with wastewater of irregularly changing quality. The broad popularity of ferrite treatment facilities in Japan, despite their relatively difficult operation and management, can be attributed to the fact that we have not only designed and constructed the facilities, but also have continued to cooperate with clients to build on our knowledge of treating various types of wastewater and also of advising them on operating and managing the facilities. In fact, of two facilities with a treatment capacity of 2000 l per batch supplied to two different universities, the one designed and constructed by a contractor having no knowledge of technology on or experiences with wastewater treatment is now not operating and is in a state of closure, while the other one, for which we provided instructions, is reported to be operating normally.

Following this example, we instructed the Korean engineering company to send their engineer for one year of training in the operation and management

Table 1 Classification of Inorganic Wastewater for Separate Collection

Symbol	Wastewater	Classification for storage
Hg	Mercury	1. Organic mercury
		2. Inorganic mercury
CN	Cyanide	3. Cyano-complex
		4. Cyanide
P	Phosphoric acid	5. Phosphate
F	Fluorine	6. Fluoride
M	General heavy metal	7. General heavy metal
		8. Acid
		9. Alkali

of a wastewater treatment facility at some of the universities in Japan. During this training, he was instructed in the separate collection system of wastewater, operation of the system, maintenance procedures, safety management, research systems, analyzing techniques, and other factors. Shown in Table 1 are the classifications of wastewater for separate collection adopted at the universities where the trainee was actually instructed.

7.2.2.4 On-Site Training

Unlike conventional neutralizing coagulation and sedimentation, ferrite treatment technology is not discussed in technical books dealing with ordinal wastewater treatment methods. Thus, the training program for the Korean engineer was started with the learning of the basis of ferrite reaction as depicted in Figure 1.

Then, the trainee learned: (1) the difference of products — Fe_3O_4, Fe_2O_3, $FeOOH$, $Fe(OH)_2$, etc. — formed by oxidation of solution, depending on oxidizing conditions; (2) how to check the progress of the reaction by pH meter and oxidation-reduction potential (ORP) meter; (3) the crystal structure of ferrite; (4) the effects of coexisting ions on ferrite reaction; and (5) the principle, type, and feature of magnetic separators.

In parallel with the learning mentioned above, the trainee studied the following treatment methods for CN wastewater, F wastewater, and Hg wastewater, all of which are included in inorganic wastewater: (1) two-step degradation of CN ion by hypochlorous acid; (2) calcium treatment of F ion; and (3) removal of Hg ion by ion exchange resin.

For organic mercury wastewater, the trainee also learned oxidizing degradation treatment using $KMnO_4$ and how to confirm the progress of the oxidation reaction by using a pH meter and ORP meter. During the training program, we demonstrated the progress of a reaction using a 1-liter glass reaction cell in our laboratory, whenever necessary, so that the trainee could visually realize the necessity for the separate wastewater collection.

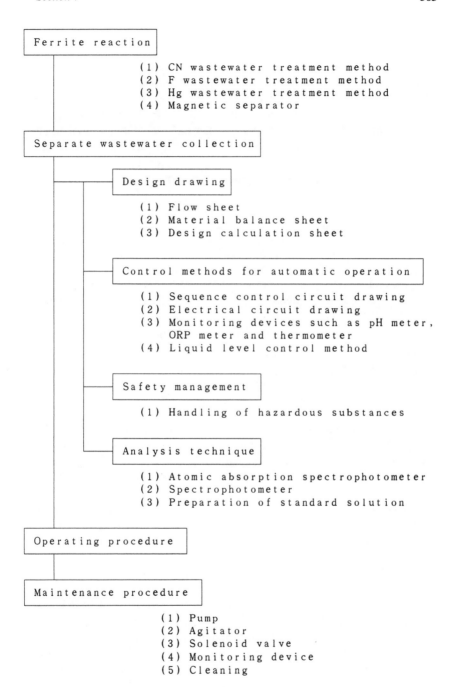

Figure 1. Training program (items learned) for operation and management of wastewater treatment facility.

Then, as hands-on training, the trainee learned the full-scale operation of facilities actually in use in our plants and universities. The training was conducted in the following sequence: (1) observation of all the operations required at the facility, with explanation given by the operators of the facility; (2) actual charging of iron sulfate and wastewater from polyethylene containers to the designated tanks with instructions given by the operators; (3) manual operation of the facility, confirming the status of the treatment through all steps of each process (the trainee was able to understand all the processes after repeating the operation several times); (4) automatic operation of the facility so that the trainee could develop his sense of actual system operation; (5) maintenance procedure required for the long and stable operation of the facility; and (6) cleaning, disassembling and reassembling of pumps and automatic valves, and replacement and calibration of the pH meter electrode, ORP meter electrode, and other instruments.

To deal with possible troubles involved with electrical circuits, it is also important to be able to read sequence control circuit drawings and to be knowledgeable about monitoring devices such as a pH meter, ORP meter, thermometer, and level switch. Understanding the whole system requires studying the flow sheet, material balance sheet, and design calculation sheet. Essential for the safety of the operators is education in the prevention of accidents, safety management, and hygiene, which includes learning about dangerous substances, hazardous substances, poisonous substances, relevant (Japanese) regulations for environmental protection, and the preparation of a checklist of mechanical and electrical equipment.

The trainee learned the analysis methods specified by the Japanese Environmental Agency using most types of analysis equipment, including an atomic absorption spectrophotometer and a spectrophotometer. With about one year of such training, an individual is capable of operating a facility by himself and also able to propose modifications for the facility and component equipment. However, even for NEC personnel, it is difficult to determine the cause of unsatisfactory treatment results that may occur when a facility is started up with actual wastewater. Also, on such an occasion, we realize clearly the great importance of the separate wastewater collection.

7.2.2.5 Situation in South Korean Subsequently

Thanks to the efforts of those concerned, the number of facilities in Korea has been increasing. The number of universities provided with a facility is 4 of 16 technical colleges and 8 among 23 universities, which account for 30% of the total number of colleges and universities. Even aside from the basic problem of whether the treatment should be conducted by each university individually, it is felt that further efforts must be made in the future. The current number of facilities, including those under construction, is 9 using the ferrite treatment method, 2 using the hydroxide method, and 1 using a method with heavy metal remover. Two of the above facilities utilizing the hydroxide method and one using the ferrite method are currently closed. Nevertheless, the ferrite treatment method has been the primary means adopted.

The Korean engineer who received training in both the designing and operation of the treatment system put his experience to good use after his return to South Korea by advising people concerned at the university on how to set up the organization prior to delivery of the facility and then by giving instructions in the operation, maintenance procedure, and analysis procedures after the facility had been delivered. The Korean design engineer played a leading role in the system design of the other facilities, for which the Korean engineering company received the orders, and also helped raise the nationalization index in the facilities in South Korea. For example, even in the first facility constructed at the University of Seoul in 1982, the construction was carried out with the definite policy that Korean products were to be used as much as possible if parts and equipment could be locally designed and manufactured. Only monitoring equipment such as the pH meter and a ferrite reactor, requiring complex special design know-how, and a magnetic separator, requiring advanced technology in its manufacturing, were imported from Japan.

As is generally known, now that South Korea has advanced in its technical development, everything can be locally produced except the magnetic separator. This is how the first wastewater treatment facility at a Korean university was materialized through technology cooperation.

7.2.2.6 The Case of University C in South Korea

Fortunately, we had a second chance to engage in technology cooperation for a facility at University C, which was completed in December, 1990. The fact that the newly constructed treatment facility building houses an analysis room, a laboratory, and a resting room testifies to the strength of the university's commitment to environmental protection.

During this period of cooperatively working together, the Korean engineering company made significant progress in hardware manufacturing technology; thus, all the design and construction work could have been carried out by Korea alone except for the magnetic separator and some program controllers which would still have been difficult to manufacture in South Korea. All NEC did was to check on the engineering flow sheet prepared by its Korean counterpart.

As employees often move from company to company in South Korea, however, it is difficult for South Korean firms to transfer technologies and know-how of hardware and software to replacement engineers. In fact, when the facility at University C was due to be started, the engineer who was in charge of the operation and management of the facility at the University of Seoul had resigned from the company, so that we were asked by South Korea to dispatch an engineer to oversee the running test and general checkup of the facility again. We sent a trained engineer who spent 10 d in this activity.

The treatment facility in University C requires relatively less strict separate wastewater collection because the facility is designed to treat wastewater by incineration followed by ferrite treatment of waste flue gas scrubbing water, no matter what the type of wastewater may be. The facility is also designed to allow the conversion of treatment from the current method to one by which ordinal organic wastewater and inorganic wastewater are independently treated, provided the separate wastewater collection system works well.

The workmanship demonstrated in the facility at University C was fairly good except for several minor problems believed to have been caused by lack of information resulting from the resignation of the operation and management engineer of the engineering company. The problems included: (1) dead space in the ferrite reactor (including piping) resulting in the presence of unreacted wastewater; (2) lack of an agitation system even though the feed water to the magnetic separator must be homogeneous; and (3) a design system which was inadequate for automatic operation.

Of the total days spent on-site for correction, more than half were spent in modifying the sequence circuits for the automatic operation. Since the sequence problems can be attributed to the sequence circuit design engineer not being thoroughly familiar with the treatment system, it is presumed that not all of the needed relevant information was transferred by his predecessor. However, since he was fully familiar with the operation of the programmable controller (a Japanese product), he was able to modify the circuit immediately after we pointed out what was wrong.

As mentioned above, there were very few problems with the design and construction of the facility, but it was still in an early stage of maturity due to lack of both relevant information and experienced personnel in charge of operation and management. We believe, however, that South Korea will be able to design a highly sophisticated facility after gaining further experience in the near future.

7.2.3 FUTURE COOPERATION

A review of environmental controls implemented by developing countries indicates that learning from history is as important for a country as standing alone in terms of economy and manpower.

Japan is a country with a valuable history of environmental problems, and whether the solutions have been good or not, it should disclose its history and make available its pollution control technologies to foreign countries and should also learn other things from those countries. The reason is that environmental problems are no longer merely domestic problems of a single country, but are global and are releated to energy and resource problems.

As mentioned in Section 7.2.2, we realized strongly that it was necessary for us not only to propose the appropriate treatment system and equipment but also to help establish the organization required for the smooth functioning of the facility, and also to provide technical cooperation in the domain of software for its operation and maintenance. We had developed the technologies employed through a succession of failures and successes and also through the continuous accumulation of the most recent technical information.

Naturally, the situation of each country is different. In South Korea, which is a leader among the Newly Industrialized Countries rather than being a still developing country, technical cooperation could be conducted smoothly. In other developing countries, greater difficulties would be anticipated in the cooperation necessary in the domain of both hardware and software. It is necessary to promote the international exchange of information to try and minimize such difficulties, including information on problems already experienced, those currently being experienced, those which can be expected in the future, and those unique to each country.

To resolve these hurdles, further efforts by those engaged in research and education are essential.

CHAPTER 7.3

Hazardous Chemical Control in a High Technology Environment: The Global Environmental Impact of Organometallics

Takashi Korenaga and Miyoko Izawa

TABLE OF CONTENTS

0-8493-682-5/94/$0.00 + $.50

7.3.1 INTRODUCTION

Since the beginning of recorded history man has utilized metals as conventional chemicals. Our use of manufactured chemicals has continued to accelerate in our lifetime in both quantity and number, and more than ten million types are now registered in the Chemical Abstract. Approximately 100,000 of those are produced for commercial products and used all over the world. Between 500 and 1000 new chemicals are added to the list each year.[1,2]

Chemicals, including metals, allow us to enjoy many benefits of modern living; however, we seem to continuously face an uneasy counterpoint of benefits and risks here in the chemical age. A primary example of the problem is the large quantities of pesticides used in agriculture worldwide, which make the control of parasites easy, but also release hazardous industrial chemical waste into the environment.

Chemicals are entering our environment directly and indirectly, and the threat to human health is tremendous. The level of danger to humans after the chemicals have entered the environment depends on two things: (1) how toxic the chemicals are and (2) the extent of human exposure. But regardless of these factors, the bottom line being to adequately protect human health, more advanced risk assessment is urgently needed that target control of the applied chemicals. In this area we not only need more information about the acute toxicity of each chemical, we also need to discuss the complete fate of the chemical after it has entered the environment. In addition, we need to unify the information available throughout the world. Vital reports and studies are in files and repositories in many countries, and without an adequate data base system to bring them together efficiently, they will remain scattered.

The subjects of discussion should therefore be extended to related areas — treatment technology, waste management, detection methodologies, policy-making for sustainable development, and different legal regulations of developed and developing countries. In this section, we give an overview of those subjects, with our main focus placed on global circulation and/or the fate of applied chemicals.

We have attempted to assess the present status and future plans for chemicals from the point of view of the impact of organometallics on the environment. Only when it is necessary will inorganic compounds be discussed. In addition, we will site the prediction for newly applied chemicals in the environment, which are based on data on earlier hazardous chemicals.

7.3.2 GENERAL SCOPE OF OCCURRENCE AND IMPACT OF APPLIED CHEMICALS ON THE ENVIRONMENT

7.3.2.1 Chemicals Discharged in Large Quantities

When chemicals are applied to the environment in large quantities, the risk of pollution increases proportionally. The complete history of applied chemicals is beyond the scope of this paper, but a few examples of organometallics are justified.

One of the oldest chemicals, organomercury or mercury, has long been utilized in large quantities and for many purposes because of its usefulness. However, we recognized very early that it is a serious environmental pollutant.[3] Organoleads have also been used in large quantities in gasoline additives. Worldwide, about 200,000 tons of lead ethyl used as an antiknock compound was produced annually beginning around 1974,[4] and subsequently, the environmental damage began to appear.

One of the most recent concerns involves new hazardous pollutants from organotins used in antifouling paint on ships or boats. The organotins used in the paint enter the aqueous environment directly.

Nor can be overlook the effect of several types of hazardous pesticides as forms of heavy metal, organometallics, or organic chemicals used in modern agriculture in most countries. The use of chemical pesticides has been rapidly growing in developing countries, as in industrialized countries. Some of the organomercuries, leads, tins, and arsenics have also been utilized as effective pesticides. As has been reported so many times, we cannot escape the seemingly inevitable accidents involving these and other toxic chemicals.

Another sobering concern is the amount of waste disposal from consumer products: this is expected to double by the end of this century and to double again before the year 2025.[2] Land disposal of hazardous waste is still being

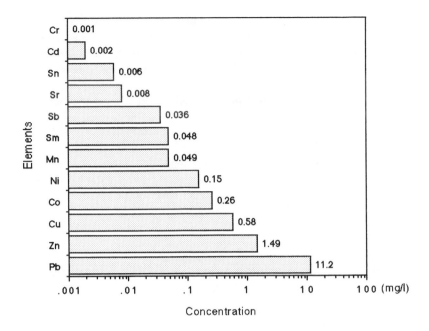

Figure 1. Elements extracted with acid from portable products.

carried out in most countries, making it important to know what kind, and how much of, the hazardous elements will shift to the environment. A primary example of this waste is used dry cells or fluorescent lights containing mercury. There is a possibility that the mercury will reenter the environment and be broadly distributed over time. Recently, Yasui[5] found that abandoned car batteries and solder (Pb-Sn alloy compounds) from electrical appliances contain a considerable amount of toxic lead or tin compounds (Figure 1). He chose the elements of Be, Mg, Al, Sc, Ti, V, Cr, Mn, Co, Ni, Cu, Zn, Ga, As, Se, Rb, Sr, Y, Zr, Nb, Mo, Ru, Rh, Pd, Ag, Cd, In, Sn, Sb, Te, Cs, Ba, La, Ce, Pr, Nd, Sm, Eu, Gd, Tb, Dy, Ho, Er, Tm, Yb, Lu, Hf, Ta, W, Re, Os, Ir, Pt, Au, Hg, Tl, Pb, and Bi and analyzed these as possible hazardous elements. The excudation of harmful chemicals from such abandoned products can be dangerous if they leak into the water sources. Yet, despite the information available and the undeniable risks, chemicals are continually being released into our environment, seemingly evading society's control.

7.3.2.2 The Legal Regulations and Legislation

Integrating environmental issues into tangible solutions to chemical waste problems will require more extensive information gathering on our part. If we expect governments to take us seriously and establish credible environmental policies, we must improve the ways we assess environmental risks and benefits. The use of hazardous chemicals has been banned in some countries, but often not completely. Moreover, many developing countries are now industrializing, and their volume of hazardous waste is growing rapidly as the use of pesticides increases. Regulations controlling chemical waste have also become more numerous, but at a much slower pace, and while technology and other methods are available to help developing countries, current methods to transfer assistance to them are inadequate.

7.3.2.3 Chemicals Discharged in Small Amounts and the Variety of Kinds from High Technology

One problem that is proving very difficult to combat originates from high technology processes, such as microelectronics, fine chemicals, and the advanced chemicals industry where chemicals are being applied in combination. This mix of chemicals produces a substance made up of a variety of different compounds, each one in only a small amount.

Conventional laws (water pollution control laws, etc.) were not designed to cope with a mix like this; they only target specific chemicals discharged in large quantities. Moreover, regulations are only concerned with the acute toxicity of newly applied chemical substances, and information and data on the toxic effects are not available on 79% of the more than 48,000 chemicals approved by the U.S. Environmental Protection Agency (EPA).[2]

7.3.2.4 Treatment Technology Toward the High Technology Industry

In view of the need for environmental assessment of high technology waste, we have already worked out various behaviors of complexed and mixed wastewaters.[6] We also researched and summarized current treatment and reuse of the wastewaters in highly advanced technological firms in Japan.[7] About 27,000 specified factories were investigated, and the results showed that the effluent character of advanced technology waste resembles that of hospitals and universities.

The main points of the problem are as follows:

1. Variety of chemicals, various forms and small quantities, irregular and seasonal discharging of waste
2. Many original effluents (various manufacturing processes, laboratories located over wide areas, etc.)
3. Multiple, random uses of various types of chemicals
4. Synthesis of newly developed material and mutagens, disease germs
5. Difficulty in continuous monitoring of discharges
6. Increase in use of chemicals not easily decomposed (organometallics involved)
7. Wastewaters mixed with organic substances (making treatment difficult)
8. Methylation of heavy metals might even occur in the manufacturing or the treatment process (authors' speculation)
9. Complex items

Organometallics are one of the chemicals difficult to decompose. We earlier reported a new method of treatment for cacodylic acid (dimethyl arsonic acid) as an example of a persistent substance.[8] Chemical oxidative degradation of organoarsenics is very difficult, and if one of these (such as cacodylic acid) is intermixed into wastewater, a conventional treatment method like coagulation sedimentation cannot be used.

It was proven that cacodylic acid degrades completely to inorganic arsenic ion by the application of Fenton's reaction;[9] using this reaction as an effective pretreatment, a process to degrade organoarsenics has been developed. We previously reported that oxidation treatment of photographic wastewaters was effectively achieved; Fenton's reaction easily and completely degraded the organic compounds which were usually difficult to decompose.

It is therefore obvious that an optimized new treatment technology and reuse system should be developed for wastewaters from high technology industry.

7.3.3 ASSESSMENT METHOD FOR CONTROLLING APPLIED CHEMICALS DISCHARGED INTO THE ENVIRONMENT

7.3.3.1 Impact of Organometallics

In this section we focus on limiting the use of organometallics and environmental pollution and discuss some aspects of assessing the control of applied chemicals.

In general, once chemicals are released directly or even indirectly into the environment, they become available for final distribution into the human environment and living cells. The occurrence of organometallics in, or under, environmental conditions can be traced to two areas: (1) when they are directly or indirectly used as commercial products and (2) when they are formed in, or under, environmental conditions. Derivatives formed are often higher in toxicity than their parent inorganic metal or ion, and their solubility in lipid tissue means a much larger residual period than metallic ions in organisms. The microbial synthesis of organometallics from inorganic precursors has been discovered for the metals Hg, Pb, Tl, Pd, Pt, Au, Sn, Cr, and As.[10] Furthermore, the decomposed products of organometallics (by methylation) is an additional way of formation.

Without a doubt, knowledge of an applied chemical's pattern of mobility is needed, including accurate information on the fate or global circulation (water, sediment, air) of the chemicals. The process moves in steps: input, persistence, and degradation. Even the possible formation of new compounds in the environment as the result of a chemical's release has to be considered.

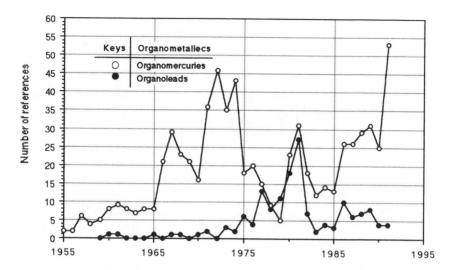

Figure 2a. Changes in reference material for organometallics and pollution (Hg, Pb) in the years 1955 to 1991.

7.3.3.2 Analysis of the Work on Organometallics and Pollution

Basically, organometallics, by definition, are those compounds containing a metal carbon bond. However, when necessary, inorganic compounds (original metals) will also be discussed here.

We earlier reviewed more than 1000 pieces of representative literature on this theme, dating from 1955 to 1988, and discussed the circulation, control, and formation of organometallics.[11] As shown in Figures 2a and 2b, we have now added approximately 400 of the latest reports from 1988 to 1991. Figure 2a charts the changes in the amount of literature produced on organomercuries and organoleads as "old" hazardous pollutants. Figure 2b follows the quantity of literature available on organotins, organoarsenics, and other organometallics concerned with pollution (in water, sediment, air).

Of the old chemicals, mercury and lead are the organometallics that have been the greatest environmental concern, and therefore, have received the most attention. Organomercuries and organoleads have been reviewed and discussed in detail,[4,11-13] and, as shown in Figure 2a, more than 160 reports are available on organomercuries; over 100 on organoleads were covered in our previous review.[11]

Our review of the literature shows that investigations have spotlighted environmental concentrations of mercury and other compounds, in addition to their commercial uses, toxicity, persistence, and methylation. Organometallics and their fate in the environment has also been the recent subject

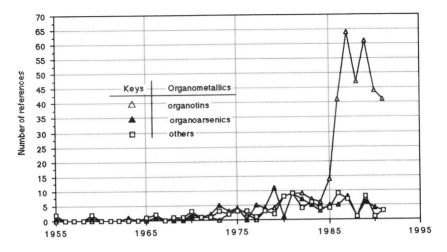

Figure 2b. Changes in reference material for organometallics and pollution (Sn, As, others) in the years 1955 to 1991.

of several independent studies,[14,15] and the World Health Organization (WHO) has published the environmental health criteria for mercury and lead.[16,17] In the West, the U.S. National Academy of Sciences has made a detailed assessment of mercury and lead in the enviornment,[3] and Figure 2a shows that much research is still being carried out.

Figure 2b charts the amount of literature produced on organotins from 1955 to the present. Organotins are recognized as a more recent source of environmental pollution, and their synthesis and application have greatly expanded since 1940. The main uses of organotins have been as stabilizers for polyvinyl chloride (PVC) and biocides; however, significant direct entry of organotins to the aqueous environment is now coming from antifouling paints which we discuss in a later section.

As shown in Figure 2b, a total of more than 210 work accounts were completed on organotins between 1955 to 1988. They were also the subject of an American Chemical Society Symposium,[18] and in 1980, WHO produced an "Environmental Health Criteria for Organotins."[19] Recently, the EPA issued an administrative report on tributyltins.[20]

In the literature we listed from 1988 to 1992 a total of more than 170 of the works were on organotins, easily leading the list of organometallics covered. A number of reviews were attempted on the organotins.[21-25] About 30 of the pieces looked at organomercuries and 17 focused on organoleads; however, very little has been attempted on organoarsenics and other organometallics.

In the latest high technological processes, organometallics consisting of a variety of elements are being used. It is important to note that their use is not limited to the well-known elements of Hg, Pb, Sn, and As, but an extensive list of other elements are also being utilized — Al, B, Ba, Be, Bi, Ca, Cd, Co, Cr, Cu, Ga, Ge, Fe, I, In, La, Li, Mg, Mn, Mo, Ni, P, S, Sb, Se, Si, Sr, Te, Tl, V, Y. Zn, etc. These and other elements are expected to be used as organometallics or as organometallic forms. More than 10,000 organometallics are now available. As yet there does not seem to have been any significant effect on the environment made by the high technology industry. In Figure 3 the metal ions of organometallics known to date are shown.

7.3.4 IMPACT OF ORGANOMERCURIES ON THE ENVIRONMENT

7.3.4.1 Assessment of Organomercuries

Mercury has long been recognized as one of the most toxic heavy metals and identified as a serious pollutant. Organomercuries are also recognized as one of the oldest hazardous chemicals in the environment, especially methylmercury. As a result, intensive research and discussions on both organomercuries and mercury metal have been carried out and reviewed numerous times since the 1950s.[3,7,11,12,15,16,26] The mobility and the fate of global organomercury circulation have also been heavily discussed in detail.[11,12,15,16] Figure 4 illustrates the way in which mercury circulates in the environment (air, water, sediment), Hg(0) is oxidized to Hg(II) under natural conditions.[27] Furthermore, Hg(II), whether discharged directly or produced from Hg(0), can be methylated by both aerobic and anaerobic bacteria.[3,11,15,16,26]

In this chapter, when reference is made to organomercuries, we mean the original inorganic metal in the form of mercury(0) and mercury(II), because the mercury discharged into the environment came from an inorganic origin. The degradation process is also cited, and it is important to note that degradation to Hg(0) is not the final step for mercury.

As shown in Figure 2a, research on organomercuries is ongoing. Organomercuries and metal have been well documented by assessments of impact studies (including environmental methylation) and detection methodologies; governmental control policies have also been introduced. However, in spite of all the documentation available, environmental risks have not been predicted well, and unfortunately, we have not yet learned our lesson from what happened in the Minamata case in Japan. Worldwide mercury contamination after the Minamata incident is shown in Table 1.

7.3.4.2 Occurrence and Impact of Organomercuries on the Environment

7.3.4.2.1 Uses, Toxicity, Input

The main uses of mercury have been as an inorganic metal in electrolytic and soda industries, or for catalysts, medicines, and pesticides. In the 1970s, indicates elements known as orgnometallics at present

Figure 3. Trends of organometallics in the periodic table of elements.

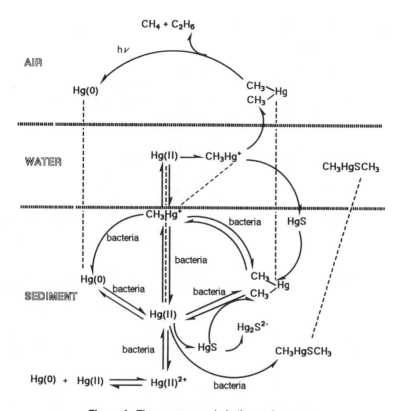

Figure 4. The mercury cycle in the environment.

Japan, Germany, Italy, England, Belgium, and other countries manufactured soda primarily using the mercury method. Several thousand of tons of mercury were estimated to have been used in that period, and most of it now found in the environment has probably come from those sources.

Japan has recently seen a reduction in the use of mercury, following the introduction of a closed system under administrative guidance. In the soda industry, introduction of the diaphragm method obviated unnecessary use of mercury. Restrictions on organomercuries as fungicides (seed dressing) and bacteriocides have also led to the reduction of their direct release into the environment. Today, with modern technology and methods, combined with a ban on usage and increasingly stringent laws, it seems the problem of these chemicals as a pollutant has been solved.

This is much the same throughout the developed world, where large quantities of mercury are no longer a major source of pollution due to strict regulations. In Japan, the effluent standard for total mercury is 0.005 μg/l, and the environmental quality standard is 0.0005 μg/l. However, even though the use of mercury has been heavily restricted and, in some cases, entirely banned in industrial nations, in developing countries many older chemicals

Table 1 Mercury Pollution in the World

Year	Place	Information
1956	Minamata, Japan	Fish contaminated with methylmercury from acetaldehyde process
1960~	Caribbean Sea	Coconut, fish, lemon contaminated from soda industry
1963	Pakistan	100 Poisoning deaths, bread contaminated with ethylmercury
1965	Niigata, Japan	Fish contaminated with methylmercury from acetaldehyde process
1966	Guatemala	45 Poisoning deaths, seeds treated with methylmercury
1967	Gahana	144 Poisoning deaths, corn contaminated with ethylmercury
1969	New Mexico and U.S.	4 Poisoning deaths, pork contaminated with ethylmercury
~1970	Sweden, Finland	Fish, birds contaminated with mercury from chlorine or pulp industry
1970	Quebec, Canada	Poisoning contaminated with organomercury from pulp, soda industry
1970~	Elbe River, Germany	Mercury pollution
1971	Iraq	4375 Poisoning deaths, wheat treated with organomercury
1974	Venezuela	16 Poisoning deaths, 200 poisoning with mercury at soda industry
1975	Bankok, Thailand	Fish or water plant contaminated with mercury from soda industry
1977	Jacarta Bay, Indonesia	Pollution in Jacarta Bay, Minamata disease symptom found
1977	Colombia	Pollution from soda industry, fishermen seek compensation
1980~	Amazon River, Brazil	Mercury contamination from purifying gold
1980	Rivers in Chirin and Heironchan, China	Minamata disease symptom found, contaminated with mercury from acetaldehyde industry
1987	Philippine	Large numbers of deaths, poisoning, and deformity of fish and domestic animals contaminated with mercury from purifying gold
1988	Dnepru River, Ukraine	Mercury pollution from light bulb industry
1988	Indian Ocean	Suspicion of ocean dumping including mercury
1990	Hancho Bay and Uencho Bay, China	Fish contaminated with mercury
1990	New Zealand	Fish contaminated with mercury
1991	Syanhai, China	Mercury pollution from medical instrument or electronics industry

including mercury are still being widely used (Table 1). There is also another possible new source for mercury pollution — mercury from used batteries or fluorescent lights, which is a threat if it reenters the environment and circulates again.

The symptoms of methylmercury poisoning are severe: sensory disturbance, tremor, ataxia, constriction of the visual field, and impaired hearing. At Minamata, mental retardation and motor disturbance developed in infants through prenatal exposure. Data are available on the acute toxicity of Hg(II) to many types, and methylmercury is the most chronically toxic compound tested,[26] but its elimination is slow.

7.3.4.2.2 Persistence

Methylmercury and other organomercuries, such as alkyl mercury or phenyl- and alkoxyalkyl-mercury, break down into organic materials and inorganic mercury,[11] but methylmercury degrades at a slower rate than phenyl or alkoxyalkyl derivatives. Since mercury has a low affinity for oxygen, and the bonds are of lower polarity than Zn–C and Cd–C, degradation is not as easy as low bond energy. The half-life of methylmercury in the human body is 70 d, compared to 4 to 5 d for Hg(II) salts.[28]

7.3.4.3.3 Environmental Methylation

Methylation may occur by an enzymatic methyl transfer process within the cell of a living organism (biomethylation). Many research projects have defined the abiotic transfer from Me[Co] (methylcobalamine) to Hg(II), and many conquer on the electrophilic transfer of Me$^-$ from Me[Co]. However, the formation of Me$_2$Hg is 6000 times slower than MeHg.[29] Where methylmercury is formed, a very low level of micrograms per liter has been involved. Therefore, several authors have suggested that there has been no real increase in organomercuries and metal concentrations in human tissues and fish during the 20th century, compared to nonpolluted locations.[11] Despite these assertions of low concentrations, however, continuous formation of organomercuries in the environment may still result in their entering the food chain, leading to much higher concentrations in organisms. The bioconcentration factor for Hg(II) is 5000, but for methylmercury it ranges from 4000 to 85,000.[26]

7.3.5 IMPACT OF ORGANOTINS ON THE ENVIRONMENT

7.3.5.1 Assessment of Organotins

Organotins are recognized as a new hazardous chemical in the environment, especially TBTs (tributyltins). TBTs are a well-documented example of the problems faced in assessing impact, introducing control policies, and predicting environmental risk. As shown in Figure 2(b), a great deal of intense research on organotins has been conducted since 1985. Some of it reflects the present status of organotins in the environment, while other reviews focus on other aspects.[25,30-36] In 1980, WHO demonstrated that low levels of TBTs (micrograms or nanograms per liter) in aquatic environments had adverse effects on marine vertebrates and invertebrates.[37] The toxicity of TBTs in aquatic organisms is well documented.[19,35,38] The observed nontoxic levels are generally 1 μg/l.[33] Despite the numerous studies, however, more are needed. The research to date has focused on the various uses and toxicity of organotins, while other facts, such as input, persistence (degradation, formation), and detection methodologies have only recently come to light.

Much work has concentrated solely on the fate of TBTs in the aquatic environment. Environmental surveys have measured levels in many areas: the Great Bay, San Diego Bay, Chesapeake Bay, Ontario and throughout Canada, the oyster culture area of the Atlantic Coast and France, the Swiss lakes, the southwest coast of England, Mediterranean ports, and the coast and bays throughout Japan.[32] There have also been studies of their persistence in water and sediment. Since TBTs exist at toxic levels in the aquatic environment, their persistence and their effect on abiotic and biotic degradation processes in aquatic systems have also been evaluated.[11]

Until recently, we were limited in our degradation experiments and attempts to determine breakdown products by the lack of analytical detection techniques for TBTs and their related compounds, which are at a low level (nanograms per liter) but still toxic. Recently, more sensitive analytical methods will allow the more thorough investigation of distribution and the persistence of TBTs, and their fate in the environment will be better established.[34] We must keep in mind, though, that the new information and data is still not enough to completely understand the levels and fate of organotins in the environment, and further studies are essential.

To assess how best to control applied organotins in the environment, we must consider the limited amount of information available on actual production and use, and the law of each nation. We must also accumulate and unify data on organotin behavior, and it will be essential to improve the data base very soon.

7.3.5.2 Occurrence and Impact of Applied Organotins on the Environment

7.3.5.2.1 Uses, Toxicity, Input

A number of organotins ($R_nS_nX_{4-n}$) have been synthesized and applied since the early 1960s. Production was at 5000 tons in 1965, and by 1987, more than 36,000 tons of organotins were being produced annually throughout the world.

The use of these chemical is broad. Dialkyltins ($R_2S_nX_2$) are used as stabilizers in PVC or catalysts, and the use of triorganotins (R_3S_nX) has increased dramatically over the past 30 years. TBTs — bis(tributyltin) oxide (TBTO), tributyl fluoride, tributyltin chloride, tributyltin hydroxide, tributyltin naphthenate, and tris(tributylstanyl) phosphate — have long been used in agricultural and industrial biocides. They are used in antifouling paints on ships and boats, or as slimicides in cooling towers.[29,40] Their toxic effect on organotins is generally estimated on the order of tri->tetra->di->mono-organic types. The major feature is the nature of the organic group (R), rather than of the anionic group (X).

In the 1970s, it was found that TBTs caused deformities in oysters and killed their spat in addition to causing damage to other bivalve mollusks, especially mussels. TBTs were found to bioaccumulate in biota and to enter the human food chain by this means.

TBTs in antifouling paint were first introduced in Europe in the beginning of the 1960s. By 1985, 20 to 30% of the vessels in the world were estimated to be using antifouling paints containing TBTs. Now, with more effective formulations of copolymer paints, and because of the hydrolysis of paint, TBTs are entering the water slower and at a more controlled level, but, nevertheless, they are still continually moving into the environment. Some formulations are effective for five years or more.[30]

When TBTs are distributed in the water, they appear in the form of hydrated cations and chlorides,[39] and the equilibrium of these different species is influenced by pH.[40] Zuckerman et al. reported the solubility of TBTO in the sea water was 51 mg/l.[41] Many studies show that TBTs adsorb strongly onto soil.[42]

It seems that the frequency of organotin entry into the environment has risen to some extent recently. Though they are no longer ''new'' chemicals, they currently appear as a new and serious pollutant in the environment. There is great concern about them because of the great potential they have of reaching the sea by uninterrupted direct entry.

Another concern about organotins as pollutants is their long persistence in the environment. They accumulate in sewage sludges, sediments, and biota. Moreover, their indirect entry is possible through other organotin products, such as plastic articles or from high technology applications.

7.3.5.2.2 Persistence

The degradation of organotins in the environment is of great interest and was recently reviewed in detail by Blunden[22] and Muller et al.[25] The S_n–C bond can be cleaved by chemical cleavage, photolysis, thermal cleavage, irradiation cleavage, and the abiotic degradation process.[43] Photochemical cleavage and the abiotic degradation process in the environment are also possible.

Numberous investigations have confirmed the photolytic degradation of TBTs, though their photolysis and biodegradation in the environment are rather slow.[11] Several studies have attempted to identify the mechanism of biodegradation, and all suggest that biodegradation occurs more rapidly in water than in sediment. Half lives for TBTs are on the order of 20 d for water, and 16 weeks to several years under anaerobic conditions; however, further studies are needed to elucidate the breakdown mechanism.

Bioaccumulation of TBTs has been shown in several aquatic species in the numbers ranging from 1000 to above 10,000 for microorganisms and 30,000 for algae and bacteria.[35]

7.3.5.3 Legal Regulations and Legislation

In 1982, France banned the application of antifouling paints containing TBTs for some boats, and the UK followed in 1985.[28] Within several years there was a significant decrease in organotin concentration in the waters of the Arcachon Bay (France), accompanied by an increase in oyster production. In 1987, the target concentration was lowered to 2 ng/l in the U.K. The U.S. prohibited or restricted the use of TBTs with the Organotin Antifouling Paint Control Act of 1988. However, various states in the U.S. have enacted their own legislation on TBTs, and the antifouling act does not cover them.

The Japanese government began calling for a reduction in the use of TBTs in 1972, and in 1987 the Japanese fishermen's union agreed to discontinue using antifouling paint containing TBTs. West Germany, Ireland, and Switzerland have also passed legislation restricting TBTs. Though there are still loopholes that allow shipowners to coat their vessels in another country's waters, many nations that have legislation pending are also moving to control TBTs, and international action is expected. Another helpful measure came from the EPA's effort, which prescribed a maximum tin release rate for any TBT used in antifouling of 4 $\mu g/cm^2$ a day.

When we consider the decomposition difficulty and the cumulative nature of organotins, however, we must realize that it is still too early to tell to what degree restrictions have reduced TBT levels in the world's waters.

7.3.6 CONCLUSION

The United Nations Conference on the Environment in Rio de Janeiro, dubbed "The Earth Summit", turned out to be the largest of its kind ever held. Scientists, political leaders, and the general public of the world all seemed to be in agreement that economic activities are having a deadly disruptive effect on the earth's natural environment and that such activities are now at a critical point. It seems an awakening in consciousness is occurring, with a move from economic gains to environmental necessity.

With this new unified platform in place it is important that we not try to fool ourselves. Serious help toward nurturing the earth back to a healthy state will be neither easy nor cheap. Some have suggested that environmental conservation should be introduced, with some of the cost reflected in a tax system. For the individual it is a time for each person to think seriously about changing his or her lifestyle for the sake of environmental protection. At the same time, industries are expected to promote more efficient clean production, including increasing recycling and reuse of residues, as well as reducing the quantity of wastes discharged.

Under the current regulations and technologies, no hazardous chemicals are being discharged into the environment in toxic amounts; however, they are still being discharged at low levels, and these low concentrations are of major concern, because they are hazardous and can cause chronic disease.

To control such chemicals in a high technology environment, new treatment methods are needed, and more advanced treatment technology is needed for complexed wastewater.

Developed nations have been reaping the comforts of industrialization, and now must be held accountable for environmental damage and its cost. While the industrialized world begins to acknowledge its responsibility, it is also hoping that developing nations will avoid the same mistakes, but hope is not enough. It remains a fact that very few hazardous waste management systems are in place in developing countries. Most of these developing nations also have no regulations governing toxic waste and no facilities capable of adequately treating and disposing of such materials. Helping developing nations industrialize in an environmentally sound way is possible, but like the answers to current world environmental problems, it is only possible when unified action is taken.

REFERENCES

1. The Earth Summit-Agenda 21, Rio de Janeiro, June 3 to 14, 1992.
2. Brown, L.R., *State of the World,* W.W. Norton, New York, 1988.
3. An Assessment of Mercury in the Environment, National Research Council, National Academy of Sciences, Washington, D.C., 1978.
4. *The Biogeochemistry of Lead in the Environment,* Noriagu, J.O., Ed., Elsevier/North-Holland, Amsterdam, 1978.
5. Yasui, I., Annual Reports of Grant-in-Aid for Research Program on Formulation and Management of Man-Environmental System, G061-N10, Ministry of Education, Science and Culture, Japan, 1992, 161.
6. Takahashi, T. and Korenaga, T., Advances in Environmental Resources I: Research in Treatment and Reuse of Waste Waters with Multiple Random, and Small Effluent Characteristics, Involving Developments of Highly Advanced Technology, Okayama University, Okayama, Japan, Administration Center for Environmental Science and Technology, 1987.

7. Shinoda, S., Korenaga, T., Kaseno, S., et al., Advances in Environmental Resources II: Assessment of Environmental Impact and the Development of Effluent Containing Mercury at Low Level, Okayama University Administration Center for Environmental Science and Technology, Okayama, Japan, 1989.
8. Korenaga, T., Takeuchi, F., Jyo, Y., Kurose, S., Izawa, M., and Takahashi, T., *Suishisu Odaku Kenkyu,* 12, 736, 1989.
9. Korenaga, T., Takeuchi, F., Jyo, Y., Kurose, S., Myodo, T., and Takahashi, T., *Suishitsu Odaku Kenkyu,* 12, 233, 1989.
10. Wood, J.M., *Environ. Health Perspect.,* 63, 115, 1985.
11. Korenaga, T., Izawa, M., and Haraguchi, H., *PPM,* 22, (10) 49, (11) 56, and (12) 34, 1991; ibid, 23, (1) 69, 1992.
12. *The Biogeochemistry of Mercury in the Environment,* Noriagu, J.O., Ed., Elsevier/North-Holland, New York, 1978.
13. Guibt, J.-C. and Degobert, P., *Rev. Inst. Fr. Pet.,* 37(6), 823, 1982.
14. Organometals and Organometalloids-Occurrence and Fate in the Environment, ACS Symposium Series No. 82, Brinckman, F.E. and Bellama, J.M., Eds., American Chemical Society, Washington, D.C., 1978.
15. Craig, P.J., *Spec. Publ. R. Soc. Chem.,* No. 44, 277, 1983.
16. Environmental Health Criteria 1, Mercury, World Health Organization, Geneva, 1976.
17. Environmental Health Criteria 3, Lead, World Health Organization, Geneva, 1980.
18. Craig, P.J., Comprehensive organometallic chemistry, in *Environmental Aspects of Organometallic Chemistry,* Vol. 2, Abel, E.W., Stone, F.G.A., and Wilkinson, G., Eds., Pergamon Press, Oxford, 979, 1982.
19. Environmental Health Criteria 15, Tin, World Health Organization, Geneva, 1980.
20. U.S. Environmental Protection Agency, *Ambient Water Quality Criteria for Organotins—1987,* 52, 37515, 1987.
21. Craig, P.J., *Environ. Tech. Lett.,* 1, 225, 1980.
22. Blunden, S.J. and Champman, A.H., *Environ. Tech. Lett.,* 3, 267, 1982.
23. Cardwell, R.D. and Sheldon, A.W., *Oceans,* No. 4, 1117, 1986.
24. Rohbock, E., Int. Conf. Manage. Control Heavy Metals Environ., 386, 1979.
25. Muller, M.D., Renberg, L., and Ripen, G., *Chemosphere,* 18, 2015, 1989.
26. PB Rep. No. PB-85-227452, 145, 1985.
27. Wood, J.M., *Science,* 183, 1049, 1974.
28. Mercury Contamination in Man and his Environment, International Atomic Energy Agency, Vienna, 1972.
29. Simone, R.E.De, Penley, M.W., Charbonneau, L., Smith, S.G., Wood, J.M., Hill, H.A.O., Pratt, J.M., Ridsdale, S., and Williams, R.J.P., *Biochim. Biophys. Acta,* 304, 851, 1973.
30. Anders, C.D. and Dalley, R., Oceans '86 Organotin Symposium, Vol. 4, Institute of Electrical and Electronics Engineers (IEEE), Washington, D.C., 1986, 1108.
31. Stebbing, A.R.D., *Mar. Pollut. Biol.,* 16, 383, 1985.
32. Clark, E.A. and Sterritt, R.M., *Environ. Sci. Technol.,* 22, 600, 1988.
33. Hall, L.W., Jr. and Pinky, A.E., *CRC Crit. Rev. Toxicol.,* 14, 150, 1985.

34. Thompson, J.A.J., Sheffer, M.J., Chau, R.C., Cooney, Y.K., Cooney, J.J., Cullen, W.R., and Maguire, R.J., National Research Council of Canada, Environmental Secretariat, Publ. No. NRCC 22494, 1985, 1.

35. Laughlin, R.B., Jr. and Linden, O., *Ambio,* 16, 252, 1987.

36. Maguire, R.J., Tkacz, R.J., and Sartor, D.L.J., *J. Great Lakes Res.,* 11, 320, 1985.

37. World Health Organization (WHO), Tin and Organotin Compounds, A Preliminary Review, Environ. Health Criteria 15, Geneva, 1980.

38. Morita, M., Annual Reports of the Grant-in-Aid for Research in Human Environmental System, G031-N12, Ministry of Education, Science and Culture, Japan, 1990, 13.

39. Maguire, R.J. and Tkacz, R.J., *Agric. Food Chem.,* 33, 947, 1985.

40. Clark, E.A., Sterritt, R.M., and Lester, J.N., *Environ. Sci. Technol.,* 22, 600, 1988.

41. Zuckerman, J.J. *Organometals and Organometalloids: Occurrence and Fate in the Environment,* Brinkman, F.E. and Bellama, J.M., Eds., A.C.S. Symp. Ser. No. 82, American Chemical Society, Washington, D.C., 388, 1978.

42. Maguire, R.J., Carley, J.H., and Hale, E.J., *Agric. Food Chem.,* 31, 1060, 1983.

43. Blunden, S.J., Hobbs, L.A., and Smith, P.J., *The Environmental Chemistry of Organotin Compounds-Environmental Chemistry,* Bowen, H.J.M., Ed., The Royal Society of Chemistry, London, 1984, 49.

CHAPTER 7.4

Reuse Concept in Metallic Wastewater Treatment

Taneaki Okuda and Izuru Sugano

TABLE OF CONTENTS

0-8493-682-5/94/$0.00 + $.50

© 1994 by CRC Press, Inc.

7.4.1 THE "FERRITE PROCESS" AND FERRITE BY-PRODUCT CHARACTERISTICS

The "ferrite process" has, since its development by the NEC Corporation in 1973, been improved and widely applied to metallic wastewater treatment.[1-3] Figure 1 shows its process flow. Its reaction principle is as follows:

$$xM^{2+} + (2 - x)Fe^{2+} + 60H^- = M_xFe_{(2-x)}(OH)_6 \tag{1}$$

$$M_xFe_{(2-x)}(OH)_6 + O_2 = M_xFe_{(2-x)}O_4 + 3H_2O, \tag{2}$$

where the first reaction is neutralization, followed by oxidation. In the process, wastewater containing heavy metal ions is mixed with the ferrous salt, ferrous sulfate, in most cases, with alkali added to adjust the pH value to about 10. It is heated at about 60°C and oxidized by aeration to form ferrite precipitates which have a spinel crystallographic structure and into which most of the heavy metal ions are incorporated in the form of a solid solution.

Some substances strongly inhibit ferrite formation in the aqueous solution.[3] Figure 2 shows the relative ferrite process treatability of individual elements. Elements indicated in the chart with black bars (A) are those metals for which wastewater treatment by the ferrite process is effective.[3] The number under individual element symbols indicates the residual concentration after a solution containing 1000 mg/l of the element has been treated with the standard ferrite process. Those elements indicated with an X sign (D), e.g., Si or P, strongly inhibit ferrite formation in the aqueous solution.

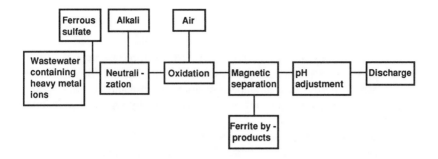

Figure 1. Ferrite process.

1	Ia	IIa		A	B	C	D		1H			Semi - metal		Non metalllic			2He
---	----	-----	--	---	---	---	---	--	----	--	--	IIIb	IVb	Vb	VIb	VIIb	----
												IIIb	IVb	Vb	VIb	VIIb	
2	3Li	4Be					Metallic					5B 97	6C	7N	8O	9F	10Ne
3	11Na	12Mg 0.3	IIIa	IVa	Va	VIa	VIIa	VIII	Ib	Ib	IIb	13Al 82	14Si	15P	16S	17Cl	18Ar
4	19K	20Ca 600	21Sc	22Ti 4>	23V 0.5>	24Cr 0.01	25Mn 0.01	26Fe 0.02>	27Co 0.02>	28Ni 0.05>	29Cu 0.2	30Zn 0.03	31Ga 79	32Ge 26	33As 0.01	34Se 17	35Br	36Kr
5	37Rb	38Sr 20	39Y 15>	40Zr 50>	41Nb (50)	42Mo 860	43Tc	44Ru	45Rh	46Pd	47Ag 0.17	48Cd 0.02	49In 5.6	50Sn 1>	51Sb (05) 67	52Te 850	53I	54Xe
6	55Cs	56Ba 1.1	*L	72Hf	73Ta 50>	74W 820	75Re	76Os	77Ir	78Pt	79Au	80Hg 0.01	81Tl	82Pb 0.02>	83Bi 0.2>	84Po	85At	86Rn
7	87Fr	88Ra	**A															

57 ₹ 71	*L	57La 50	58Ce 0.5	59Pr	60Nd	61Pm	62Sm 76	63Eu	64Gd 14	65Tb	66Dy	67Ho	68Er	69Tm	70Yb	71Lu
89 ₹	**A	89Ac	90Th	91Pa	92U	93Np	94Pu	95Am	96Cm	97Bk	98Cf	99Es	100Fm	101Md	102No	103Lr

A : Ferrite process effective.
B : Ferrite formation moderately inhibited and removal incomplete.
C : Removal impossible.
D : Ferrite formation strongly inhibited.

Figure 2. Element treatability with the ferrite process.

The "ferrite process" is extremely well suited to the treatment of complex mixtures having a variety of different heavy metal ions. Wastewater which contains single metallic ions in higher concentrations might better be treated by some simple process other than the ferrite process, and the metal might then be recovered.

Ferrite sludge has the following characteristics:

1. It is chemically stable.
2. It is uniform in consistency and made up of fine particles, the average size of which is about 50 nm in diameter.

3. It is magnetic.
4. Its density is high, about 5 g/cm³.

Ferrite sludge is referred to as a "ferrite by-product" because it can be used as a raw material. Most applications are safe because the ferrite by-product is chemically stable and does not normally dissolve in water. When exposed to highly acidic rain, however, some ferrite by-products containing harmful metal ions may indeed become soluble.

Generally, waste recycling requires special precautions when the waste contains harmful substances. In Japan, the dissolution test, using neutralized water, is the only officially sanctioned test for determining sludge stability. This test is not, however, recognized as being valid for recycling, only for landfill activities. So far, many ferrite by-product applications have been carried out with special precautions or using magnetite, ferrite which is free from any harmful substances, for safety.

Further investigations must be carried out for wider and safer waste reuse.

7.4.2 SOME FERRITE BY-PRODUCT APPLICATIONS

7.4.2.1 Vibration Damping Materials

Ferrite by-products have a high density (about 5 g/cm³), and a composite formed with an appropriate resin can be used as an excellent vibration damping material[4] for cars, ships, or any machines which generate mechanical vibration. It can also be used to isolate a precision machine from adverse vibration. Figure 3 shows a vibration-proof pedestal for microscopy. Figure 4 shows a vibration-proof base for a magnetic disk evaluation system. The mask aligner system, furnished with the vibration damping materials, has played an important role in the lithography process used in semiconductor production, enabling the precise adjustment of mask patterns during optical exposure. Vibration damping paint, made from a ferrite by-product, can be applied to complexly shaped places.

7.4.2.2 Ferrite Magnetic Markers

Ferrite by-products retain large magnetic permeability in the high frequency region of electromagnetic waves, where steel loses its magnetic permeability. When ferrite by-product resins or concrete composite materials are used as magnetic markers,[5,16,19] in connection with suitable magnetic sensors, those markers can be detected and distinguished from steel. Figure 5 shows an A-type sensor used for an automated guided vehicle. Figure 6 shows a B-type magnetic sensor which can be mounted on a white sensor cane for use by the sight-impaired. Figure 7 illustrates a mobility support system for use with such a cane. Figure 8 shows an example of an automated guided vehicle for

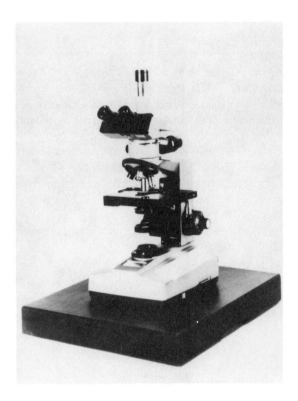

Figure 3. Vibration-proof pedestal for a microscope.

factory use. Several such systems are in actual operation in Japan today, in such public places as libraries, railroad stations, and welfare facilities for handicapped persons.

7.4.2.3 Electromagnetic Wave Absorbers

Ferrite by-products are a kind of ferromagnetic material which can be made into an effective electromagnetic wave absorber in the MHz-GHz region.[6] NEC Corporation has developed an excellent absorber of this type, "NEBOA" (NEC Broad-Band Absorber). Figure 9 shows its structure. In the figure, the first layer transforms electromagnetic waves to prevent them from reflecting of its surface, and the second layer absorbs them.

Radar navigation is often disturbed by ghost images, which can be caused by the reflection from large bridges or by such shipboard objects as poles or masts located direclty in front of radar antenna. Figure 10 shows three examples of types of ghost images which such shipboard obstructions can produce. These can be eliminated by applying NEBOA to the obstructing surfaces.

Figure 4. Vibration-proof base for magnetic disk evaluation system.

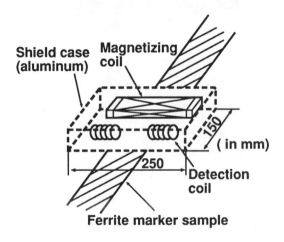

Figure 5. A-type sensor structure.

7.4.2.4 Ferrite Algae Reefs

Certain kinds of calcareous algae often cover the surface of reefs and interfere with the lives of other types of algae. One such phenomenon, that of pink rocks, is widely found in Japan's coastline areas. In such places, ferrite algae reefs[7] are expected to be very effective for the development of new reefs on the surfaces where seaweed breeds. Artificial reefs made of ferrite offer the following important advantages:

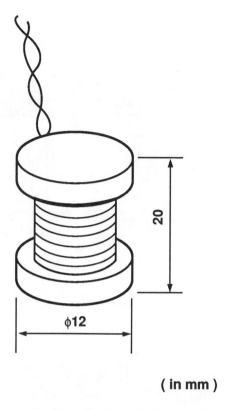

(**in mm**)

Figure 6. B-type sensor structure.

1. Ease and speed of fabrication
2. Strength and durability
3. Location easily determined by magnetic sensor
4. Attractive environment for abalone

Figure 11 shows a ferrite block for use as an artificial reef.

7.4.2.5 Carbon Dioxide Decomposition and Hydrogen Recycling

This idea, which Tamaura et al. have reported recently,[8-10] is expected to be one of the most important methods for suppressing global warming. Figure 12 shows a model for carbon dioxide decomposition and hydrogen recycling. In the cycle, carbon dioxide is converted either to carbon monoxide or methane (raw material for C–1 chemistry). Energy required for the cycle to circulate will be supplied as heat by waste at a power station. In the figure,

1. Fe_3O_4 is magnetite.
2. Fe_3O_{4-d} is oxygen deficient magnetite.
3. The C_r in $Fe_3O_4C_r$ is reactive carbon fixed on the surface of magnetite.

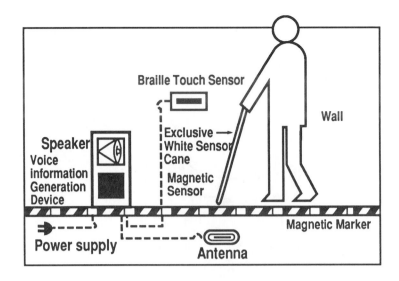

Figure 7. Mobility support system for the sight impaired.

Figure 8. Automated guided vehicle for FA.

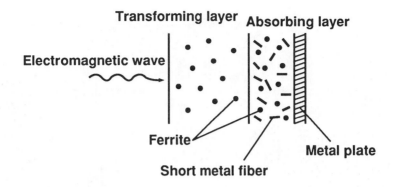

Figure 9. NEBOA structure.

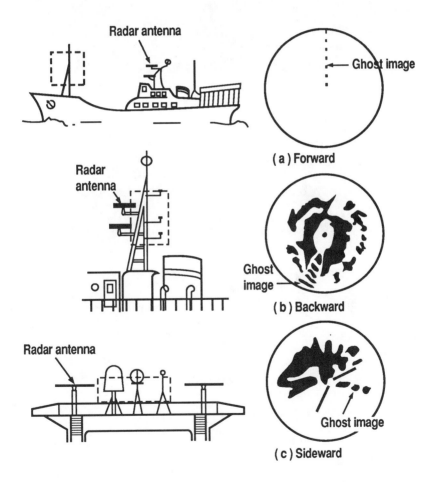

Figure 10. Ghost image examples caused by shipboard obstructions.

Figure 11. Ferrite block for use in an artificial reef.

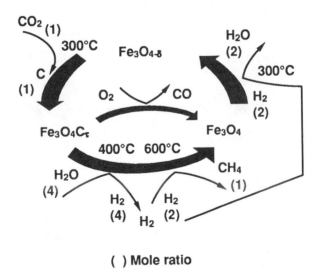

( ) **Mole ratio**

Figure 12. CO_2–CH_4 carbon cycle system.

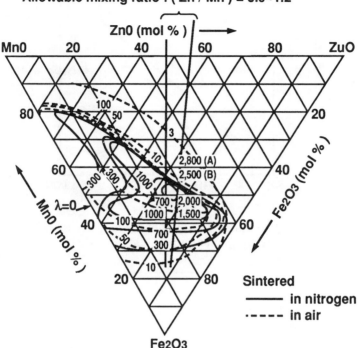

Figure 13. MnO–ZnO–Fe₂O₃ compositional region for allowable magnetic permeability.[11]

It should be noted that excess hydrogen is produced by this model from water. In the future, this technology will add to the importance of ferrite by-products.

7.4.2.6 Used Dry Battery Recycling

Although dry batteries contain a number of useful resource materials, such as manganese and zinc, recycling is often seen to be impractical because they also contain small amounts of harmful mercury, and the expense of removing that mercury may be greater than the potential value of the materials recovered.

Such recycling can be made economically viable, however, by using the manganese and zinc in combination with a ferrite by-product to produce highly valuable ferrite magnetic markers. Specifically, when a ferrite by-product is mixed with manganese and zinc and sintered under appropriate conditions, ferrite magnetic markers with much higher magnetic permeability than ferrite resin composite materials can be obtained (see Figure 13 and Reference 11).

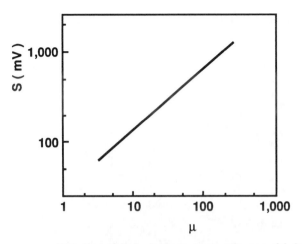

Relationship between magnetic sensitivity and magnetic permeability.

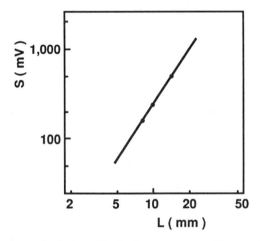

Relationship between magnetic sensitivity and sensing distance

Figure 14. Magnetic sensitivity vs. magnetic permeability and sensing distance.[11]

The magnetic markers thus obtained can be applied to high performance automated guidance systems which work well even outdoors, where the sensing distance may change significantly. The markers can also be used in greatly improved guidance systems for the sight impaired, in which large distances between magnetic sensors and magnetic markers will still be practical (see Figure 14 and Reference 11).

7.4.2.7 Cement Tracer for Detection of Cement Content and for Ground Improvement

If cement, which includes ferrite-by-product powder as a component, is used in civil engineering fields, the cement content in a concrete structure can be precisely measured during or after construction. The flow rate for fluid cement to be injected into a construction zone can be controlled by this method. This method will be applied to check the cement content in underground soil, which is to be solidified by the cement. Magnetic sensors for these purposes have already been developed.[12-14]

7.4.3 MINE DRAINAGE TREATMENT AND RECOVERED FERRITE BY-PRODUCT APPLICATION

Some unused mines produced large amounts of toxic drainage. Reduction or reuse of the waste or sludge, resulting from mine drainage treatment, is becoming an important subject because of the shortage of appropriate landfill places. Matsuo Mine[15] was Asia's largest mine that produced sulfur. It was located on the side of Hachimantai Mountain, on the uppermost branch of the Kitakami River in Iwate prefecture. In its prime, during the middle 1950s, it produced 80,000 tons of sulfur a year. Since 1972, however, the mine fell into disuse, because cheap sulfur has become available from a crude oil desulfurization process since about 1965. Then a strongly acidic (pH 2) mine drainage, including ferrous ions in high concentration and arsenic began to leak from the unused mine. The phenomenon can be explained as follows.

$$\text{(iron sulfide)} \quad \text{(oxygen)} \quad \text{(rain, ground water)}$$
$$2FeS_2 \quad + 7O_2 \quad + 2H_2O$$
$$\text{(sulfuric acid)} \quad \text{(ferrous sulfate)}$$
$$= \quad 2H_2SO_4 \quad + \quad 2FeSO_4$$

In November, 1981, the New Neutralization Plant at Matsuo and a dam were constructed[15] by the Metal Mining Agency of Japan (MMAJ). The dam life was estimated as about 20 years. Reduction in the amount of sludge and its recycling has been examined. Application tests on the "ferrite process" for the mine drainage at Matsuo was one of several countermeasures.

7.4.3.1 Mine Drainage Treatment[16,19,20]

Figure 15 shows the basic flow diagram used to treat the mine drainage and recover ferrite by-products. Because the mine drainage contains silicate

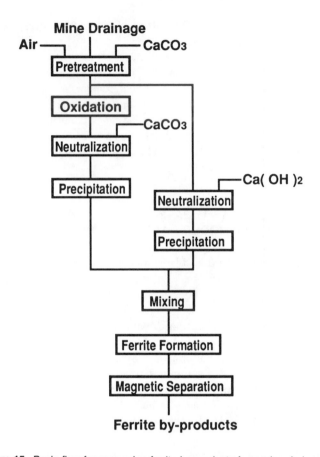

Figure 15. Basic flow for recovering ferrite by-products from mine drainage.

and aluminate, both of which inhibit ferrite formation in the aqueous solution, ferrite formation is different from normal ferrite process. Therefore, the basic flow consists of four main processes: silicate elimination (pretreatment in Figure 15), ferric hydroxide formation (neutralization in the left branch of Figure 15), ferrous hydroxide formation (neutralization in the right branch of Figure 15), and ferrite formation.[16]

7.4.3.1.1 Silicate Elimination Process

It is generally very difficult to obtain ferrous hydroxide that is free from silicate. The precipitation of a part of iron ions, at pH values between 5.5 and 6.7, proved to be very effective. Even by this process, however, the silicate could not be eliminated completely. The SiO_2/Fe ratio could be minimized, as shown in Figure 16.

The larger the SiO_2/Fe ratio, the smaller the magnetization for the final ferrite by-product, as shown in Figure 17. In this process, aluminate and

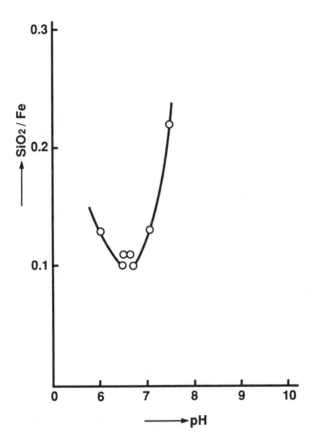

Figure 16. Relationship between SiO$_2$/Fe ratio and ferrous hydroxide precipitation pH.

arsenic could be completely eliminated. The best pH value was 5.8. This condition could minimize the iron consumption, too.

7.4.3.1.2 Ferric Hydroxide Formation

Ferric hydroxide, free from silicate, could be obtained by drainage oxidation, using iron oxidation bacteria at pH 3.8 to 4.0, as shown in Figure 18. Oxidation, using iron oxidation bacteria, has been carried out ever since the Neutralization Plant at Matsuo was established.

7.4.3.1.3 Ferrous Hydroxide Formation

Repeating precipitation from the silicate eliminated drainage at pH 9 to 10 is effective to concentrate ferrous hydroxide suspension.

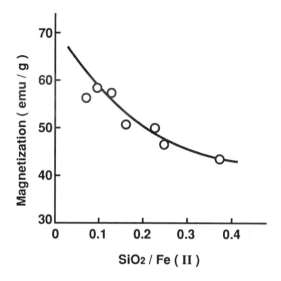

Figure 17. Residual SiO_2 concentration effect on ferrite by-product magnetization.

7.4.3.1.4 Ferrite Formation

Ferrite formation is advanced by effectively mixing ferrous and ferric hydroxides. Characteristics for the typical ferrite by-product, thus obtained, are shown in Figure 19.

7.4.3.2 Ferrite By-Product Application to Magnetic Fluid

The ferrite by-product recovered from mine drainage had smaller particle size (around 20 nm in diameter) than conventional magnetitite which is recovered from wastewater treatment in the titanium industry or the steel-making industry.

Utilizing such a characteristic, magnetic fluid was prepared from the recovered ferrite by-product and its applicability to the sink- and float-separation for nonmagnetic scraps was investigated.

Figure 20 shows a preparation procedure of water-base magnetic fluid from ferrite by-product.[16-19] The apparent density for magnetic fluid changes with the applied magnetic field strength, according to the following equation:

$$d' = d + (M \, \nabla H/4\pi g)$$

where

H = magnetic field strength (Oe), d' = apparent magnetic fluid density (at $H = H$), d = intrinsic magnetic fluid density (at $H = 0$), M = magnetic

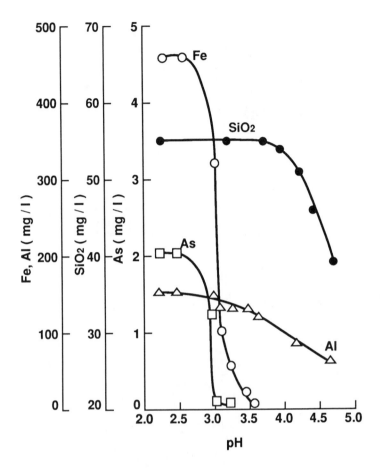

Figure 18. Oxidized mine drainage neutralization curves.

Saturation Magnetization		65 emu / g
Specific Surface Area		51 m^2 / g
Chemical Composition	Fe^{2+}	7.7 %
	Fe^{3+}	47.0 %
	SiO$_2$	3.94%

Figure 19. Characteristics for ferrite by-products, recovered from mine drainage.

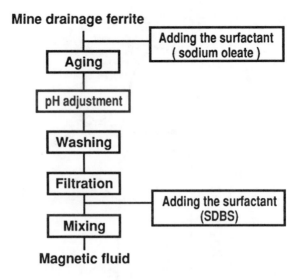

Figure 20. Water-base magnetic fluid preparation flow chart from ferrite by-product recovered from mine drainage.

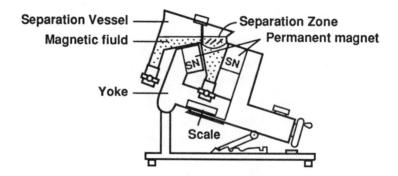

Figure 21. Experimental apparatus for sink- and float-separation for nonmagnetic scraps.

fluid magnetization (at H = H), ∇H = magnetic field gradient, g = gravity acceleration constant.

In some experiments. individual magnetic fluid samples, having 1.15, 1.21, and 1.18 g/cm^3 intrinsic density values, showed apparent density changes at 2.55 to 4.09, 2.55 to 5.51, and 2.55 to 4.51 g/cm^3, respectively, in accordance with the change in magnetic field strength, up to 2300 Oe, using the apparatus shown in Figure 21. Figure 22 shows results obtained from a sink-and-float-separation test, where the magnetic fluid, prepared from mine drainage ferrite, is compared with the fluid prepared from the reagent.

It was confirmed that the ferrite by-product compared favorably with the ferrite from the reagent and could be applied to magnetic fluid for separating

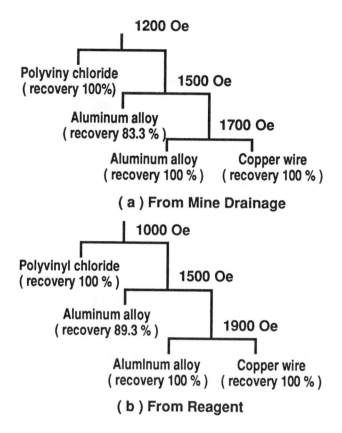

1200 Oe

Polyviny chloride
(recovery 100%)

1500 Oe

Aluminum alloy
(recovery 83.3 %)

1700 Oe

Aluminum alloy
(recovery 100 %)

Copper wire
(recovery 100 %)

(a) From Mine Drainage

1000 Oe

Polyvinyl chloride
(recovery 100 %)

1500 Oe

Aluminum alloy
(recovery 89.3 %)

1900 Oe

Aluminum alloy
(recovery 100 %)

Copper wire
(recovery 100 %)

(b) From Reagent

Figure 22. Sink- and float-separation flow diagram: (a) mine drainage ferrite magnetic fluid and (b) reagent ferrite magnetic fluid.

nonmagnetic scrap mixture. Actually, the scrap mixture for vinyl chloride, aluminum, and copper could be separated into its components.[16-19]

7.4.4 RELEVANT APPLICATION TECHNOLOGIES

Even if all of the iron in the mine drainage at Matsuo was to be converted to ferrite by-products, the quantity would amount to, at most, 3800 tons per year, while the total expenditure of iron oxide powder in Japan amounts to about 94,000 tons per year, as shown in Table 1.[19] Nevertheless, the present market could not accept the safe ferrite by-product from mine drainage because it contains a small amount of SiO_2 and CaO as impurities. Therefore, a new market or usage must be developed for useful recycling of ferrite by-product. As mentioned above, relevant application technologies, such as cement tracer, used dry battery recycling, ferrite algae reefs, and carbon dioxide decomposition catalyst, have been developed, or some of them are presently in the

Table 1 Uses of Iron Oxide Powder

Use	Type	Price (¥/kg)	Production	Anticipation Rate of Production Expansion
			(t/month)	
Raw material powder for electronic devices	α-Fe_2O_3	100	2800	?
Pigment			2770	
Red	α-Fe_2O_3	200	1860	+4%
Black	Fe_3O_4	300	210	+4%
Yellow	α-FeOOH	250	700	+4%
Recording material	γ-Fe_2O_3	700	2200	+4% (Audio)
	Coγ-Fe_2O_3	2000		~ +15% (VTR)
Magnetic fluid				
Vacuum seal	Fe_3O_4	10^7		
Magnetic tape	Fe_3O_4	5×10^5	?	?
Combustion catalyst	Fe_3O_4	$2\text{~}3 \times 10^4$		
Toner	Fe_3O_4	300	300	$-\alpha$
Soil improvement	Fe_3O_4	$1\text{~}2 \times 10^4$	?	
Magnetic marker			(t/year)	
Ferrite tile	Fe_3O_4	560	7.5	25 t/year ⎛ 3 years
Ferrite block	Fe_3O_4	300	125	375 t/year ⎝ from now ⎠
Microwave absorber	Fe_3O_4	290	1	?

Data gathered in March, 1991.

development stage. All these technologies will be very important, not only for ferrite by-product recycling, but also for solving big environmental problems.[19,20] Small amounts of impurities, such as SiO_2 and CaO, cannot be any obstacles to the cement tracer, used dry battery recycling, and ferrite algae reefs.

7.4.5 THE FUTURE OF "FERRITE PROCESS" AND FERRITE BY-PRODUCTS

The ferrite by-product contains impurities. It is not for use in electronics materials, but can be used in large amounts and as cheap ferrite for various methods, which will be useful for human life in the near future. The ferrite technology has contributed, not only to environmental protection, but also to the effective reuse of waste, and is now expanding the application area to cover the civil engineering field. The "ferrite process" and ferrite by-product will become much more important for solving large scale environmental problems, such as disruption caused by mining, ecological reef destruction, global warming, and rising sea level, in the future.[19,20]

ACKNOWLEDGMENTS

The authors would like to express their sincere thanks to Dr. T. Tsuji (President of NEC Environment Engineering, Ltd.) for his continuous encouragement over a long period of time. It is a pleasure also to thank Mr. S. Asada, Mr. T. Yoshida (presently a research assistant at the Tokyo Institute of Technology), and Dr. T. Tsutaoka (presently a research assistant at Hiroshima University) for technical support in the mine drainage treatment. The authors further wish to acknowledge gratefully the valuable suggestions received from Dr. T. Inui and Mr. F. Yamauchi.

Our study of mine drainage has been accomplished by contract with the Metal Mining Agency of Japan.

REFERENCES

1. Tsuji, T., New process for removing heavy metal ions from wastewater using ferrite manufacturing technology, *Electron. Parts Mater.,* 12(9), 70, 1973.
2. Okuda, T., Sugano, I., and Tsuji, T., Removal of heavy metals from wastewater by ferrite co-precipitation, *Filtr. Sep.,* September/October, 472, 1975.
3. Okuda, T. and Ishihara, T., Removal of heavy metals from wastewater by "ferrite process", *NEC Tech. J.,* 37(9), 35, 1984.
4. Yamauchi, F., Emoto, S., Yokoyama, K., and Saijo, H., Ferrite-resin composite material for vibration damping and its application, *NEC Tech. J.,* 37(9), 21, 1984.
5. Kondo, Y., Miura, T., and Ito, A., Magnetic guide system, *NEC Tech. J.,* 37(9), 14, 1984.
6. Inui, T., Hatakeyama, K., Harada, T., and Yoshiuchi, S., Electromagnetic absorber — application of ferrite by-product, *NEC Tech. J.,* 37(9), 2, 1984.
7. Mayama, M., Applications of artificial inorganic materials to alga reef, *Denki Kagaku,* 58(1), 15, 1990.
8. Tamaura, Y. and Tabata, M., Complete reduction of carbon dioxide to carbon using cation-excess magnetite, *Nature,* 346, 255, 1990.
9. Tamaura, Y. and Yoshida, T., New material; oxygen-deficient magnetite — application for high efficient decomposition reaction of CO_2, *Funct. Mater.,* 11(1), 38, 1991.
10. Abe, H., Yoshida, T., Tamaura, Y., Ootsuki, H., Sugimoto, M., Tabata, M., and Niahizawa, K., Digest 3rd Symp. Soc. Environ. Sci., Japan, 46–48, 1990.
11. NEC Environment Engineering, Ltd. and the Industrial Pollution Control Association of Japan, A technical development in search of the advanced use of ferrite sludge from treatment equipment of effluent with heavy metal contaminants, *Ind. Pollut. Control,* 25(5), 363, 1989.

12. Tsutaoka, T., Ema, S., Yamauchi, F., Imai, T., and Yoshiwara, M., Cement content measurement technique using ferrite as cement tracer, *Dig. 14th Annu. Conf. Magn. Jpn*, 8pG-12, 122, 1990.
13. Japan Patent Application No. 1-492781, (Jpn, Patent Kokai, H02-167464).
14. Japan Patent Application No. 1-221676, (Jpn, Patent Kokai, H03-87411).
15. The Metal Mining Agency of Japan, *Kyu Matsuo Kozan Kohaisui-shorijigyo no Gaiyo*, MMAJ, March, 1988.
16. Ema, S., Ferrite recovery from mine drainage, *Dig. Symp. Metal Recovery Mine Drainage*, Metal Mining Agency of Japan (MMAJ), Oct. 2, 1990, 9.
17. Sato, H., The characteristics of magnetic fluid made from mine drainage ferrite and solid waste scrap separation, *Dig. Symp. Metal Recovery Mine Drainage*, MMAJ Oct., 1990, 23.
18. Tsutaoka, T., Ema, S., and Sato, H., Physical and chemical properties of magnetic fluid using ferrite material from mine drainage and possibility for some applications, *Proc. Int. Conf. Ferrites, 1989*, Bombay, 1989, 1113.
19. Okuda, T., Ema, S., Ishizaki, C., and Fujimoto, J., Mine drainage treatment and ferrite sludge application, *NEC Tech. J.*, 44(5), 4, 1991.
20. Okuda, T., Ema, S., Ishizaki, C., and Fujimoto, J., Mine drainage treatment and ferrite by-product application, *Proc. 6th Int. Conf. Ferrites, 1992*, Tokyo, 224, 1992.

CHAPTER 7.5

Control Strategy of Biohazardous Waste

Kengo Kurahasi

TABLE OF CONTENTS

7.5.1 WASTE GENERATED IN INSTITUTIONS INCLUDING MEDICAL FACILITIES

7.5.1.1 Risk of Infection Accompanying the Handling of the Waste

Hospitals house patients suffering from different kinds of disease, some of which are infectious. There also are many individuals who are carrying some type of disease, perhaps a communicable disease, but who are not aware that they are ill. Contact with their discharge has the potential of direct transmission of their diseases. Most of the waste they generate is handled in the same way as normal waste without any special precautions. Once their condition is diagnosed as infectious, however, the waste is also considered infectious and must be handled and treated in a special way.

Even in a clinical setting, confirmed instances of infection of employees due to the mismanagement of generated waste are very rare. In fact, Japanese regulation requires that needles, absorbent cotton, and gauze contaminated by a visible amount of blood and used for injections or in surgery must be handled as infectious, whether it is actually infectious or not.

It is logical that the waste generated by an unknown outpatient should be handled as potentially infectious, but to the author, waste from a patient professionally diagnosed as noninfectious by a physician would not appear to have to be considered infectious. This concept could also apply to an inpatient or one undergoing a surgical operation. In fact, the Japanese Ministry of Health and Welfare explains in their "Guidelines on Medical Waste Management" that waste containing blood need not necessarily be handled as infectious waste when judged unnecessary by a specialist with professional knowledge, such as a medical doctor.

Identifying all medical waste as potentially infectious seems not only impractical, but rather expensive, because too much medical waste currently is being specially treated by steam sterilization or incineration. It also seems unnecessary to treat all waste displaying a trace of blood as potentially infectious.

Identification of waste as infectious or noninfectious should be based on scientific grounds, and at the same time the method of treatment should be technically and economically practicable. The objective confirmation that certain medical waste bears no risk of transmission, however, is not only practically impossible in an institution, but requires the expenditure of considerable labor, time, and expense that may be beyond practical limits. Scientific methods, such as culturing waste in order to examine it for infectious agents, assures no reliable results; even if an infectious microorganism is present in a culture, it may not reveal sufficient virulence to consider it infectious. It may be more practical to make decisions on the basis of the potential risk inherent in the waste. If the waste could be infectious, it is prudent to manage it as infectious. It is also important that aesthetic and perceptual considerations or perceived health risks override the actual risk of disease transmission.

In Japan, the concept of the universal precautions proposed by the Center for Disease Control (CDC) applicable from 1983 through 1988 have been introduced to determine whether or not particular wastes requires special handling.

7.5.1.2 Universal Precautions[1-5]

The concept of universal precautions was originally recommended in 1983 by the CDC.[2] Appropriate precautions were suggested for the blood and body fluids of patients infected with potential blood-borne pathogens. These precautions were broadened to include all patients in 1987.[3] In a 1988 document,[4] the specific body fluids were defined more precisely: the precautions do not apply to feces, nasal secretions, sputa, perspiration, tears, urine, and vomitus unless they contain visible blood. The main elements of the universal precautions stressed by the CDC are: (1) care to prevent injuries when using needles, scalpels, and other sharp instruments and devices; (2) appropriate use of protective barriers, such as gloves or face shields, to prevent contact with blood, body fluids containing visible blood, and other specific body fluids such as cerebrospinal fluid, synovial fluid, pleural fluid, peritoneal fluid, pericardial fluid, amniotic fluid, cervical secretions, and semen; and (3) immediate and thorough washing of hands and other skin surfaces that inadvertently become contaminated with blood or any of the special fluids. Under the updated precautions, blood and certain body fluids of all patients are considered to be infectious for human immunodeficiency virus (HIV), hepatitis B virus (HBV), and other blood-borne pathogens. Hepatitis C virus (HCV) should be added because recent research clarified that the virus also causes blood-borne hepatitis.

7.5.2 BIOHAZARDOUS WASTE

7.5.2.1 The Waste Generator and His Liability

There is ambiguity as to the exact point of time after use that medical supplies and material become waste. For example, when a nurse accidentally sticks her finger with a disposable syringe after it was used for injection but before she discarded it in a waste container, there may be some argument as to whether she was injured by a waste needle or not. It may be necessary to trace the origin of such a needle injury, for example, when classification of the origins of injuries caused by medical waste are under investigation.

A nurse is allowed to give an injection subcutaneously to a patient under the direction of the physician in charge, and the responsibility for the medical treatment, including the injection, lies with the physician. The generator of the waste is the physician, and he is considered liable for any potential hazard caused by any worker's mistake at any stage during the successive steps of treatment until treatment has been concluded; this is true regardless of the number of people and stages involved. This is what is meant by the term "cradle-to-grave responsibility".

7.5.2.2 Biohazardous Waste

The definition of the term "infectious waste" is discussed to some extent in publications of the World Health Organization (WHO),[6] the Environmental

Protection Agency (EPA),[7] and the Japanese Ministry of Health and Welfare,[8] and a paper by Steuer.[9] There is relative agreement on the use of this term for the type of waste which is discussed next.

Biohazardous waste, although occasionally used as a synonym for infectious waste, is not discussed much within the range covered by this terminology. First of all, the meaning of biohazardous waste and the range covered by this term should be considered. Biohazardous waste is that originated in a living creature which contains potentially hazardous substance and which has the potential of being a biohazard to human health upon exposure or damage to the environment if released. A biohazard is a hazard caused by (1) microorganisms such as bacteria, viruses, yeast, protozoa, fungi, etc., (2) their body components such as proteins and nucleic acids, and/or (3) their products such as toxins and immunogenic substances. Workers dealing with a microorganism and its products are, therefore, exposed to risks of biohazards, and the waste generated is potentially biohazardous.

The most important and principal sources of biohazards are microorganisms, especially bacteria and viruses. Various bacteria are not only ubiquitous in man's environment, but on and in his body in enormous numbers, while he is in the best health. Documented microbiological groups which have caused infectious disease by needle sticks, however, are restricted to a relatively small 21 groups. *Staphylococcus aureus* was the only bacteria causing a disease as a result of puncture injury of a needle contained in waste during housekeeping, although publications of the CDC and others list a number of pathogenic microorganisms.[10-13,26]

Biohazards encountered in the course of handling waste are mainly caused by living organisms and affect the health of personnel. Most of the waste considered biohazardous is synonymous with microbiologically hazardous waste or infectious waste. Principal sources that generate infectious and potentially infectious waste are health care establishments, including medical institutions, and the homes of patients suffering from infectious disease. The term biohazardous waste can then be replaced by infectious medical waste, infectious hospital waste, infectious health care waste, or simply infectious waste. Here, medical waste refers to that from a medical faculty, a medical institute, and/or their affiliated hospital. Infectious and potentially infectious wastes that should be handled in a special way are generated in these facilities in relation to medical treatment and care. Of course, not all wastes generated in medical institutes and their hospitals are infectious. The amount of infectious waste is estimated to be 15 to 40%[14-17] of the total waste generated in a hospital. This high ratio of infectious waste to total waste may be ascribed to the fact that general waste is contaminated with infectious waste when the two are mixed and when noninfectious waste is inadvertently discarded in the container meant for infectious waste.

Waste associated with research in intoxication, allergy, oncogenicity, and gene technology are considered by Dr. Shinoda in the introduction of this book, but in this chapter, the term "infectious waste" is used synonymously with "biohazardous waste" in the sense described above.

7.5.2.3 Generation Sources of Infectious Waste

Research laboratories and hospital divisions where infectious or potentially infectious waste is generated are principally the following:

1. Research laboratories — microbiological, pathological, and immunopathological laboratories
2. Hospitals — operating room, treatment room, artificial dialysis room, clinical pathology room (microscopical, cytological, histological, and bioptical examinations), clinical laboratories (blood, biochemical, microbiological, serological, and immunological examinations), ward kitchen, and laundry

7.5.3 ROUTE OF TRANSMISSION AND MANIFESTATION OF INFECTIOUS DISEASE

7.5.3.1 Route of Transmission and Portal of Entry to the Human Body

Most infectious waste is not generated by transmission of a microbial agent to medical waste, rather the waste has been contaminated before the point when it became waste, and the source of the agent may be the patient or animal suffering from communicable disease. Therefore, the waste is a mediator having no life activity of its own which may nonetheless passively transmit pathogenic agents from the patient to the waste handler.

In an infected body, microorganisms and their products are contained in blood, tissue fluid, cerebrospinal fluid, synovial fluid, tissues, and mucus, which protects the surface of mucous membranes of the digestive tract, airways, and the excretory and genital organs. Any article coming in contact with these microorganisms in the course of medical treatment and care could be a possible mediator of disease transmission. Waste contaminated by blood is known to be the primary mediator of infection.

If a waste handler has an injury on the surface of his body, pathogens may enter his body while he works. There are three different routes of bodily entry for infectious agents: the parenteral, enteral (oral or ingestive), and respiratory routes. Infectivity may differ depending on the route. For example, *Salmonella* is highly infectious if ingested, but shows no infectivity on contact with skin.

The portal of entry of an infective agent by the parenteral route is a skin wound made by a cut or stab with a used needle, a discarded scalpel blade or other sharp, and broken skin or mucous membranes of the eye, nose, or mouth. This is particularly critical for workers involved in waste handling, because blood-borne type B and type C hepatitis viruses are the main pathogens causing waste-mediated infection. The acquired immunodeficiency syndrome (AIDS) virus is known to rapidly weaken in infectivity, especially in a moist atmosphere, so the risk of infection posed by waste containing this virus is considered to be minimal.

The oral route is the portal of entry for agents such as *Salmonella* which cause typhoid fever or food poisoning in the waste handler via the gastrointestinal tract. This bacteria can be transmitted from the contaminated hand of the handler to his mouth. The respiratory route is the portal of entry for agents such as tubercle bacillus which causes lung infection. A waste handler working in a hospital may run the risk of inhaling aerosol or the scent of droplets of patient body fluid released when a red bag containing contaminated wet waste is improperly handled, such as compacting more waste into the bag.

7.5.3.2 Preventing Pathogens from Entering the Human Body

As mentioned in an earlier section, care should be taken to prevent injuries when handling needles, scalpels, and other instruments. Appropriate use of protective methods, such as gloves or face shields, in a clinical setting prevents contact with blood, body fluids containing blood, and other specific body fluids. Immediate and thorough washing of hands and other skin surfaces even when contaminated with blood or any potentially harmful fluid may reduce infections.

7.5.3.3 Manifestation of Infectious Disease

The presence of an infectious agent in waste is not synonymous with the transmission of the agent or with the capacity of the agent to produce an infectious disease. The risk of infection for the waste handler depends on the likelihood of contact with the infectious waste, and an infected human body does not always manifest the disease.

Different conditions contribute to disease manifestation: virulence and the dose of infecting agent, host resistance and susceptibility, and portal of entry in the host. The risk of manifestation of disease increases with an increase in the virulence and in the dose and decreases with an increase in resistance in the host.

A few cases of infectious disease caused by contaminated waste have been documented. Only one case of injury by needle puncture during infectious waste handling has been reported.[10,11] This seems to imply that conditions do not really favor contracting a disease; one reason may be the difficulty of pathogens surviving in an unfavorable waste environment. This is fortunate for workers involved in the management of infectious waste.

For the last decade, nosocomial infection (hospital infection) caused by methicillin-resistant *S. aureus* (MRSA) has been found. Inhabitant bacteria in hospitals sometimes transmit an infectious disease, especially to inpatients whose resistance is remarkably reduced by old age or by any number of other diseases. Administration of too much medication in the form of antibiotics, antineoplastics, and immunosuppressors may also reduce resistance to infection.

Most of the large university or institutional hospitals in Japan have a prudent infection management program. They treat infectious waste by high pressure steam sterilization, incineration, or spraying of a chemical disinfectant (for example, 2% glutaraldehyde solution) before temporal accumulation* for waste collection in a restricted area under the regulation for prevention of nosocomial infections. This action may also contribute to minimizing potential infection due to waste.

Therefore, the risk of nosocomial infection through waste could be said to be almost zero, at least in the large hospitals in Japan. Weakly pathogenic microorganisms which cause nosocomial infection include *S. aureus, S. pyrogenes, S. viridans, Pseudomonas aeruginosa, P. cepatica*, and *Klebsiella pneumonia*.

7.5.4 CATEGORIES AND CONSTITUENTS OF MEDICAL AND INFECTIOUS WASTES

Medical institutions produce a variety of waste associated with their research and medical treatment, each of which may be in rather small amounts; this is in addition to the waste emanating from domestic life. WHO[6] classifies eight main categories of waste: general waste, pathological waste, radioactive waste, chemical waste, infectious waste, sharps, pharmaceutical waste, and pressurized containers.

According to WHO, infectious waste is that which contains pathogens in sufficient concentration or quantity that exposure to it could result in disease. This includes inherently infectious waste such as pathological waste originating from an infected patient and waste items which have been in contact with pathogens on their use.

Pathological waste is tissues, organs, body parts, human fetuses, and most blood, serum, urine, feces, sputa, gastric juice, bone marrow, pleural fluid, semen, secretions, and discharges. The carcasses of laboratory animals inoculated with a sufficient amount of infectious agent could pose a potential risk of disease transmission, in other words could potentially be infectious.

Cultures and stocks of infectious agents become infectious waste after use. Waste from infected patients in isolated wards has at least the potential risk of carrying infection.

Disposable items which could become infectious waste by contact with pathogens include equipment for transfusion, infusion and dialysis, syringe,

---

* Accumulation refers to temporary holding of the waste at the point of generation.[18]

blood collectors, needles, tubing, probes, vessels, cotton, gauze, towels, gowns, aprons, gloves and sheets for surgical treatment, and culture vessels, test tubes, petri dishes, pipettes, and tubing used in laboratory and research work.

Sharps include needles, syringes, scalpel blades, broken glass, saws, and any other items that could cause a cut or puncture.

Research work involving highly virulent or concentrated pathogenic organisms, or recombinant DNA molecules is performed under strictly controlled conditions in an isolated room, and the safe management of the waste may be rather easy to control.

Reusable items which are cleanable (such as glass syringes, scalpels, surgical instruments, and lab clothing) should not be included in waste materials.

The Committee for Medical Waste Management (Counterplan and Examination), a consulting organ of the Japanese Ministry of Health and Welfare, listed in its "Medical Waste Management Guideline" published in 1991, six categories of infectious waste and their main constituents:

1. Blood, blood products, and other fluids (blood, blood serum, blood plasma, whole blood products, blood component products, body fluids such as semen, interstitial fluids)
2. Pathological waste produced by surgical operation (organs, body tissues)
3. Sharp items contaminated with blood (hypodermic and intravenous needles, scalpel blades, test tubes, petri dishes, broken glass)
4. Test implements and cultures used for test and inspection of pathological microorganism (test tubes, cultures, and petri dishes used for experiments and examinations)
5. Implements used for artificial dialysis (drainage tubing, filters)
6. Other clinical supplies contaminated with blood disposables such as gloves used in experiments and surgery, cotton, gauze, bandages)

After consideration of the various guidelines of CDC and EPA for infectious waste and optional categories, and EPA's list of regulated medical wastes, Reinhardt and Gordon[19] recommended that the following types of waste be classified and managed as infectious waste:

- Human blood and blood products
- Cultures and stocks of infectious agents
- Pathological wastes
- Contaminated sharps*
- Contaminated laboratory wastes
- Contaminated wastes from patient care
- Discarded biologicals
- Contaminated animal carcasses, body parts, and bedding

---

* From the management perspective, it is best to manage all sharp items uniformly, without differentiating between contaminated and noncontaminated items.

- Contaminated equipment
- Miscellaneous infectious wastes

Miscellaneous infectious waste includes general types of infectious waste that are not readily assigned to another specific categories. These are usually generated in the handling of infectious materials, for example, waste that is generated during cleanup of spills of infectious material (rubber gloves, masks, aprons, lab coats, absorption materials, towels, mops, etc.).

Their idea is that this is a more logical grouping of waste types that is suitable for practical application in a hospital, laboratory, and other places where infectious wastes are generated.

7.5.5 IDENTIFICATION OF INFECTIOUS WASTE

The most important activity for safe conduct of a total infectious waste management program with time and cost effectiveness is the segregation of infectious and noninfectious wastes at the site where it is generated. The generators of waste (physicians, nurses, and laboratory workers) know best whether or not their waste should be handled specifically as infectious waste. One of the best ways is for the generator himself to discard the waste into designated containers placed close to the site of generation. This may reduce the possibility of trouble arising in subsequent stages of the waste processing. It is desirable that an infection control manager, who has sufficient knowledge and experience about infectious waste management, serves in a certain area such as a ward or surgical theater and makes the final decision on whether or not the waste needs to be handled as infectious. All wastes generated in isolation wards and clinical laboratories should be handled as infectious.

At the stage of segregation, microbiological identification of any pathogen potentially present in the waste from an unknown patient is practically impossible, so it is prudent that waste containing a visible amount of blood be handled as infectious, based on the universal precautions and on aesthetic, as well as perceptional, considerations. It is important for the waste workers to be protected from exposure to contact with a potentially infectious waste in the course of their work. Accordingly, each segregated waste should be placed into an appropriate container in a way that it will not leak or spill any infectious agent.

7.5.6 CONSIDERATION OF PRETREATMENT REQUIREMENTS

Regulations are required on handling infectious waste between the site of its generation and its transport to the on-site or off-site location for the next treatment (such as steam sterilization, incineration, or disposal). The following are the requirements for segregation, packaging, collection, accumulation, storage,* labeling and marking, and transport.

7.5.6.1 Segregation

Adequate segregation is required to protect the waste handler from the risk of exposure to potential disease transmission and from injury by sharps. Preparation of an appropriate and durable container in which to discard the segregated waste is also necessary, as is a correct form of primary packaging and loading on the transport vehicle safely.

7.5.6.2 Container and Packaging

Use of an appropriate container is an important factor in minimizing risks of the attendant handling the infectious waste.

It is recommended that a selected type of container be placed close to the site of waste generation so that infectious waste can be discarded promptly and directly after its generation. Various types of containers have been designed for different infectious wastes. The differences are in potential pathogenesis (infectious or noninfectious) and physical nature (liquid, solid, muddy, sharp, etc.). One of the most important is one container for contaminated sharps and one for solid waste, including blood-soaked material. Essential requirements for the former are that it be puncture resistant, impermeable, and rigid. It is recommended that needles be removed from disposable syringes with the aid of a loosening device and discarded into a metal container with a lid for heat treatment in an incinerator at 800°C. This treatment will completely consume the pathogens, and the sharp iron material will be oxidized making it easily destructible. Infectious solid waste containing wet material (such as blood-soaked gauze) may be discarded in a plastic bag. The bag must be tightly closed to prevent leaking and spilling of any residual liquid and strong enough to hold the waste without tearing or bursting. Requirements for a plastic bag to be used for steam sterilization are discussed in a later section.

Use of an impermeable and rigid container at the site of waste generation will help in transporting off-site to the next treatment facility because it will be easy to pack and load such a container on a transport cart, reducing the total dimensions and the risk of leakage or spillage of the waste material.

Liquid infectious waste to be steam sterilized should be pooled in a metal container with a tightly closed screw stopper. Liquid infectious waste to be combusted in an incinerator should be discarded into a plastic container with thick walls. Important in avoiding waste worker contact with infectious

---

* Storage[18] refers to rather long-term holding of the waste prior to transport, treatment, or disposal in a dedicated facility or centralized area.

material is the pouring of liquid waste into the container in a careful way so that the outside wall of container is not contaminated.

7.5.6.3 Labeling and Marking

Proper marking is needed to indicate that the container contains infectious sharp items or other waste. A proper color code, the universal biohazard symbol, and wording such as "contaminated sharps" are in common use. Plastic bags that are used for infectious wastes are usually red in color, so that the term "red bags" refers to the plastic bags that contain infectious waste. One problem after the steam sterilization of infectious waste contained in red bags is the difficulty in differentiating between treated and untreated bags since the bags usually do not change color. Labeling by color or other marking which denotes the bag and its contents have been treated with steam may be necessary on the bag. One solution may be to put the sterilized bag into a second bag with a color which indicates it has been sterilized, for example, into a black which is usually used for general waste. Development of a red bag which would change in color following the steam sterilization, or an ink which changes color with steam, is anticipated.

7.5.6.4 Collection, Accumulation, and Storage

It is recommended that infectious wastes be collected as soon as possible after they are generated for transport to the location of steam sterilization or incinerator on-site or off-site. However, temporary holding (accumulation) of waste prior to collection is unavoidable, and there should be an area set aside for this to which only authorized personnel are allowed access. This area should be separated in some way from other areas and must be easily cleaned. A patrol that collects infectious waste at frequent intervals or immediately upon the request of the generating individual would improve the circumstances.

Storage of the waste prior to transport is acceptable only when no further processing is available; the limited number of employees concerned with this often brings about such a situation in national and public institutions in Japan. Storage at room temperature of waste containing pathogenic microorganisms, tissues, or organs from surgery may allow bacteria to multiply or surgical substances to putrefy, resulting in an increased hazard, a putrid odor, or unpleasant handling. Storage area should be built to exclude small wild animals (such as rats, cats, or dogs) from entering and should have a lock, a refrigeration device for putrefiable waste is desirable.

7.5.6.5 Carts for Waste Collection and Transport

Each facility should have carts which are used solely for the purpose of collecting and transporting waste not yet disinfected. Different types of carts may be necessary to carry the different kinds of waste containers. Regardless of its type, the cart should be able to be cleaned and disinfected, have a bottom flange or groove all around it to provide adequate containment of any spill of waste liquid that may occur, and have casters allowing the cart to move easily in any direction.

A sloped walkway or elevator, preferably exclusively used for the movement of such carts, should be provided between upper and lower floors, and the exit from the first floor should also offer a gradual incline to ground level.

7.5.7 RISK AND LIABILITY

One of the most common risks to individuals handling infectious waste is the exposure to potential infection by direct or indirect contact, and another is injury by sharp items. These occur primarily through incorrect handling at some step of the pretransport stage described in Section 7.5.6.

The public and the environment are at risk if exposed to water from a facility working with hazardous waste. This is mainly attributable to the improper handling of infectious waste in one of the steps during the course of waste processing, including pretransport, on-site or off-site treatment, and intermediate treatment. Environmental risks are also present if there is a release of waste contaminants into groundwater, surface water, or the air. Release of infectious agents into the air through the steam exhausted from a sterilizer and the smoke from an incinerator that has not been equipped with a smoke cleaner may create an environmental hazard. Infectious liquid waste is allowed to be discarded in a sewer only after it has been treated in a wastewater treatment facility.

A clear hierarchy of responsibility and a written management program of waste treatment procedures, warnings for special precautions, and a record of actual practices are required to assure the safety of those working with waste. Ultimate responsibility for any disadvantage to workers, the public, or the environment due to the waste itself should fall to the generator of the waste, whether an individual or an organization. This concept is known as "cradle-to-grave" responsibility, that is, the generator may be liable even for a mistake committed by a commercial waste management facility.

Accidents will be reduced if appropriate guidance is given on the proper steps to be followed in a waste management system; this guidance and instruction should be provided by specialists including the risk manager, safety director, biosafety officer, biohazardous material manager, infection control staff, or the institution's attorney. In this way, economic risk can also be decreased.

7.5.8 ON-SITE AND OFF-SITE TREATMENTS

7.5.8.1 On-Site Treatment

Waste, especially infectious waste generated in an institution conducting research and medical activities, must be processed on-site to remove its infectiousness and other potentially hazardous qualities before it is moved off the premises for final disposal such as in a landfill.

7.5.8.1.1 Incineration

Infectious waste is usually treated in an incinerator on the premises of the institution; the resulting ash and other residue is then transported to the final disposal site.

Requirements of an incinerator to assure complete combustion of infectious wastes containing various plastic materials are: (1) that it have sufficient capacity to combust an increasing amount of waste under optimal conditions with an excess quantity of oxygen and at high temperature, whatever the type of incinerator is used and (2) that it have a gas washer and a scrubber to wash out and collect acidic gases (such as hydrogen chloride and sulfur dioxide) and hazardous chemicals which may be products of incomplete combustion formed under less than optimal conditions. Toxic products include polychlorinated dibenzo-*p*-dioxins (dioxins), polychlorinated dibenzo-furans (furans), polyaromatic hydrocarbons, and polyaromatic organic substances. A device for the recovery of mercury should be installed because medical waste often contains very small quantities of mercury originating from broken mercurial thermometers, and mercury sphygmomanometers, or mercurial disinfectants. Waste plastics produce a great deal of heat during incineration, which shortens the life of the incinerator, and halogen-containing gas may pollute the air when emitted from the stack. The cost may be extremely high for a single institution to own an incinerator meeting these high standards. On-site treatment of infectious waste also has several other disadvantages: for effective operation, top-notch maintenance, and good management of the incinerator is essential, as is the ongoing training of the operator to assure a high grade of skill. Joining other institutions in the cooperative use of an incinerator may help to reduce the expenses. As stated, one advantage of on-site treatment is that it avoids having to remove potentially infectious materials from the grounds of the institution where they were generated. Other advantages are that it provides a better chance for immediate incineration and shortens the transport of waste to the incinerator, thus reducing the risk of exposure and spilling, making a manifest unnecessary, and increasing control over the handling of the process. Furthermore, use of an on-site incinerator may provide a opportunity for heat recovery.

7.5.8.1.2 Steam Sterilization

Steam sterilization is another option for the handling of infectious waste. This has long been practiced for reusable medical supplies and implements (such as clothing, glass syringes, scalpel blades, and other surgical instruments) and has also been proven to be a reliable method to treat infectious waste.[20-22] Steam sterilization is based on the fact that saturated steam is itself a powerful sterilizing agent. In addition to time and temperature, a critical requirement is that the saturated steam must come into direct contact with the infectious agents contained in the waste. Factors which impede this process are factors which interfere with the steam sterilization. These include the physical characteristics of the material composing the infectious waste, overly compact stacking of the waste, and the use of ordinary plastic bags on the market to contain the waste. Water material with low heat conductivity (such as plastics) requires more heat and a longer time to reach a sterilizing temperature. Overpacking waste in a container hinders the permeation of steam throughout the waste, as well as preventing sufficient evacuation; plastic bags impede the air displacement by saturated steam through the wall of the bag; the resulting is retardation of direct contact between saturated steam and infectious agents. One means of effectively overcoming this interference is to prolong the exposure period of the infectious waste or to use higher temperature. Specific conditions must be determined by a rigorous test of each steam sterilizer installed.

A container to be used in steam sterilization must allow the penetration of air and steam. The entry and exit of gases through the opening of a plastic bag do not ensure direct and complete contact of the saturated steam with pathogens adhering to the waste. A fine mesh bag made of heat-stable plastics allows the complete passage of air and saturated steam at the stages of air removal and steam charging in the process of sterilization. Following sterilization, placing the treated bag in another plastic bag with a mark, word, or color indicating the contents have been sterilized will facilitate subsequent handling.

Types of the waste appropriate for steam sterilization include cultures, stocks of infectious agents, associated biologicals, and contaminated labware.[23,24] When pathological wastes are steam sterilized, additional treatment rendering them unrecognizable is required before disposal for aesthetic and perceptual reasons.[25] Disposable plastic implements should be broken after sterilization to prevent reuse.

7.5.8.2 Off-Site Treatment

In off-site treatment, untreated infectious waste is transported off-site to another location where it is treated.

Advantages of off-site treatment are that the institution need not have a treatment facility. Precautions for the generator of infectious waste involve proper segregation of the waste, its packaging, and its collection for off-site transport by a vendor. Disadvantages of this treatment may include the difficulty in contracting with a credible infectious waste management facility, especially in local regions or the countryside. There may also be problems in controlling the entire pathway of the waste management system, which is composed of several stages, each of which might be managed by a different firm in conjugation with each other and may result in an increased risk of liability to the generator. One other possible disadvantage is that the waste may have to be packed in a second sturdy container to be picked up by a commercial service for long distance transportation.

7.5.8.3 Options

Selection of the appropriate management alternative will be based on the evaluation of various aspects of the institutional situation. Whether on-site or off-site treatment is decided on, of greatest importance is to manage infectious waste safely, properly, effectively, efficiently, and inexpensively. In some instances, the employment of both options together may best meet these objectives. One idea, for example, is to install an incinerator owned by a commercial facility on the premises of an institution and to have it operated and maintained by the commercial service on a pay-for-use basis. This may also reduce the necessity for second packaging, off-site transport, a return trip of manifest, and long-time storage.

7.5.9 STRATEGY OF INFECTIOUS WASTE MANAGEMENT

7.5.9.1 Planning Program for Waste Management Practice

The proper management of infectious waste involves a variety of individual steps: generation, identification, segregation, collection, accumulation, storage, packaging, transport, disinfection, steam sterilization, incineration, and disposal. Today, these individual steps are regarded as only one stage of an integrated waste management system. A program of infectious waste management should control all stages from generation through disposal successively and integrally. Because facilities differ and the management program planned by one institution may not necessarily be appropriate for another, it is prudent for each institution to prepare a plan commensurate with factors like the scale of the institution, the type and amount of waste generated, the budget allocated to waste management, conditions of location of the institution, and the availability of commercial service for waste management. Some of these items may also be parameters to be considered environmentally by the executives making the decisions.

An evaluation of the current system is a starting point for strategic planning: examination of the type and amount of waste, potential risks of transmission of infectious agents and injury for those handling the waste, and documented accidents to date caused by mismanagement of waste in the institution. A review of the current waste management program, practices, and administration should be made, and the problematic areas need a further improvement targeted in the development of the best possible system.

The treatment of medical waste under the universal precautions is not necessarily the best way because generation of an increased amount of infectious waste might result, as seen in Japan. For the present, leaving the identification standards of infectious waste as they are may be advisable, while discussions continue on developing standards and procedures to reduce the quantity generated. The tremendous increase in the use of disposable supplies utilized under the guise of better control of infection and making good use of time is a burden to institutions both from the viewpoint of expenditure and waste management disposal. This is a serious problem and should be settled without delay.

It is regrettable, however, that the reduction of infectious waste will not have the effect of reducing overall waste, but instead will increase the generation of noninfectious material which is another type of medical waste.

7.5.9.2 Program for Safety Supervision

Systems and circumstances must be established under which those who work with waste can do their jobs safely, effectively, and efficiently. Precedence must never be given to economy over safety. Safety is the foremost concern for workers engaged in waste management, the public, and for the environment in which we live.

7.5.9.3 Program of Education and Training

A comprehensive education program should be offered to employees of each institution: those who generate infectious waste, those who work with waste, and the supervisors and managers. The institution's waste management philosophy, technology, safety practices, and overall administration should be set forth in detail. A program to educate intended specialists who would be devoting themselves to management of infectious waste is also necessary.

7.5.9.4 Emergency and Contingency Planning

The most important and common emergencies may be spills of infectious materials or injuries by contaminated sharp instruments. To assure prompt reaction to such emergencies, institutions and hospitals should have a program of care for the individuals involved, procedures for decontamination, cleanup of the spill area, and recovery of the spilled material. Employees should have knowledge and understanding of the institution's program. Personal protective equipment (clothing, gloves, masks), spill control supplies (paper towels, forceps, broom, mop, bucket, etc.), disinfectant (bleach solution), and appropriate medication (including vaccine and immunoglobulin) should be available promptly if an accident should occur.

A contingency plan is also important in the event of a failure in the waste handling stream, for example, the breakdown or malfunction of a steam sterilizer or incinerator treating infectious waste. It is prudent to determine in advance and to post the commercial firm to contact in an emergency. Establishment of a command hierarchy and lines of communication in an emergency are also important.

7.5.9.5 Preparation of Guidance

In addition to the guidelines on medical waste management issued by legal authorities, a comprehensive guide for the management of infectious waste exclusively for use by institutional faculties and staff members will be beneficial. A table showing a necessary pretreatment at the point of generation of infectious waste and a flow chart of the entire waste management system will aid generators of waste in their understanding and practice of proper waste handling.

Planning should be flexible enough to accommodate future change in the law, regulations, available methods, and other factors.

REFERENCES

1. Walter, W.B., Universal precautions, in *Proc. Natl. Conf. Management Medical and Infectious Waste — Practical Considerations,* Hazardous Material Control Research Institute, Silver Spring, MD, 1989, 45.
2. Garner, J.S. and Simmons, B.P., Guidelines for isolation precautions in hospitals, *Infect. Control,* 4, 245, 1983.
3. Centers for Disease Control, Recommendation for prevention of HIV transmission in health-care settings, Morbidity and Mortality Weekly Rep., Public Health Service, U.S. Department of Health, Education, and Welfare, Atlanta, 36 (Suppl. 2S), 1S, 1987.
4. Centers for Disease Control, Update: universal precautions for prevention of transmission of human immunodeficiency virus, hepatitis B virus, and other blood-borne pathogens in healthcare settings, Morbidity and Mortality Weekly Rep. Public Health Service, U.S. Department of Health, Education, and Welfare, Atlanta, 37, 377, 1988.
5. Walter, R.D., Infectious waste: how do we decide?, in *Proc. Natl. Conf. Management of Medical and Infectious Waste — Practical Considerations,* Hazardous Materials Control Research Institute, Silver Spring, MD, 1989, 3.
6. World Health Organization — Regional Office for Europe, Management of Waste from Hospital and Other Health Care Establishments, EURO Reports and Studies 97, 1983.

7. U.S. Environmental Protection Agency, Office of Solid Waste, EPA Guide for Infectious Waste Management, EPA/530-SW-86-014, Washington, D.C., 1986.

8. Kousei-sho (The Ministry of Health and Welfare, Japan), *Iryou Haikibutsu Shori Gaidorain (Guideline for Medical Waste Management)*, Chuuoh Houki Pub. Tokyo, Japan, 1989.

9. Steuer, W., Hygienic aspects on hospital refuse in consideration of pollution control, in *Recycling International*, EF-Verlag, Berlin, 1982, 858.

10. Collins, C.H. and Kennedy, D.A., Microbiological hazards of occupational needles stick and sharps injuries, *J. Appl. Bacteriol.*, 62, 385, 1987.

11. Shirato, S., Iryou Gyoumuto Baiohazaado (Medical practice and biohazard), in *Iryou Haikibutsu (Medical Waste)*, Tanaka, M. and Takatsuki, H., Eds., Tyuuou Houki Shuppan, Tokyo, Japan, 1990, 66.

12. Shinoda, S., Okuda, J., and Shirato, S., Iryou gyoumuto kansensei haikibutu (Medical practice and infectious waste), *Iryou Haikibutsu (Medical Waste)*, Tanaka, M. and Takatsuki, H., Eds., Tyuuou Houki Shuppan, Tokyo, Japan, 1990, 32.

13. Iwata, K., *Biseibutsuni yoru Baiohazaadoto sono Taisaku (Microbiological Biohazard — General Considerations and Control)*, Iwata, K., Ed., Sohuto Saiensu Sha (Soft Science), Tokyo, Japan, 1980, 12.

14. Murayama, N., Iryoukei haikibutsu kanrishisutemuno sekkeini kansuru kenkyuu (Medical wastes management system design), *Juuten-ryouiki Kenkyuu Houkokusho* (Study Report in Priority Area Research), GO36 N32-08, 1989, 4.

15. Allen, R.J., Breiman, G.R., and Darling, C., Air pollution emissions from the incineration of hospital waste, *J. Air Pollut. Control Assoc.*, 36, 829, 1986.

16. Takatsuki, H. and Sakai, S., Iryou haikibutsuno haishutsu doutai (Discharge of medical waste), in *Iryou Haikibutsu (Medical Waste)*, Tanaka, M. and Takatsuki, H., Eds., Tyuuou Houki Shuppan, Tokyo, Japan, 1990, 83.

17. Sammet, D. and Lausterer, W., Characteristic value of quantity and composition of hospital waste, in *Recycling International*, EF-Verlag, Berlin, 1982, 888.

18. Reinhardt, P.A. and Gordon, J.G., The search for a successful waste management strategy, in *Infectious and Medical Waste Management*, Lewis Publishers, Chelsea, MI, 1991, 3.

19. Reinhardt, P.A. and Gordon, J.G., Identification of infectious waste, in *Infectious and Medical Waste Management*, Lewis Publishers, Chelsea, MI, 1991, 31.

20. Rutala, W.A., Stiegel, M.M., and Sarubbi, Jr., Decontamination of laboratory microbiological waste by steam sterilization, *Appl. Environ. Microbiol.*, 43, 1982, 1311.

21. Lauer, J.L., Battles, D.R., and Vesley, D., Decontaminating infectious waste by autoclaving, *Appl. Environ. Microbiol.*, 44, 690, 1982.

22. Cooney, T.E., Techniques for steam sterilizing laboratory waste, *AMSCO Waste Processing Technical Report DB-3014*, AMSCO, Pittsburgh, PA, 1988, 7.

23. U.S. Environmental Protection Agency, Office of Waste and Emergency Response, National Technical Information Service, EPA Guide for Infectious Waste Management, PB 86-199130, 1986, (5)2.

24. National Research Council, Biosafety in the Laboratory — Prudent Practices for the Handling and Disposal of Infectious Materials, National Academy Press, Washington, D.C., 1989, 39.

25. U.S. Environmental Protection Agency, Office of Waste and Emergency Response, National Technical Information Service, EPA Guide for Infectious Waste Management, PB 86-199130, 1986, xiv and (5)1.

26. Laboratory Safety at the Center for Disease Control: U.S. Department of Health, Education, and Welfare, Public Health Service, Center for Disease Control, Atlanta, May, 1979.

**Section
8**

Index

INDEX